Spectroscopy in Biochemistry

Volume II

Editor

J. Ellis Bell, Ph.D.
Assistant Professor of Biochemistry
University of Rochester Medical School
Rochester, New York

CRC Press, Inc.
Boca Raton, Florida

Library of Congress Cataloging in Publication Data

Main entry under title:

Spectroscopy in biochemistry.

Bibliography: p.
Includes index.
1. Spectrum analysis. 2. Biological chemistry—Technique. I. Bell, John Ellis.
[DNLM: 1. Biochemistry. 2. Spectrum analysis. QH324.9.S6 S741]
QP519.9.S6S63 574.19'285 79-26879
ISBN 0-8493-5551-6 (v. 1)
ISBN 0-8493-5552-4 (v. 2)

Direct all inquiries to CRC Press, Inc., 2000 N.W. 24th Street, Boca Raton, Florida 33431.

International Standard Book Number 0-8493-5551-6 (v. 1)
International Standard Book Number 0-8493-555-2-4 (v. 2)

Library of Congress Card Number 79-26879
Printed in the United States

PREFACE

Over the years spectroscopy has played an important role in the development of modern biochemistry. Once the sole domain of spectroscopists with access to highly specialized and expensive equipment, most biochemistry departments now either have, or have access to, facilities for many forms of spectroscopy. This is reflected in the increased uses spectroscopy has found in biochemistry over the last few years, and the increasing number of articles in journals describing work based on one spectroscopic technique or another. This book was conceived with the idea of providing a text which would describe and illustrate the many ways spectroscopic techniques are being, and will be applied to biochemical problems. We have sought to illustrate the potential of various spectroscopic techniques rather than review all that has been achieved. Since there are many theoretical texts describing the basis of the various spectroscopic techniques we discuss here, we have, in general, not extensively discussed the phenomena described, choosing rather to discuss applications, giving only theoretical treatments where we feel it is necessary to understand the principal as related to the application described.

This idea for the aims of this book originated in discussions with Drs. Charles Tanford, Jackie Reynolds, and David Richardson several years ago when we all were involved in teaching spectroscopy to graduate students in biochemistry. This book has been written in part with the aim of providing a text which will be useful in teaching the biochemical applications of spectroscopy. We have tried in addition to write a book which will be of use to the biochemist or biologist who does not have a background in spectroscopy, but desires to find out what sort of information spectroscopy can provide. This book will hopefully fill this need. We have limited our attention to those techniques most frequently used, and which at present have the widest applications. We have stressed both the applications and the limitations of the techniques.

THE EDITOR

J. Ellis Bell, Ph.D., was born in London, England in 1948, and educated in England. He received the B.A. degree (with honors) from Oxford University in 1972, M.A. (oxon) 1974, and his doctorate in 1974. His doctoral work at Oxford (with Keith Dalziel) was concerned with negative cooperativity in oligomeric enzymes, and involved kinetic, equilibrium, and spectroscopic approaches. In 1974, he first came to the United States to work with Dr. Robert Hill in the Department of Biochemistry at Duke University Medical Center. In 1978 he was appointed Assistant Professor of Biochemistry in the Medical School at the University of Rochester, New York, where he is currently working on structure-function studies of membrane proteins and the basis of cellular interaction.

CONTRIBUTORS

Robert D. Bereman, Ph.D.
Professor
Department of Chemistry
North Carolina State University
Raleigh, North Carolina

Lawrence J. Berliner, Ph.D.
Associate Professor
Department of Chemistry
The Ohio State University
Columbus, Ohio

Don Alexander Gabriel, M.D., Ph.D.
Assistant Professor
Department of Internal Medicine
University of North Carolina
School of Medicine
Chapel Hill, North Carolina

Carole L. Hall, Ph.D.
Research Scientist in Chemistry
School of Chemistry
Georgia Institute of Technology
Atlanta, Georgia

Charles Sidney Johnson, Jr., Ph.D.
Professor
Department of Chemistry
University of North Carolina
Chapel Hill, North Carolina

Daniel J. Kosman, Ph.D.
Associate Professor
Department of Biochemistry
State University of New York at Buffalo School of Medicine
Buffalo, New York

Joseph R. Lakowicz, Ph.D.
Associate Professor
Department of Biological Chemistry
University of Maryland School of Medicine
Baltimore, Maryland

Thomas Nowak, Ph.D.
Associate Professor
Department of Chemistry
University of Notre Dame
Notre Dame, Indiana

Elizabeth R. Simons, Ph.D.
Professor of Biochemistry
Boston University School of Medicine
Boston, Massachusetts

ACKNOWLEDGMENTS

We would like to acknowledge the help and contributions of our colleagues to the preparation of the chapters in this book. The editor would in addition like to acknowledge the assistance and forbearance of his coauthors in the preparation of their chapters; Drs. Keith Dalziel, George Radda, and Raymond Dwek at Oxford for introducing him to the ways and wonders of spectroscopy; his students, who have helped in the preparation of his chapters, and Evelyn O'Keeffe, who has been of invaluable assistance at every stage in the preparation of this book.

DEDICATION:

to Evelyn

TABLE OF CONTENTS

Volume I

Volume II

Chapter 1

USING THE SPIN LABEL METHOD IN ENZYMOLOGY

Lawrence J. Berliner*

TABLE OF CONTENTS

* Established Investigator, American Heart Association.

I. WHAT IS IT ABOUT? — ESR LABEL OR PROBE APPROACH

A. Reporter Groups

We have seen how the use of specifically incorporated radioactive nuclides in substrates or inhibitors allows us to gain mechanistic information about enzymes and metabolic pathways. The "labeling" technique may be adopted in physical studies of biological systems by incorporating specific physical probes (or "reporter group") at an intended target site in, e.g., an enzyme or protein. In optical spectroscopy the reporter group may be a chromophore or fluorophore; in NMR a molecule containing a unique nucleus (e.g., ^{19}F). With the ESR technique we call this group a spin label or spin probe, which is a molecule containing a stable paramagnetic group. While some proteins contain their own intrinsic spin probe, such as the paramagnetic Fe(III) heme of met-myoglobin or met-hemoglobin, they are a minority of all proteins and enzymes for the general application of ESR to biochemistry.

A reporter group must possess some physical property of the protein under study in order to (1) "report" changes or details of its immediate environment, (2) be distinct from the remainder of the protein with respect to the physical property under measurement, and, most important, (3) not contribute any significant perturbations to the biomolecular conformation or catalytic properties.

Those physical methods that are amenable to reporter group techniques are summarized in Table 1. A reporter group molecule placed strategically at a specific site in a biomolecule may yield a wealth of information about the macromolecule's structure and function.

B. Nitroxide (Spin Labels — Spin Probes)

The ESR reporter group that comprises the majority of most spin labeling research is shown as several structural examples in Figure 1. The common paramagnetic moiety of all of them is the nitroxide group ($\rangle N-0$), a stable organic free radical. Since some of these molecules are intended for covalent labeling of the biomolecule while others as noncovalent probes, we have strictly defined the two cases with the nomenclature: (1) *spin labels* — covalently binding nitroxides designed for modification of specific amino acid residues and enzyme active sites; and (2) *spin probes* — noncovalently binding ligands that bind to proteins and membranes by ionic, hydrogen bonding, and especially hydrophobic forces. Some examples of each are shown in Figure 2.

Table 1
REPORTER GROUP TECHNIQUES AND THEIR SENSITIVITY TO THE PHYSICAL ENVIRONMENT[a]

Technique	Instrumentation	Physical or chemical property detected
UV-visible	Optical absorption spectrophotometry	Polarity
	Polarized absorption spectrophotometry	Chemical, electronic environment
Fluorescence	Emission spectrophotometry	Polarity; some specific aspects of the chemical environment (proximity to other fluorescence acceptors or donors)
Fluorescence depolarization	Polarized emission spectrophotometry	Motion; polarity, chemical environment (as above)
NMR	NMR spectroscopy (absorption or pulsed)	Magnetic environment, motion, polarity
ESR (spin labeling)	ESR spectroscopy	Motion, orientation, polarity, magnetic environment
γ-Ray perturbed-angle correlation spectroscopy	Radioactive coincidence spectrometry	Motion
Resonance Raman	Laser Raman spectrophotometry	Chemical environment

[a] All techniques involving extrinsic labels must be checked for any perturbation to the bio-molecular system ("one must ensure that the reporter group is reporting the news, not making the news").

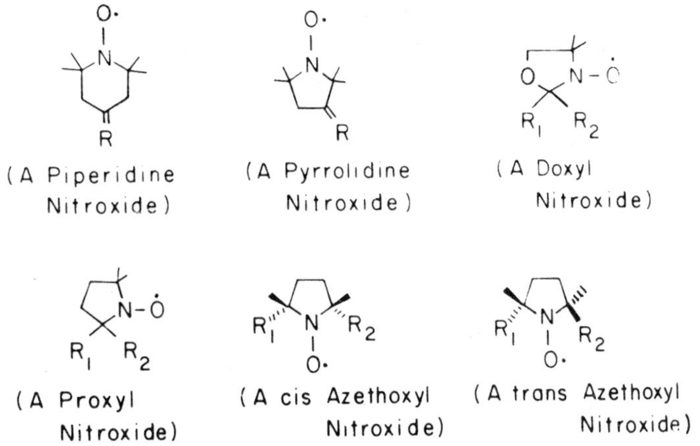

FIGURE 1. Stable nitroxide structures.

While there are, in fact, other paramagnetic species that may find even more potential usefulness as spin labels in the future, they are described briefly below for the aid of the student for future reference.

C. Paramagnetic Metals

1. Extrinsic Probes — Mn(II)

By an extrinsic probe we mean one that may be incorporated into a system where it does not normally occur. Manganese (II) is an extremely useful probe by virtue of its ability to substitute for Mg(II) in many enzymes and their nucleotide complexes; its limited, but specific natural occurrence in a number of metalloproteins; and its ability

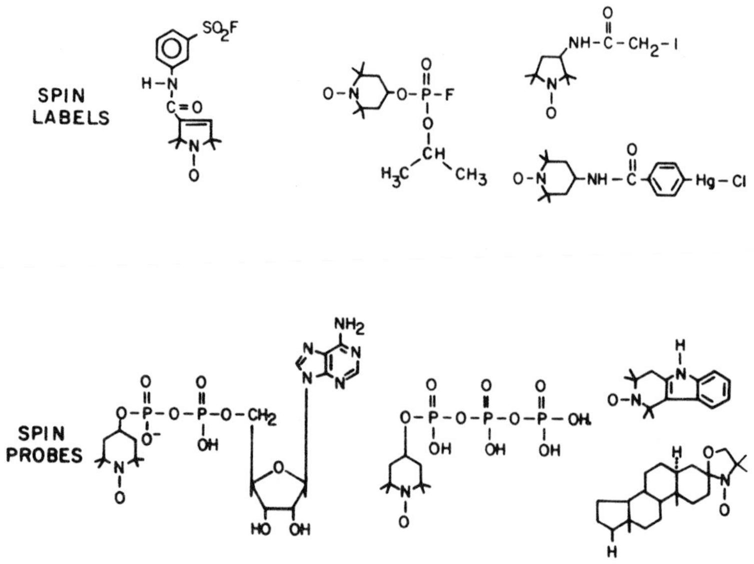

FIGURE 2. Examples of nitroxide spin labels and spin probes. A spin label contains a reactive functionality for covalent modification. A spin probe is designed to bind reversibly to a macromolecule.

to relax neighboring nuclei by virtue of its large magnetic moment ($S = 5/2$). From an ESR viewpoint, an aqueous Mn(II) solution gives a six-line spectrum of slightly different intensities and linewidths arising from hyperfine coupling with the Mn(II) nuclear spin. Upon complexing with a nucleotide or protein, the new asymmetric environment induces an axial distortion of the Mn(II) zero field splitting yielding a significantly broadened, more complex spectrum as exemplified in Figure 3. In certain cases these "bound" spectra may be analyzed theoretically to detect changes in the Mn(II) environment upon addition of ligands.[1] Use of the apparent diminution of the Mn(II) aquo ESR signal may be exploited as a monitor of the concentration of Mn(II) bound. Figure 4 shows an example of such a titration for the binding of UDP-galactose and two Mn(II) per mole of bovine galactosyl transferase.

2. Intrinsic Probes — Iron Proteins

Many proteins (particularly those in the respiratory cycle) contain tightly bound heme coordinated Fe(II) or Fe(III). Besides this class of hemeproteins, most commonly exemplified by the oxygen carrier in the blood, hemoglobin, there is a large group of nonheme iron cases, such as the iron sulfur protein, ferridoxin. Many of these systems exhibit characterizable ESR spectra usually, at 4° or 77°K, which contain a high degree of information content as to the iron protein environment, ligand complexation, magnetic state, and other parameters. As a purely qualitative example, Figure 5 shows the enormous variation in spin state and g-factor sensitivity to Fe(III)-met hemoglobin ligand complexes at 77°K.[4]

D. Other Organic Radicals — The First Spin Probe

Actually the first spin probe experiment involved a nitrogen centered radical, but not a nitroxide, the chlorpromazine radical cation. In their experiments, Ohnishi and McConnell[5] confirmed by ESR techniques that this phenothiazine derivative, used pharmaceutically to treat schizophrenia, intercalated DNA as depicted in Figure 6. The proof was based on the fact that the nitrogen hyperfine coupling in the oriented flow-

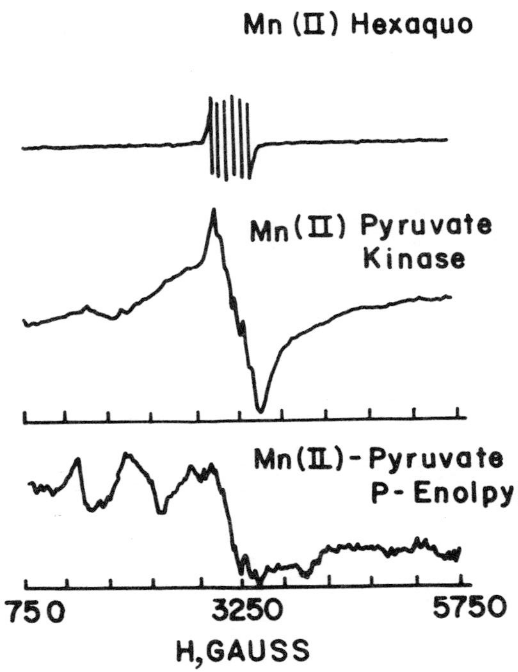

FIGURE 3. X-band ESR spectra for Mn(II) in varying environments. Top: 0.2 m*M* $MnCl_2$, aqueous; center: 1.0 m*M* $MnCl_2$/pyruvate kinase (200 mg/mℓ); bottom: manganese-pyruvate kinase — phosphoenolpyruvate (2 m*M*) complex of binary complex in center spectrum. Note that the top sample was measured at much lower signal gain than the lower two spectra. (From Cohn, M., Leigh, J. S., Jr., and Reed, G. H., *Cold Spring Harbor Symp. Quant. Biol.*, 36, 533, 1972. With permission.)

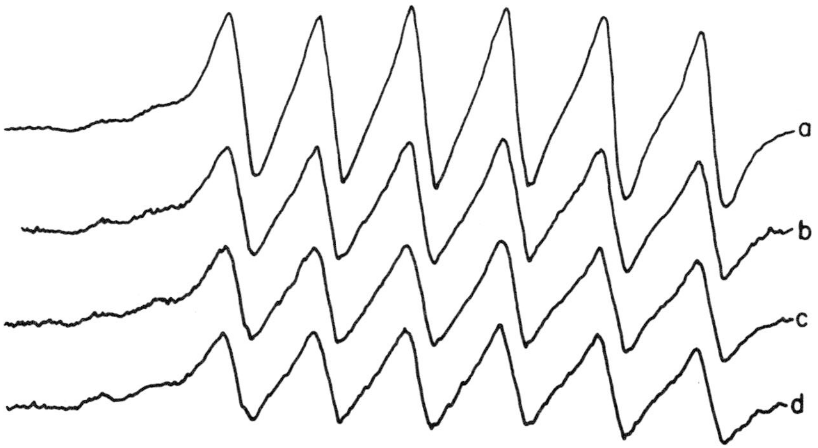

FIGURE 4. ESR spectra of Mn(II) aqueous (''free $MnCl_2$'') in the presence of bovine galactosyl transferase and increasing concentrations of UDP galactose. The apparent decrease in free Mn(II) signal height is caused by the extremely broad line shape spectra of the Mn·enzyme·UDP galactose complexes. Conditions were pH 7.4 (0.05 *M* Tris-Cl, 0.1 *M* $(NH_4)_2SO_4$, 2 m*M* ε-amino caproic acid, 26 ± 2°C); enzyme concentration was 58 μM, Mn(II) total was 251 μM and UDP galactose total varied from (a) none, (b) 1.7 m*M*, (c) 3.2 m*M* and (d) 4.5 m*M*. Thus Mn(II) free decreased by about 100 μM or two Mn(II) per mole of enzyme. (From Andree, P. J. and Berliner, L. J., *Biochemistry*, 19, 929, 1980. With permission.)

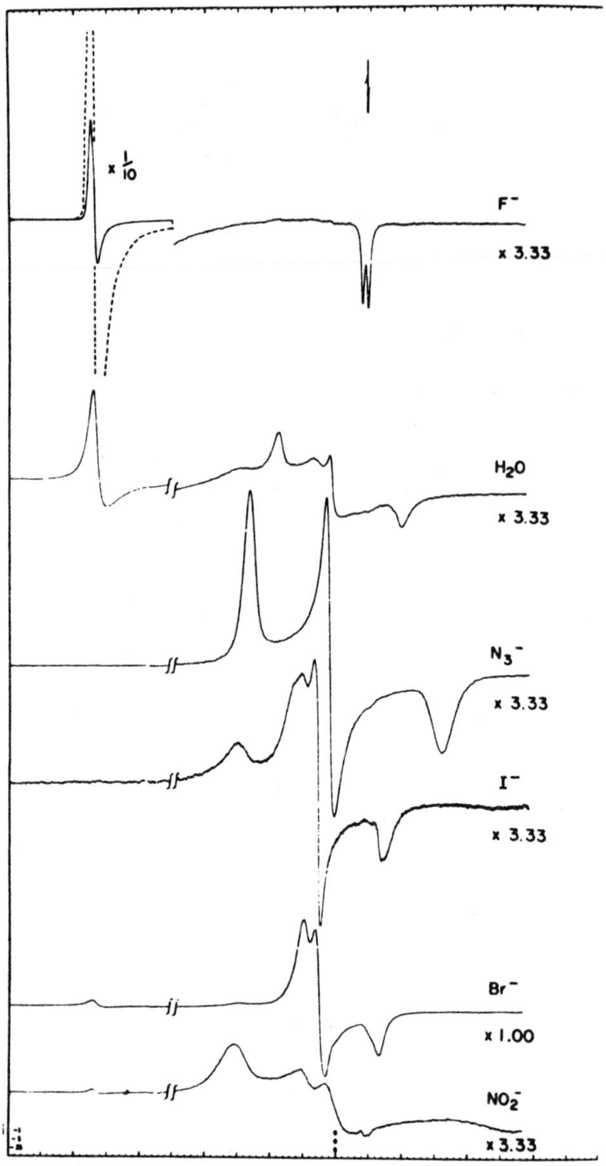

FIGURE 5. ESR spectra of met-hemoglobin complexes at X-band, 77°K. The total field sweep shown was 5000 G centered at 3020 G. The very top line at g = 2.0028 was a strong pitch reference sample. The ligand complex is designated with each spectrum. (From Uchida, H., Berliner, L. J., and Klapper, M. H., *J. Biol. Chem.*, 245, 4606, 1970. With permission.)

ing DNA corresponded to the parallel and perpendicular components of the chlorpromazine radical cation ^{14}N coupling for the magnetic field direction parallel and perpendicular to the long axis of the DNA, respectively.

II. MAKING THE LABELS — ORGANIC CHEMISTRY OF NITROXIDE SPIN LABELS

A. The Simple, Most Common Labels and Probes

The chemistry of spin labels is now quite diverse, yet much of the basic chemistry is

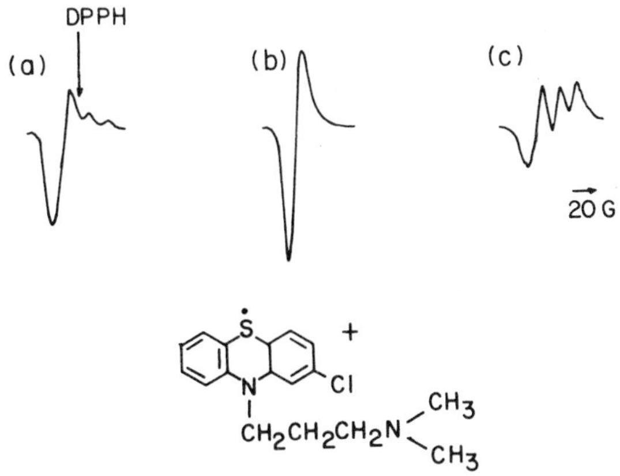

FIGURE 6. ESR spectra of the chlorpromazine radical cation: calf thymus DNA complex, pH 5.0 (a) no flow, (b) in a flow system where the DNA helices were oriented perpendicular to the applied field, and (c) parallel flow (with respect to the applied field). (Reprinted with permission from Ohnishi, S. I. and McConnell, H. M., *J. Am. Chem. Soc.*, 87, 2293, 1965. Copyright by the American Chemical Society.)

based on the three simple structures, I to III, shown in Figure 7. The diamagnetic secondary amine precursor of IV, triacetonamine, was synthesized from acetone, $CaCl_2$ and ammonia in 1927;[6] the synthesis optimized recently[7] as an economical starting material for several of the nitroxides shown in Figure 7. The oxidation to the nitroxide for piperidine (I) or pyrrolidine (II) nitroxides is accomplished in aqueous H_2O_2 with phosphotungstate or sodium tungstate/EDTA catalyst. Oxazolidine nitroxides (III) are synthesized by a condensation reaction followed by oxidation in ether with *m*-chloroperbenzoic acid (MCPA)

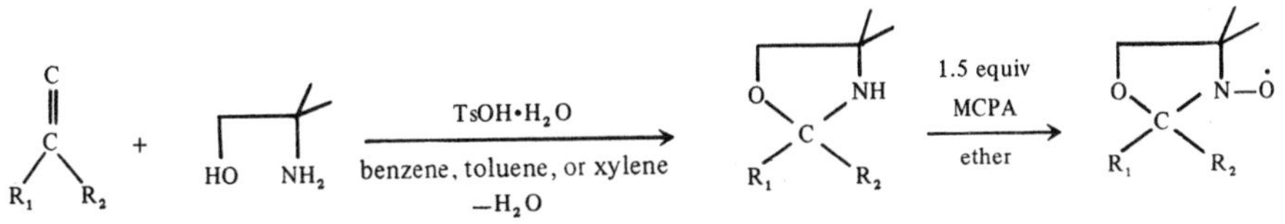

A well-documented synthetic scheme for several of the five and six membered ring nitroxides types I and II are shown in Figure 8.[9-11]

Of much general use are spin labels that are analogs of protein modification reagents. These have, in fact, been the most commonly used, and a few examples are therefore discussed below for their general importance. The two labels are of the maleimide or iodoacetamide functionality on the six membered (piperidine) or five membered (pyrrolidine) ring nitroxide:

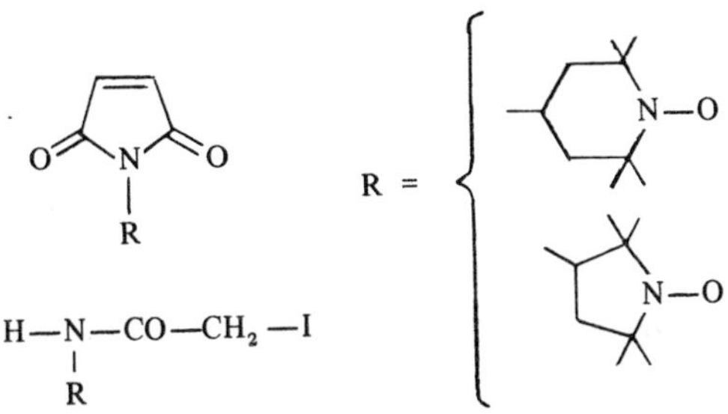

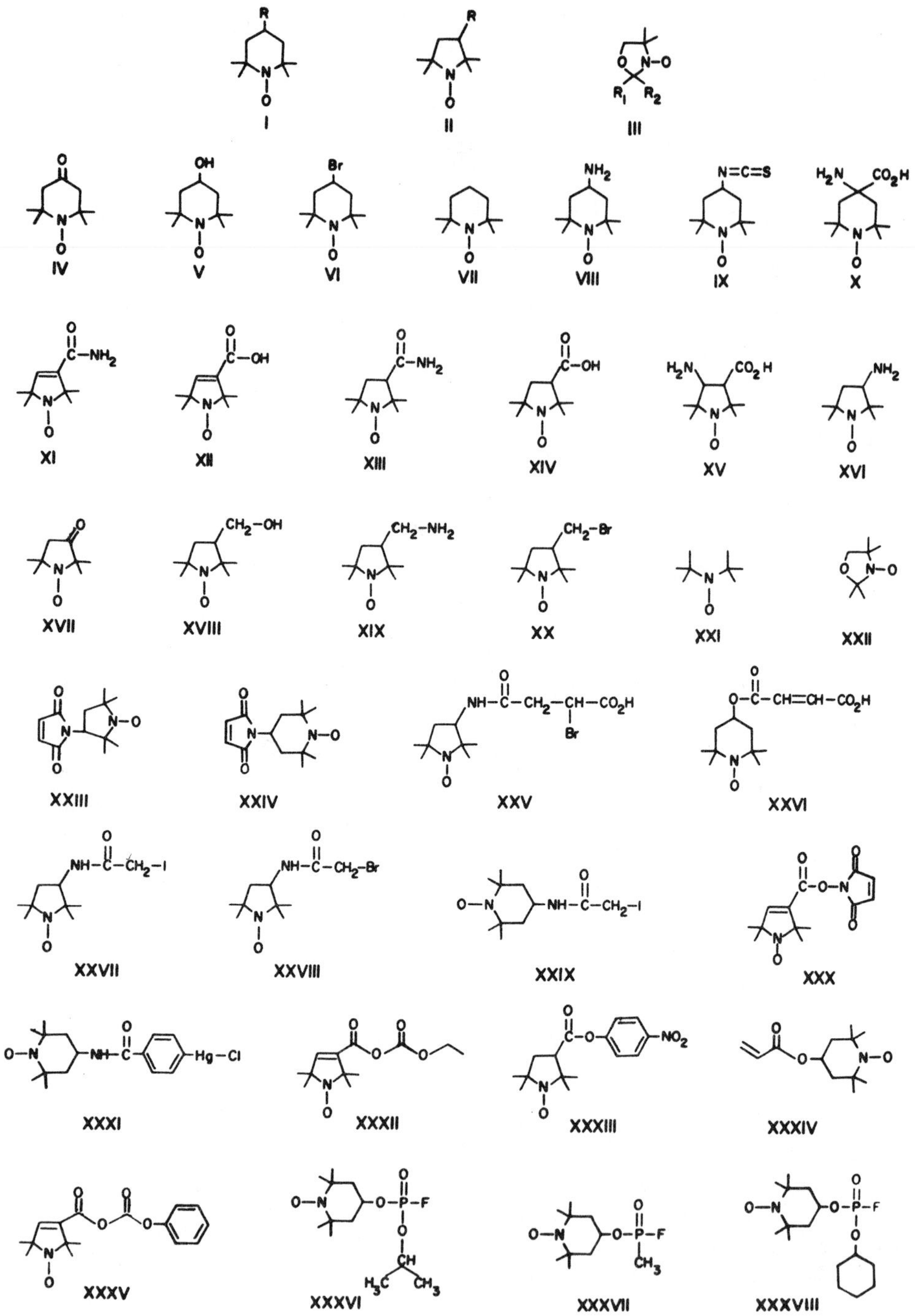

FIGURE 7. Structures of a representative sampling of protein spin labels. (From Jost, P. and Griffith, O. H., *Methods in Pharmacology,* Vol. 2, Chignell, C. F., Ed., Appleton-Century-Crofts, New York, 1972, 223. With permission.)

Both labels are quite highly reactive with cysteine side chains, but will also react with primary amino groups, the iodoacetamide being less discriminating than the maleimide. Both label types are available commercially as a series of increasing size with

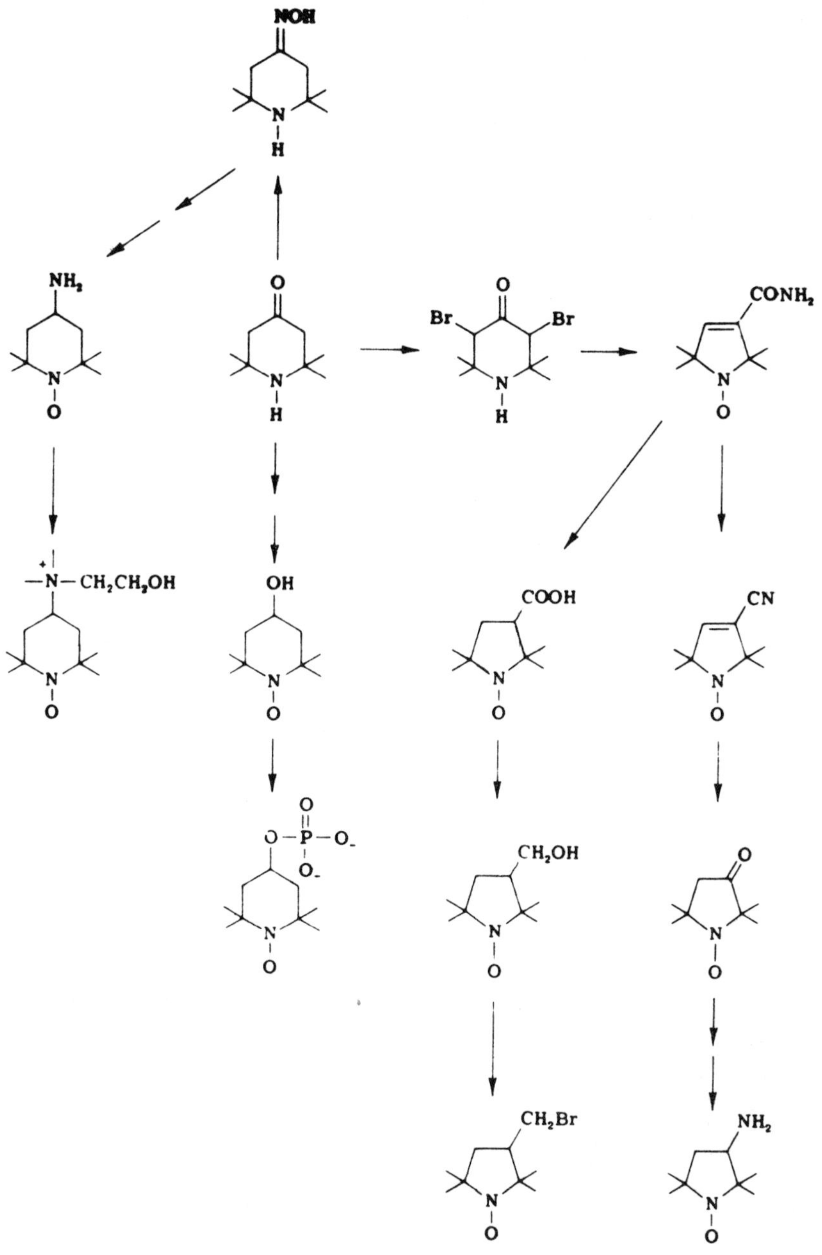

FIGURE 8. Reaction schemes leading to spin label precursors. All of these structures begin with triacetonamine (2,2,6,6-tetramethyl piperidone).

a variable methylene bridge spacer between the pyrrolidine ring and the reactive functional group.

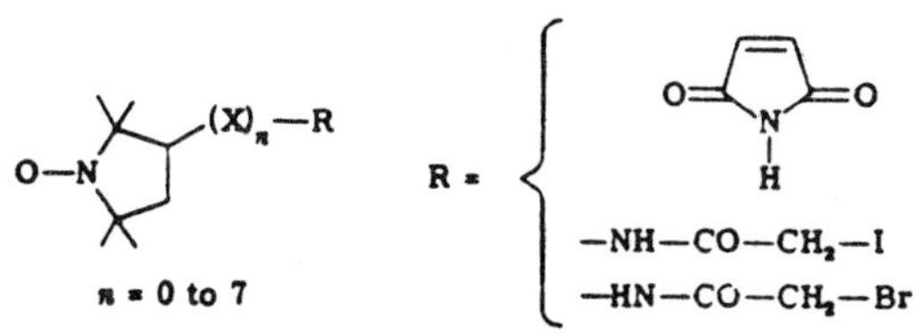

Another general reactive type, useful as an active site modification reagent with the serine esterase family of enzymes (chymotrypsin, trypsin, etc.) are analogs of fluorophosphonate protease and nervous system (acetylcholinesterase) blockers:

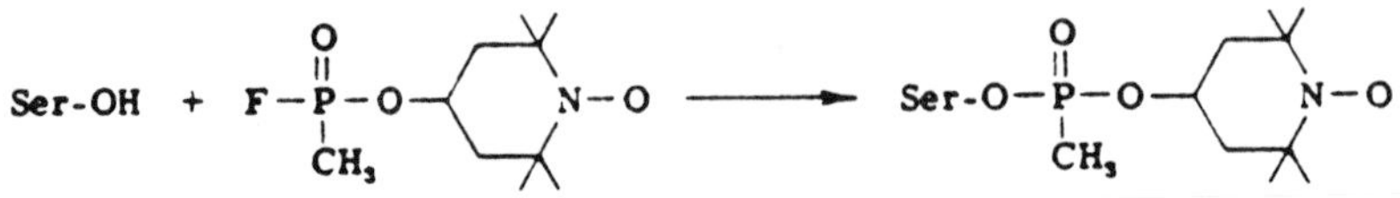

Some examples with serine protease enzymes will be described in later sections.

B. Tailored Size, Stereochemistry, and Shape—New Nitroxides

The most recent activity in nitroxide chemistry has centered around the oxazolidinyl, proxyl, and azethoxyl nitroxides (see Figure 1).[11] The oxazolidine syntheses (which were discussed earlier) allow any tailored length or functionality in the side chains R_1 and R_2 and have found their greatest potential in lipid analogs. The proxyl nitroxides, which are based on the pyrrolidine-*N*-oxyl structures, are advantageous over the doxyl nitroxides by their increased chemical stability and hydrophobicity. Their general synthetic scheme is outlined below:

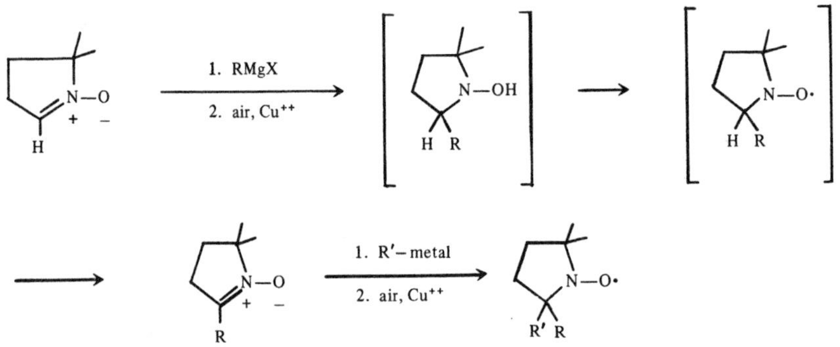

The starting material, a nitrone, is commercially available from Aldrich Chemical Company. Again, the desired spin labels may be "tailor-made" by proper choice of starting grignard reagents, RMgX.

Lastly the *cis* and *trans* azethoxyl nitroxides offer minimal steric bulk ("minimum perturbation spin labels") with choice of *cis* or *trans* stereochemistry about the nitroxide group, the highest chemical stability of all of the nitroxide types and a higher resistance to chemical reduction over that of the proxyl nitroxides. In the general synthetic schemes shown below, the *trans* isomer predominates over *cis* for steric reasons.

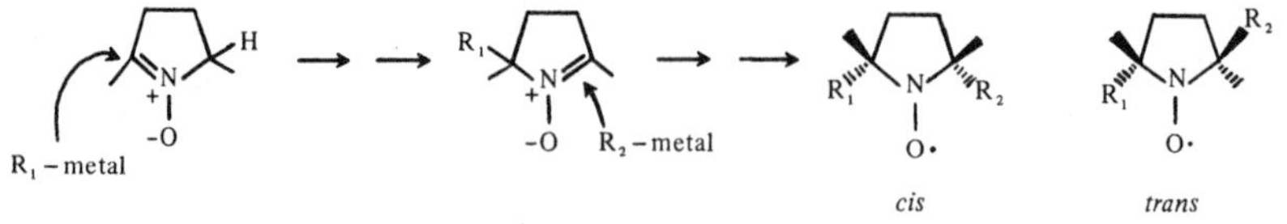

However by choosing a different starting nitrone, the *cis* isomer may be made to predominate. The reader is referred to the literature for more detailed synthetic information.[11]

C. Useful Reactions of the Nitroxide Group

1. Reduction

The paramagnetism of the nitroxide radical may be reversibly reduced by a one-electron process

$$-\overset{|}{\underset{|}{C}}-\overset{|}{\underset{\underset{O}{\cdot}}{N}}-\overset{|}{\underset{|}{C}}- \xrightarrow{e^{-}} \cdot\overset{|}{\underset{|}{C}}-\overset{|}{\underset{\underset{O_{-}}{}}{N}}-\overset{|}{\underset{|}{C}}- \xrightarrow{H^{+}} -\overset{|}{\underset{|}{C}}-\overset{|}{\underset{\underset{O}{}}{N}}-\overset{|}{\underset{|}{C}}-$$

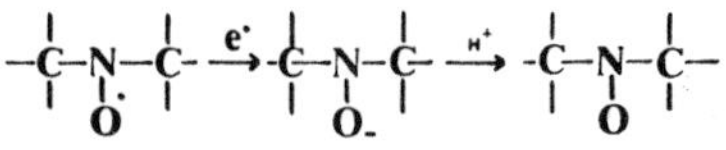

Reduction

that is accomplished in aqueous solution by ascorbate, hydroxylamine dithionite, and other reagents. All lead to the hydroxylamine, which is easily reconverted to the paramagnetic nitroxide by exposure to oxygen or other mild oxidizing agents while reduction back to the secondary amine precursor is obtained by hydrogenation on Pd/C catalyst, most aqueous treatments are of the one-electron type to give the hydroxylamine. A listing of useful reducing and oxidizing agents, both for analytical and synthetic purposes, is in Table 2. Analytically it might be useful in a protein labeling study to diminish the paramagnetism in order to discriminate between two or more labeled sites, as a control in paramagnetic relaxation or fluorescence quenching measurements, or as a method of quantitating nitroxide spin concentration on a labeled macromolecule.

2. Other Reactions (Spin Annihilation, Spin Traps)

Specific methods have now been developed for annihilating nitroxide ESR signals by photolytic methods. For example, Schwartz and McConnell have used the following reaction:[12]

$$[Co^{III}(CN)_5(CH_2CO_2^-)]^{4-} + R-\overset{}{\underset{}{\boxed{N}}}-O \xrightarrow[\lambda=350nm]{h\nu} R-\overset{}{\underset{}{\boxed{N}}}-O-CH_2CO_2^- + Co^{II}(CN)_5^{3-}$$

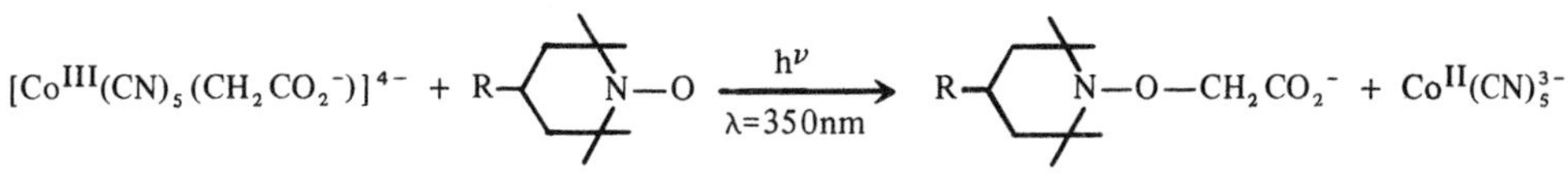

In general, in nitroxide systems that have no unsaturation the principal point of photochemical attack is the nitroxide group itself, resulting again in annihilation of the ESR signal. In the example shown below, which might take place in a very apolar biological environment, toluene photoadds to the piperidinol nitroxide with subsequent loss of signal.[13]

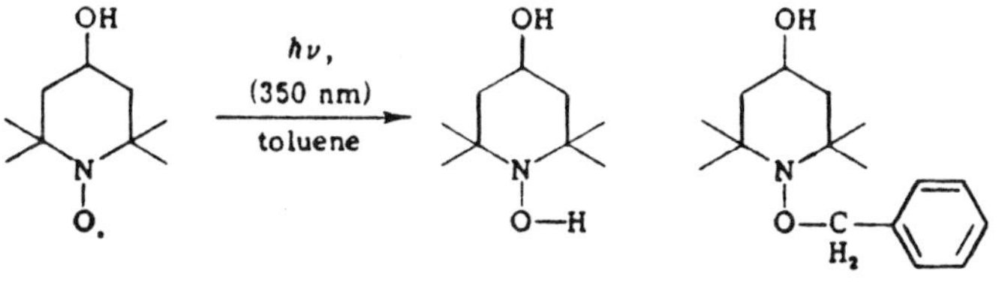

Moderately stable nitroxides may be produced at sites of radical reactions using nitrone spin traps. In the case shown below 5,5'-(dimethyl-1-pyrroline-1-oxide) reacts with an ·OH radical to yield the nitroxide shown and resultant ESR spectrum (Figure 9).[14]

III. WHAT CAN WE LEARN? PHYSICO-CHEMICAL INFORMATION FROM ESR MEASUREMENTS

One of the principal parameters one obtains from a spin label experiment related to motion is, specifically, the tumbling rate of the nitroxide group. This may or may not reflect the motion of the biological system to which the nitroxide is bound, depending upon the degree of immobilization of the label/probe with respect to the macromolecule and also to the "time window" of the experiment. For the conventional X-band ESR experiment (9.5 GHz), we show an example of the tumbling motion range by simulating a series of tumbling rates by continuously increasing the viscosity in a

Table 2
SOME REDUCING AND OXIDIZING AGENTS FOR NITROXIDES

Reducing Agents[a]	Reoxidizing agents
Ascorbate	H_2O_2
Phenylhydrazine	O_2
Hydroxylamine	MCPA (meta-chloroperbenzoic acid)
Thiols	PbO_2
Organolithium	Ag_2O
Na_2S	Cu^{++}
K_3FeCN_6	
$LiAlH_4$	
$NaHAl(OCH_2CH_2OCH_3)_2$	

[a] Doxyl nitroxides tend to irreversibly hydrolyze from the hydroxylamine intermediate.

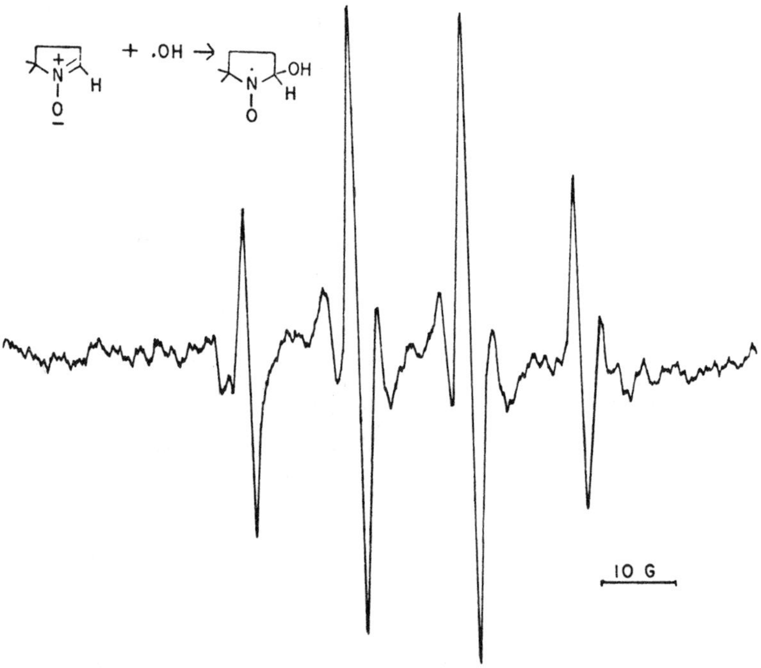

FIGURE 9. ESR spectrum of 5,5′-(dimethyl-l-pyrroline-l-oxide) after addition of an OH· radical across the double bond. The OH· radical was generated in a microsomal system. (From Lai, C. S. and Piette, L. H., *Biochem. Biophys. Res. Commun.*, 78, 51, 1977. With permission.)

water/glycerol mixture of a small nitroxide (Figure 10). The categories of motion are called (1) fast motion or "weakly immobilized", (2) intermediate motion or "moderately immobilized", (3) slow motion or "strongly immobilized", and (4) very slow motion, which is indistinguishable from "slow motion" on a conventional instrument, but attainable with new instrumental modifications.

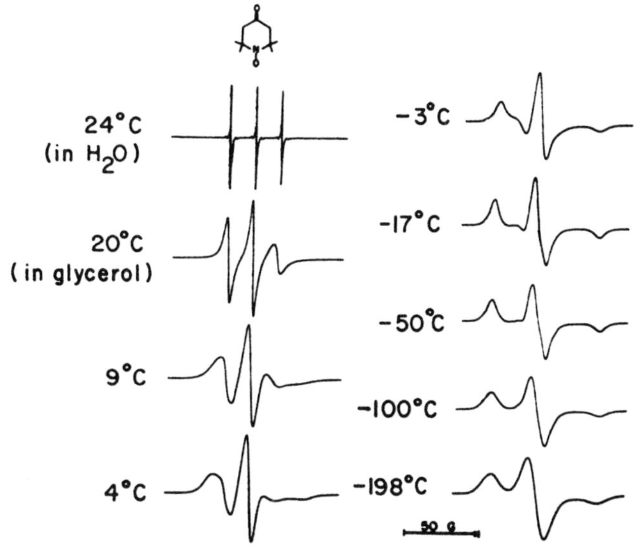

FIGURE 10. Effects of molecular tumbling (viscosity) on nitroxide ESR (X-band) spectra. The spectra in glycerol represent $5 \times 10^{-4}M$ nitroxide in reagent grade glycerol at the temperature shown. (From Berliner, L. J. and Shen, Y. Y., *Chemistry and Biology of Thrombin*, Lundbland, R. L., Fenton, J. W., II., and Mann, K. G., Eds., Ann Arbor Science, Ann Arbor, Mich., 1977, 197. With permission.)

A. Motion

1. Fast Motion

For the tumbling time range 10^{12} to 10^{10} sec^{-1} (fast tumbling, weakly immobilized), the Kivelson theory is quite applicable for a purely isotropic tumbling model; that is, where motion is completely random about all of the (principal hyperfine) axes of the nitroxide. Formally, the equation for the rotational correlation time, τ_c, is given by

$$\frac{T_2(O)}{T_2(M)} = 1 - \frac{4}{15}\tau_c b\,\Delta\gamma\,H_0 T_2(O)M + \frac{\tau_c}{8} b^2 T_2(O)M^2 \qquad (1)$$

where the longitudinal relaxation time ratios, $T_2(O)/T_2(M)$ are related by $[T_2(O)/T_2(M)]^{-1}$ = ratio of the linewidth of the center line to that of the low field ($M = +1$) or high field ($M = -1$) line; M = ^{14}N nuclear spin, H_o = applied field, b and $\Delta\gamma$ are constants related to the hyperfine and g-tensor anisotropy of the particular nitroxide under study.

The factor $T_2(O)/T_2(M)$, the line width ratio, can be approximated by the peak height ratios of the two lines $\sqrt{h(O)/h(M)}$. For the spectra of radical XVI in Figures 11a and 11b, we can simplify Equation 1 by substituting in $T_2(O)/T_2(-1)$ and $T_2(O)/T_2(1)$; respectively.

$$\frac{T_2(O)}{T_2(-1)} = 1 + \frac{4\tau_c b}{15}\Delta\gamma\,H_0\,T_2(O) + \frac{\tau_c}{8} b^2 T_2(O) \qquad (2)$$

$$\frac{T_2(O)}{T_2(+1)} = 1 - \frac{4\tau_c b}{15}\Delta\gamma\,H_0\,T_2(O) + \frac{\tau_c}{8} b^2 T_2(O) \qquad (3)$$

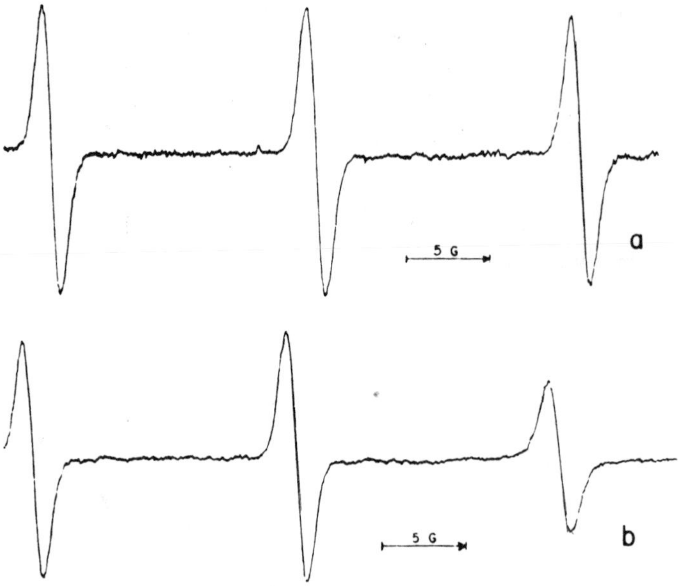

FIGURE 11. ESR (X-band) spectra at room temperature of radical XVI (Figure 7) in (a) water and (b) 60% aqueous sucrose. (From Stone, T. J., Buckman, T., Nordio, P. L., and McConnell, H. M., *Proc. Natl. Acad. Sci. U.S.A.*, 54, 1010, 1965. With permission.)

Solving the two simultaneous equations we obtain an equation *linear* in M

$$\frac{T_2(O)}{T_2(-1)} - \frac{T_2(O)}{T_2(1)} = \frac{8\tau_c b}{15} \Delta\gamma H_o T_2(O) \qquad (4)$$

and an equation quadratic in M

$$\frac{T_2(O)}{T_2(-1)} + \frac{T_2(O)}{T_2(+1)} = 2 + \frac{2\tau_c b^2}{8} T_2(O) \qquad (5)$$

Solving each equation for τ_c and substituting the square root of the peak height ratios for linewidth ratios

$$\frac{T_2(O)}{T_2(-1)} = \frac{\Delta H(-1)}{\Delta H(O)} = \sqrt{\frac{h(O)}{h(-1)}} \qquad (6)$$

and where $\Delta H(M)$ is the linewidth in gauss and $[T_2(M)]^{-1} = \Delta H(M) \sqrt{3}\, \pi\, 2.8 \times 10^6$ sec^{-1} for a Lorentzian line

$$\tau_c = \Delta H(O) \sqrt{3}\, \pi\, (2.8 \times 10^6) \left[\frac{8}{15}\, b\, \Delta\gamma\, H_o\right]^{-1} \left[\sqrt{\frac{h(O)}{h(-1)}} - \sqrt{\frac{h(O)}{h(1)}}\right] \text{sec} \qquad (7)$$

and

$$\tau_c = \Delta H(O) \sqrt{3}\, \pi\, (2.8 \times 10^6) \left(\frac{b^2}{4}\right)^{-1} \left[\sqrt{\frac{h(O)}{h(-1)}} + \sqrt{\frac{h(O)}{h(1)}} - 2\right] \text{sec} \qquad (8)$$

The terms $\Delta\gamma$ and b are expressed by

$$\Delta\gamma = -\frac{-|\beta|}{\hbar}\,[g_{zz} - \tfrac{1}{2}(g_{xx} + g_{yy})] \qquad (9)$$

where β is the electron Bohr magneton, $\hbar$ is Planck's constant and g_{zz}, g_{xx}, and g_{yy} are the components of the anisotropic g-tensor for the nitroxide under study.

$$b = \frac{4\pi}{3}(A - B) \qquad (10)$$

where $A = T_{\parallel}$ and $B = T_{\perp}$ in frequency units (sec^{-1}).

Approximating the hyperfine and g-tensor parameters for this label from the available data for di-*t*-butyl nitroxide, $(CH_3)_3C{-}\underset{\underset{O}{|}}{N}{-}C(CH_3)_3$, the following parameters may be used in the calculations

$$
\begin{aligned}
\hbar &= 1.054 \times 10^{-27}\ \text{erg sec}\\
|\beta| &= 9.27 \times 10^{-21}\ \text{erg G sec}^{-1}\\
H_o &= 3400\ \text{G for X-band}\\
g_{zz} &= 2.0027\\
g_{xx} &= 2.0089\\
g_{yy} &= 2.0061\\
T_{\parallel} &= 87\ \text{MHz}\\
T_{\perp} &= 14\ \text{MHz}
\end{aligned}
$$

then,

$$
\begin{aligned}
b &= 3.06 \times 10^{8}\ \text{rad sec}^{-1}\\
\Delta\gamma &= 4.22 \times 10^{4}\ \text{rad sec}^{-1}\ \text{G}^{-1}
\end{aligned}
$$

and we calculate

$$\tau_c = 6.51 \times 10^{-10}\,\Delta H(O)\left[\sqrt{\frac{h(O)}{h(-1)}} - \sqrt{\frac{h(O)}{h(1)}}\right] \text{sec} \qquad (11)$$

and

$$\tau_c = 6.51 \times 10^{-10}\,\Delta H(O)\left[\sqrt{\frac{h(O)}{h(-1)}} + \sqrt{\frac{h(O)}{h(1)}} - 2\right] \text{sec} \qquad (12)$$

For Spectrum 11b (see Figure 11b) the peak height ratios are 1.29 and 1.04 for

$$\sqrt{\frac{h(O)}{h(-1)}}$$

and

$$\sqrt{\frac{h(O)}{h(1)}}$$

respectively, $\Delta H = 1.22$ G, and we obtain $\tau_c = 2 \times 10^{-10}$ sec/rad from Equation 11, and 2.6×10^{-10} sec/rad from the quadratic term (Equation 12). The small discrepancy in the two terms has never been totally explained; however, Hoffman et al.[17] noted that the quadratic term (Equation 12) was more useful by its insensitivity to microwave power saturation effects. If we carry out the same calculations for Figure 11a, we obtain 2.5×10^{-11} sec/rad for either term.

A crude yet "order of magnitude" estimate of the rotational correlation time is derived from the Stokes-Einstein or Debye diffusion model for a spherical molecule of radius R, $\tau_R = 4\pi\eta R^3/3kT$ where η is the viscosity, k is Boltzmann's constant, and T is the absolute temperature. For spectrum 11a we would calculate for R = 5 Å at $\eta = 0.871 \times 10^{-2}$ poise, 299°K, a τ_R of 11×10^{-11} sec/rad, about fivefold longer than that calculated by ESR. A value of R = 3.1 Å would give perfect agreement, $\tau_R = 2.6 \times 10^{-11}$ sec/rad.

2. *Intermediate and Slow Motion*

This time range $\tau_R = 10^{-8}$ to 10^{-10} sec has been treated in some detail by Freed who has developed several computer simulation models for both isotropic and anisotropic tumbling.[18] A commercially available software is called EPRCAL®, which is suited for some Nicolet computer systems. Figure 12 gives some examples of simulated spectra for isotropic motion in the intermediate to slow motion range. For the slower part of this tumbling range, graphical methods of estimating τ_R in the 10^{-8} to 10^{-9} sec range have been tested by Shimshick and McConnell as outlined below.[20] A labeled protein of approximately <50,000 daltons of an intrinsic tumbling rate 10^8 to 10^9 sec^{-1} will yield a "strongly immobilized" or powder type spectrum if the nitroxide is rigid with respect to the macromolecule. By increasing the solution viscosity with, e.g., sucrose, the hyperfine extrema separation is plotted against viscosity and extrapolated to the infinite viscosity limit. Actually, one plots the change in hyperfine splitting, $\Delta H(\tau_2)$, where $\Delta H(\tau_2) = H_\infty - H_\eta$, and H_η and H_∞ are the position of the high field component of the spectrum at viscosity η and at infinite viscosity. From plots of $\Delta H(\tau_2)$ vs. $(\tau/\eta)^{2/3}$, the extrapolated infinite viscosity value of $\Delta H(\tau_2)$ and fitting to theoretical plots of $\Delta H(\tau_2)$ vs. $(\tau/\eta)^{2/3}$ the appropriate τ_2 is determined.

3. *Very Slow Motion*

Where the macromolecular tumbling times (and presumably that of the spin label) are in the range 10^{-7} to 10^{-3} sec, a conventional X-band ESR instrument will show a strongly immobilized powder spectrum since any tumbling time below approximately 10^{-8} sec is beyond the normal time detection "window" of the ESR experiment. However, now with the aid of a second modulation frequency (a second harmonic of a base modulation frequency) in the instrument, the very slow tumbling range (10^8 to 10^3 sec^{-1}) may be measured by a technique known as saturation transfer spectroscopy (ST-EPR).[21] In its simplest use, Hyde and Dalton[21] describe "second harmonic out-of-phase detection" as a diagnostic of tumbling motion in this range. While the student is referred to a more detailed description of the theory, methodology, and application of the method, Figures 13 and 14 depict an example with spin labeled bovine serum albumin, which compares the conventional (Figure 13) and ST-EPR spectra (Figure 14) for examples over a range of tumbling rates.[22]

B. Polarity

The nitroxide group is also sensitive to the microscopic "solvent" polarity by virtue of the polarizability of the formal change distribution in the nitroxide group:

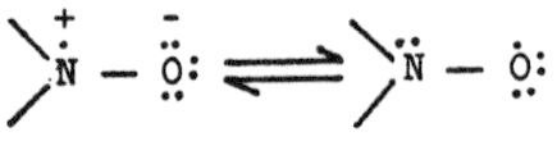

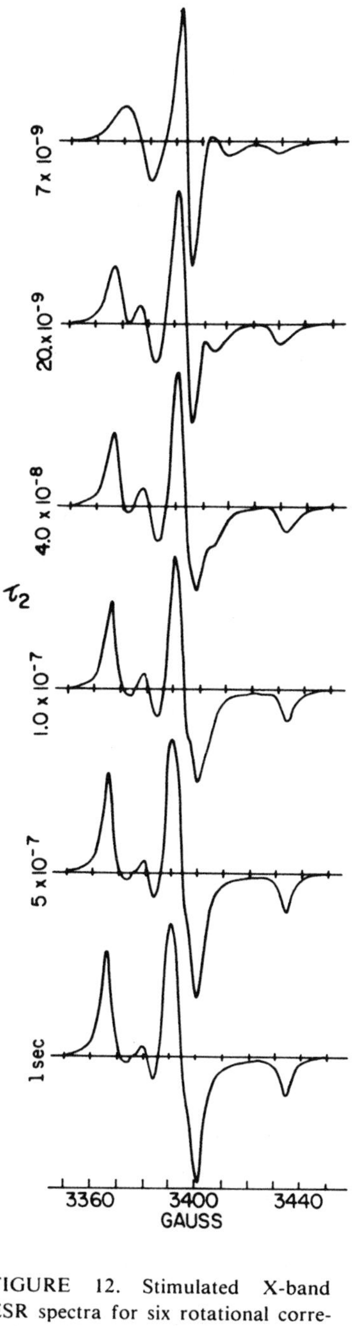

FIGURE 12. Stimulated X-band ESR spectra for six rotational correlation times, τ_2, in the intermediate to slow tumbling range. The parameters used in this diffusion-coupled Block equations calculation were g_{xx} = 2.00901, g_{yy} = 2.00601, g_{zz} = 2.00241, T_{xx} = 6.8 G, T_{yy} = 6.2, T_{zz} = 34.3 G, and T_2 (electron) = 2.4 × 10^{-8} sec. (From Berliner, L. J., Ed., *Spin Labeling: Theory and Applications,* Academic Press, New York, 1976, Appendix I. With permission.)

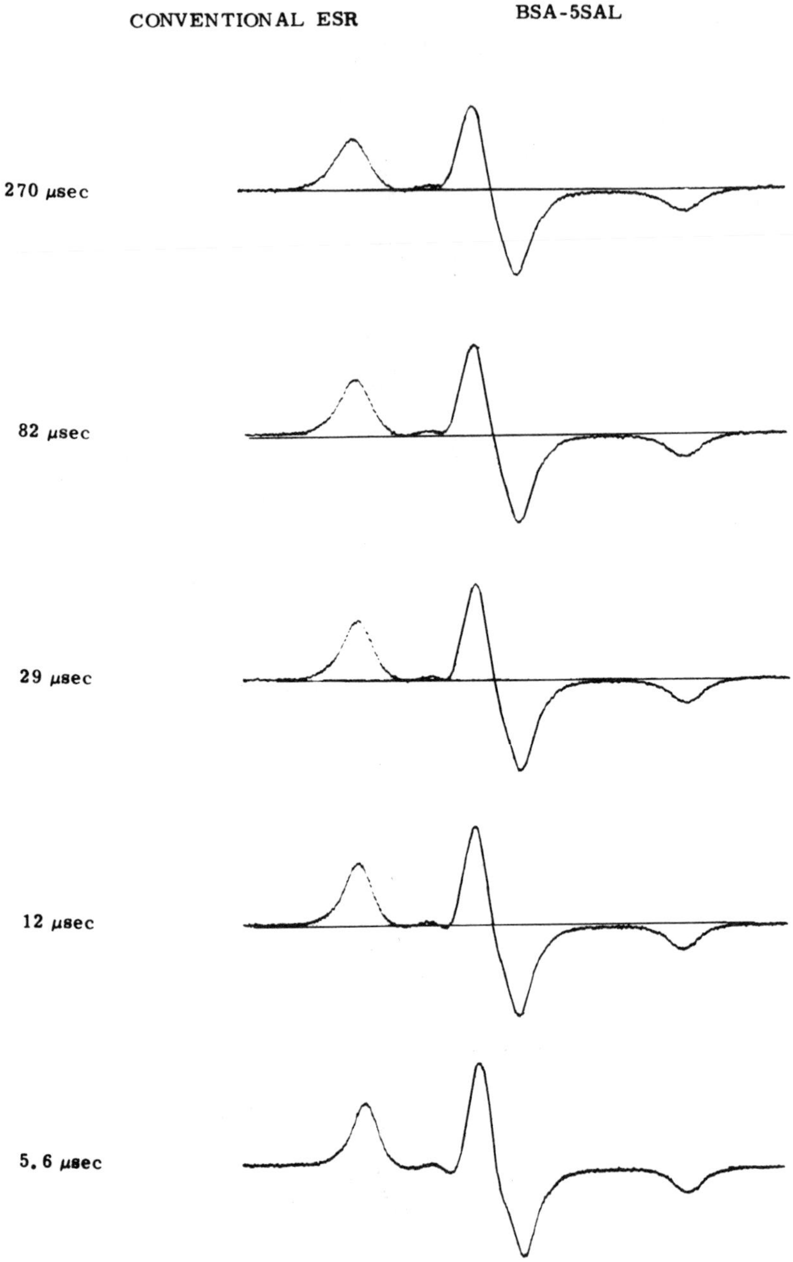

FIGURE 13. Conventional ESR spectra in the very slow time range. The spectra shown are for a 2:1 molar complex of 5-doxyl stearic acid; bovine serum albumin in 53:47 (mol/mol) glycerol/water over the temperature range 20° to −20°C. The rotational correlation times were calculated using a macromolecular radius of 35Å at the known glycerol viscosities. (Courtesy of Dr. A. Kusumi, Kyoto University.)

The species on the left, which would be stabilized in a highly polar environment, leaves a higher free electron density in the nitrogen $2p_z$ orbital than the species on the right. This results in a larger isotropic splitting, due to the electron-nuclear hyperfine coupling, for the species on the left than on the right. Thus we observe that the hyper-

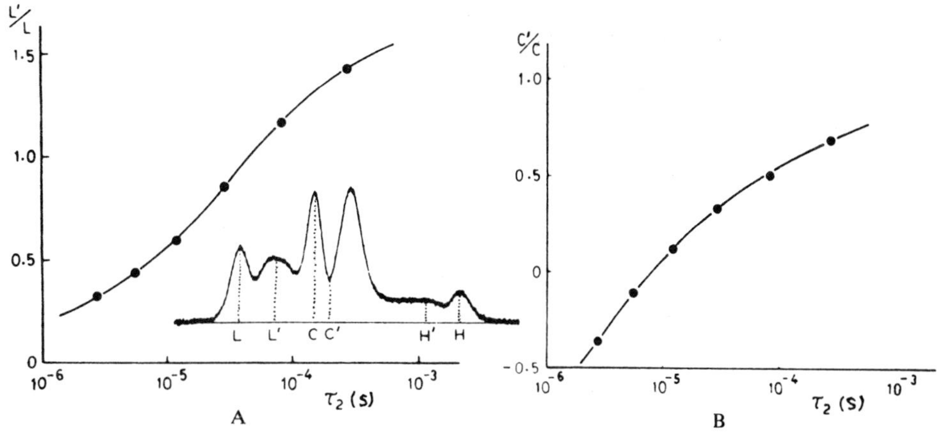

FIGURE 14. (A) Saturation transfer spectra, ESR spectra of the same samples in Figure 13 observed by second harmonic out-of-phase absorption. (B) Graphical dependence of the rotational correlation time, τ_2, on the indicated peak height ratios of the saturation transfer (ST-EPR) spectra in (a). Note the contrast in sensitivity to the very slow motion time range (10^{-6} to 10^{-3} sec) between conventional Figure 13 and ST-EPR. (From Dr. A. Kusumi, Kyoto University; and Kusumi, A. S., Ohnishi, T., Yoshizawa, T., and Ito, K., *Biochim. Biophys. Acta*, 507, 539, 1978. With permission.)

fine splitting decreases from 17.5 to 15.1 G over a solvent range from 10 *M* LiCl to hexane. For reference, Table 3 gives both hyperfine coupling constant, T_o, and *g*-factor, g_o, in 33 different solvents. As a probe of the actual polarity of the binding site we must be able to measure some absolute component of the hyperfine splitting, for example T_{zz} (or A_{zz}), the maximum component of the anisotropic hyperfine interaction which is related to the extrema separation in a ("strongly immobilized") powder spectrum as $2T_{||} = 2T_{zz}$ ($= 2A_{zz}$). There is, in general, a direct relationship between the polarity effect on this parameter and on the isotropic hyperfine splitting A_o ($= T_o$) as Griffith et al. found for fatty acid spin probes frozen (liquid nitrogen) in varying homogeneous solvent environments.[23] Figure 15 shows a plot of A_o vs. A_{max}^{-196} (liquid N_2) and with the corresponding results included for a membrane sample. For strongly immobilized nitroxides, Lassman et al.[24] had dfined a hydrophobicity parameter, h, where

$$h = \frac{A_{zz}(H_2O) - A_{zz}\ (\text{sample})}{A_{zz}(H_2O) - A_{zz}\ (\text{decalin})} \qquad (13)$$

using water and decalin as the extremes of polarity. The value of h increases from zero for an aqueous environment to above unity for environments even less polar than decalin. It should be noted that while these comparisons above are related to what we may term as homogeneous isotropic standard solvents, the possibilities of asymmetric effects, e.g., hydrogen bonding between>N–O and some protein side chain (e.g., imidazole) are conceivable. While rarely observed, Humphries and McConnell reported a hyperfine splitting of 79 G for a spin labeled antigen bound to its specific antibody, the>N–O group polarized by a strong hydrogen bond to the protein (see Figure 16).[25]

C. Quantitative Analysis — "Spin Count"

It is always necessary to determine the labeling stoichiometry in any chemical modification of a protein or enzyme. This is easily accomplished with spin labeled macromolecules by either of the following approaches. The spectrum of interest, or a specific

Table 3
ISOTROPIC ESR PARAMETERS FOR DI-*t*-BUTYL NITROXIDE[a,b]

No.	Solvent	T_o	g_o
1	Hexane	15.10	2.0061
2	Heptane-pentane (1:1)[c]	15.13	2.0061
3	2-Hexene	15.17	2.0061
4	1,5-Hexadiene	15.30	2.0061
5	Di-*n*-propylamine	15.32	2.0061
6	Piperidine	15.40	2.0061
7	*n*-Butylamine	15.41	2.0060
8	Methyl propionate	15.45	2.0061
9	Ethyl acetate	15.45	2.0061
10	Isopropylamine	15.45	2.0060
11	2-Butanone	15.49	2.0060
12	Acetone	15.52	2.0061
13	Ethyl acetate saturated with water	15.59	2.0060
14	*N,N*-Dimethylformamide	15.63	2.0060
15	EPA[d] (5:5:2)[c]	15.63	2.0060
16	Acetonitrile	15.68	2.0060
17	Dimethylsulfoxide	15.74	2.0059
18	*N*-Methylpropionamide	15.76	2.0059
19	2-Methyl-2-butanol	15.78	2.0059
20	EPA[d](5:5:10)[c]	15.87	2.0060
21	1-Decanol	15.87	2.0059
22	1-Octanol	15.89	2.0059
23	*N*-methylformamide	15.91	2.0059
24	2-Propanol	15.94	2.0059
25	1-Hexanol	15.97	2.0059
26	1-Propanol	16.05	2.0059
27	Ethanol	16.06	2.0058
29	Formamide	16.33	2.0058
30	1,2-Ethanediol	16.40	2.0058
31	Ethanol-water (1:1)	16.69	2.0057
32	Water	17.16	2.0056
33	10 *M* LiCl aqueous solution	17.52	2.0056

[a] From Reference 23.
[b] All data measured at room temperature (23° to 24°C). Estimated uncertainties are ± 0.02 G and ± 0.0001 for T_o and g_o, respectively, relative to the standard dilute aqueous solution of di-*t*-butyl nitroxide for which T_o = 17.16 and g_o = 2.0056.
[c] By volume.
[d] EPA designates a mixture of ethyl ethy (diethyl ether), isopentane (2-methylbutane), and alcohol (ethanol).

spectral component of a multicomponent spectrum comprising more than one motional state, may be digitized, double integrated, and compared with that for a known concentration of the same or similar nitroxide. For a single component spectrum, this is easily accomplished with most time averagers. If data collection or computer hardware are not available we use the "poor man's spin count technique". The spin label is chemically removed (e.g., hydrolyzed) to yield the "free", rapidly tumbling nitroxide moiety, which is compared with a standard concentration of (preferably) the identical nitroxide species. If the standard and sample are in fact identical, they may simply be related by their peak heights. Since the area under a first derivative ESR spectrum is related to the spin concentration as the (linewidth)^2 × (peak height), the linewidth

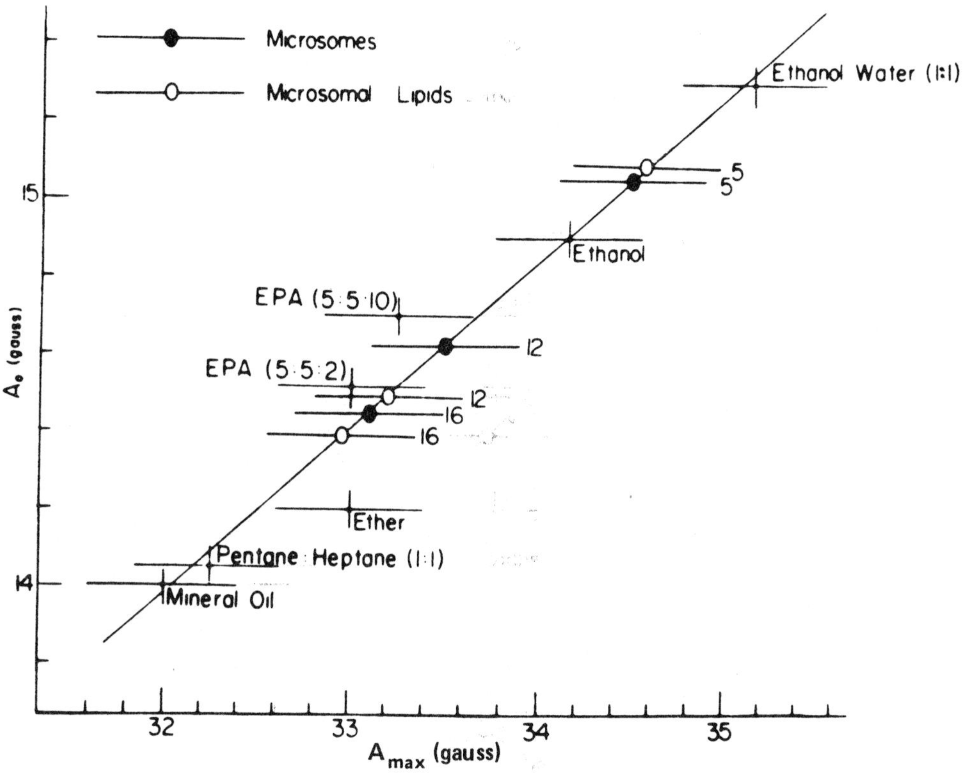

FIGURE 15. The solvent dependence of the hyperfine parameters $A_o(=T_o)$ and $A_{max}^{-196°}$ $(=T_{max}^{-196°})$. Three fatty acid spin labels 5-, 12- and 16-doxyl stearic acids or their methyl esters were measured in the solvents indicated at −196°C. EPA, ethylether:isopentane:ethanol. (From Griffith, O. H., Dehlinger, P. J., and Van, S. P., *J. Membr. Biol.*, 15, 159, 1974. With permission.)

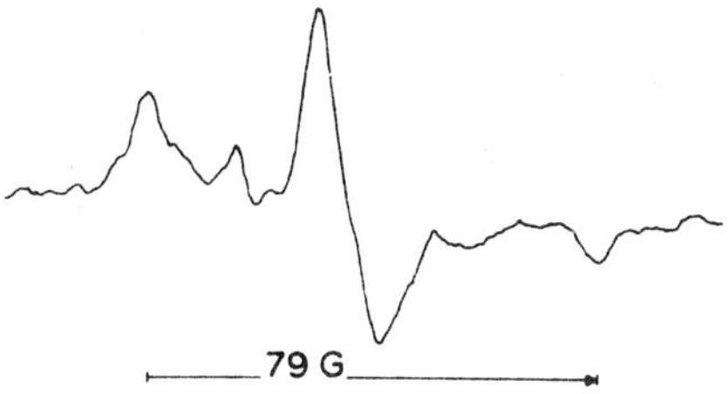

FIGURE 16. ESR spectrum of a peperidine nitroxide specific antibody (rabbit) complexed with a nitroxide dimyristoylphosphatidylcholine analog after incubation at 37° for 2 hr. The observed outer hyperfine extrema splitting of 79 G $(=2T_{zz})$ is the largest observed to date for nitroxide spin labels. (From Humphries, G. M. K. and McConnell, H. M., *Biophys. J.*, 16, 275, 1976. With permission.)

term will be identical for identical nitroxide species (sample vs. reference standard) and therefore "cancel out". Where the standard nitroxide is different (e.g., lower MW), it is best to measure linewidth as well. Thus the concentration (spin count) of the sample C_s is related to that of the reference, C_{ref} by

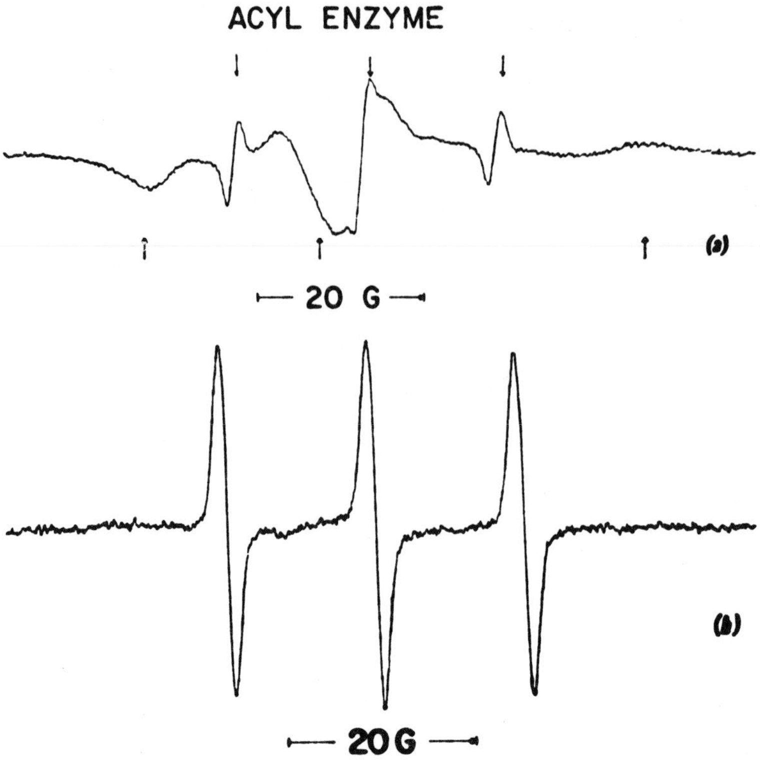

FIGURE 17. (a) ESR spectrum of α-chymotrypsin acylated with 1-carboxyl-2,2,5,5-tetramethyl pyrrolidine-l-oxide at pH 3.5. The broad resonance lines (↑) represents the immobilized nitroxide moiety while the narrow line components (↓) are due to a small amount of free label which is hydrolyzed off the enzyme. (b) After hydrolyzing the label completely free from a known concentration of enzyme. While in this specific case the enzyme will catalyze the hydrolysis (deacylation) at pH 7.0, an "irreversibly" bound label requires harsher conditions (1 *M* NaOH). This free line spectrum may be compared to a standard solution of the same compound to calculate the concentration of label originally bound to the enzyme in Figure 17a. (From Berliner, L. J. and McConnell, H. M., *Proc. Natl. Acad. Sci. U.S.A.*, 55, 708, 1966. With permission.)

$$C_s = C_{ref} \frac{(w_s)^2 \, h_s}{(w_{ref})^2 \, h_{ref}} \qquad (14)$$

where w and h denote the linewidth and peak heights, respectively.

Figure 17 gives an example of spectra of a spin labeled protein and the respective (hydrolyzed off) freed nitroxide moiety. The lower spectrum is then compared with a reference of known concentration to obtain, finally, the labeling stoichiometry. Some caution must be taken when removing the nitroxide by methods such as acid or base hydrolysis. Most nitroxide species will slowly disproportionate in acid or alkali at greater than 1 *M* concentrations. A control experiment with a reference standard in the same solvent will aid in optimizing treatment conditions and correcting for nitroxide destruction. This latter "chemical" method for obtaining the spin count is quite sensitive (10^{-6} to 10^{-8} *M*) and may be measured on samples as small as 20 $\mu\ell$ in tiny capillary sample cells.

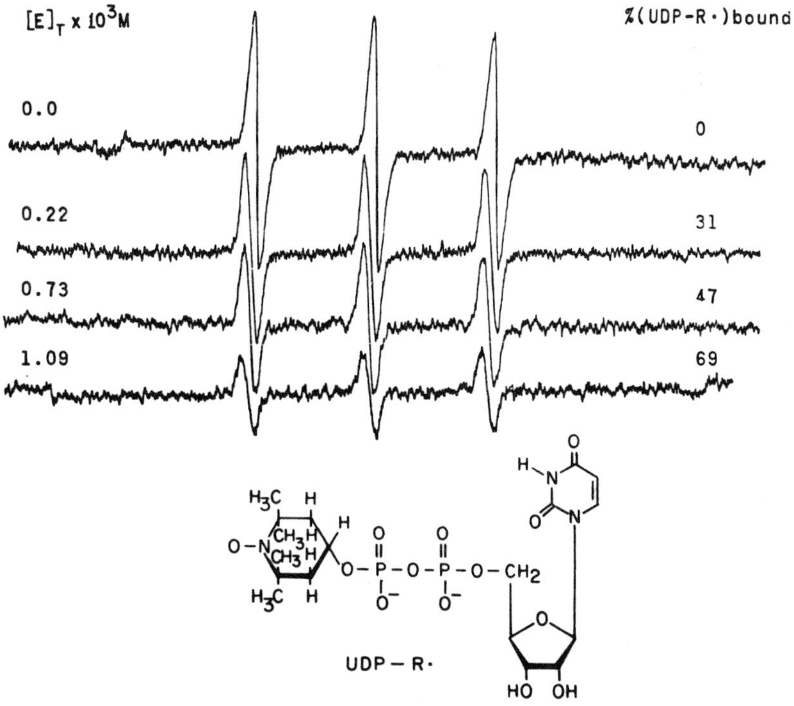

FIGURE 18. ESR spectra of UDP-R in the absence and presence of galactosyl transferase, 26 ± 1°. The total UDP-R concentration was 1.39×10^{-5} *M* in 0.1 *M* *N*-methylmorpholine (pH 8.0) containing 10% (w/v) $(NH_4)_2SO_4$. No other substrates were present. The left column gives galactosyl transferase concentration while the right column shows the percent spin label bound. Receiver gain and modulation amplitudes were identical for all spectra. (Reprinted with permission from Berliner, L. J. and Wong, S. S., *Biochemistry,* 14, 4977, 1975. Copyright by the American Chemical Society.)

D. Biochemical Data

1. Kinetics

There are many potential approaches to monitoring catalytic or other rate processes by spin label ESR spectra where the reaction is accompanied by a change in the motional state of the nitroxide. This is most obvious where the nitroxide is itself a substrate or irreversible inhibitor that is tumbling rapidly as the free label in solution and is less mobile when covalently bound to the enzyme. Conversely, the rate of release (e.g., hydrolysis, conversion from enzyme bound intermediate to products) would be detected by the reverse change in motional state.

2. Dissociation Constants

The spectral changes that occur when a small nitroxide ligand binds (reversibly) to a macromolecule are again associated with a shift to slower tumbling times and consequently broader line shape spectra. In most cases with nitroxides, the bound spectral component is broadened sufficiently that the apparent peak height approaches a negligible value compared to that of an equivalent concentration of free label. If the label is strongly immobilized the bound "powderlike" spectrum is visually distinct from the free spectrum due to the differences in splitting of the low and high field lines (~64 G vs. 30 G). An example is shown in Figure 18 for a spin labeled inhibitor for bovine galactosyl transferase. The spin probe, UDP-R, shows a decrease in its "free" spec-

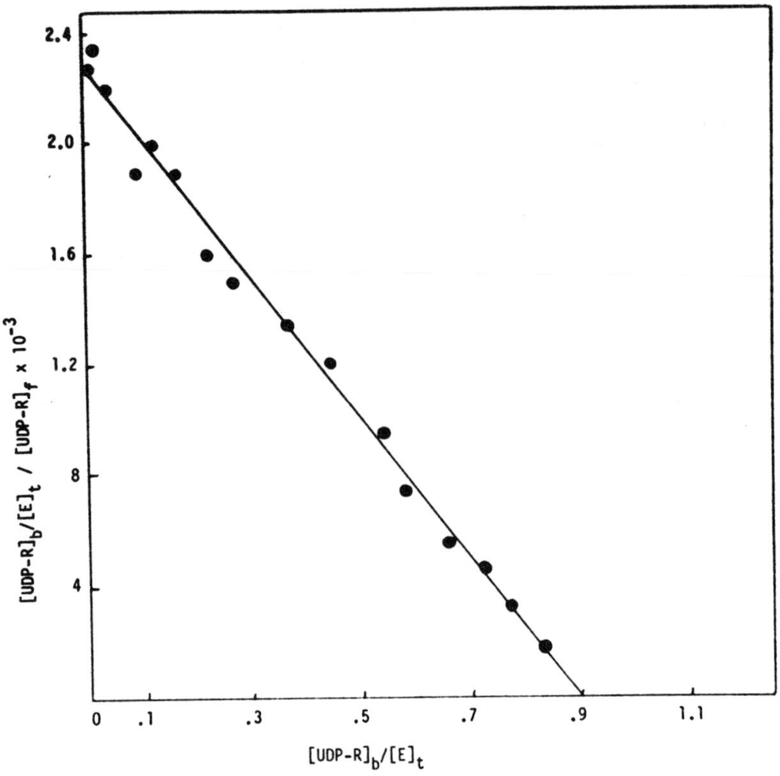

FIGURE 19. Scatchard plot of the ESR data for UDP-R binding to galactosyl transferase. Enzyme concentration was held constant at $8.4 \times 10^{-5} M$ while UDP-R concentration was varied from 7.8×10^{-5} to $5 \times 10^{-3} M$. All conditions were the same as in Figure 18. (Reprinted with permission from Berliner, L. J. and Wong, S. S., *Biochemistry*, 14, 4977, 1975. Copyright by the American Chemical Society.)

trum with increasing enzyme concentration. The fraction of UDP-R bound to the enzyme is given by

$$\frac{[h]_o - [h]_{enz}}{[h]_o} \tag{15}$$

where $[h]_o$ and $[h]_{enz}$ are the peak heights before and after addition of enzyme, respectively. These data may then be plotted to a Scatchard plot (Figure 19) to obtain both the dissociation constant(s) and binding stoichiometry. In the example shown here the dissociation constant obtained for the equimolar binding of UDP-R to enzyme was 0.40 ± 0.07 m*M*, which compared favorably with the kinetically determined competitive inhibition constant of 0.38 m*M* for this spin labeled inhibitor.[27] When the bound spectrum starts to contribute significantly to the free line peak height, i.e., where free and bound spectral components overlap, the bound contribution to the peak height must be corrected out of the calculation. While this correction is difficult to accomplish without computer graphics, it rarely introduces a severe error in these calculations until the bound label comprises approximately greater than 70% of the total spectrum.

3. *Structural Perturbations*

The most sensitive property of the spin label technique relative to protein structure-

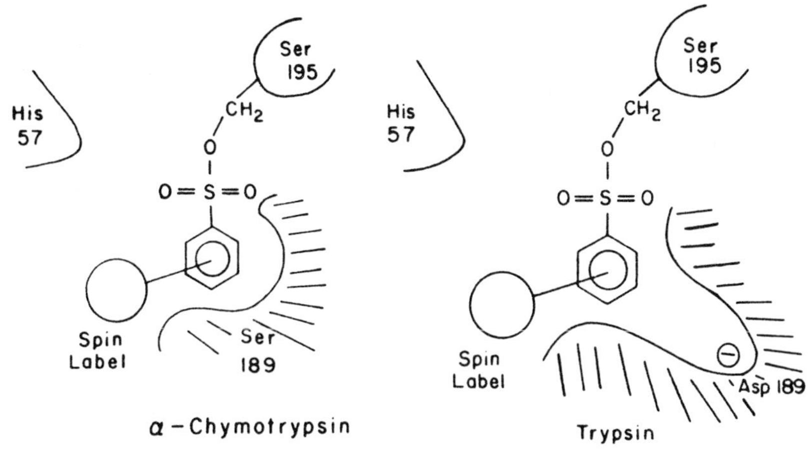

FIGURE 20. A model for the binding of spin labeled sulfonyl fluorides with α-chymotrypsin (left) and trypsin (right), respectively. The trypsin specificity pocket probably cannot accommodate any "tosyl" reagent which must bind outside this pocket. (From Berliner, L. J., *Methods Enzymol.*, 49G, 418, 1978. With permission.)

function studies is its ability to detect small (local) structural changes. Since the tumbling motion of the nitroxide group is influenced largely by its (noncovalent) interactions with the binding site environment, small changes in the macromolecular environment will likewise alter these latter interactions. Since we must always be cognizant of the fact that we are monitoring the label motion, and only indirectly the protein environment (or motion), we must also consider those cases where the label shifts its binding environment, upon, e.g., some extrinsic ligand binding nearby, without any concomitant change in enzyme structure (conformation). We class these two general cases as conformation changes where the protein structure (conformation) changes, and orientation shifts where the nitroxide shifts to a different environment, while no change occurs on the conformation of the surrounding protein structure.[28] For illustration, examples are given below that have been classified into each of the above types.

a. *Displacing Only the Spin Label — Orientation Shifts*

Figure 20 shows a schematic model summarizing a comparative spin label active site study on the two closely related serine proteases α-chymotrypsin and trypsin. The spin labels depicted were substituted fluorosulfonylphenyl compounds, which specifically inhibit the active serine 195 in each enzyme. In Figure 21 the extreme right spectra are those obtained for trypsin after reaction (and exhaustive dialysis) with each of the two spin labels shown. The extreme left spectra (Figure 21) were obtained with α-chymotrypsin. Since the phenyl moiety in the α-chymotrypsin case would tend to bind in or near the aromatic substrate specificity pocket as shown in Figure 20, the competitive inhibitor indole was added to the spin labeled α-chymotrypsin samples in order to "displace" the spin label by sterically competing for part of its binding locus. The center spectra of Figure 21 show the resultant spectrum after exposure to indole. This results from an orientation shift of the nitroxide from one binding environment (specificity pocket region) to another without any change in protein structure. Actually, the identity between the α-chymotrypsin indole spectra and trypsin spectra supported the fact that the general active site regions of each enzyme outside of their respective specificity pockets are almost identical from their respective X-ray crystal structures.[29]

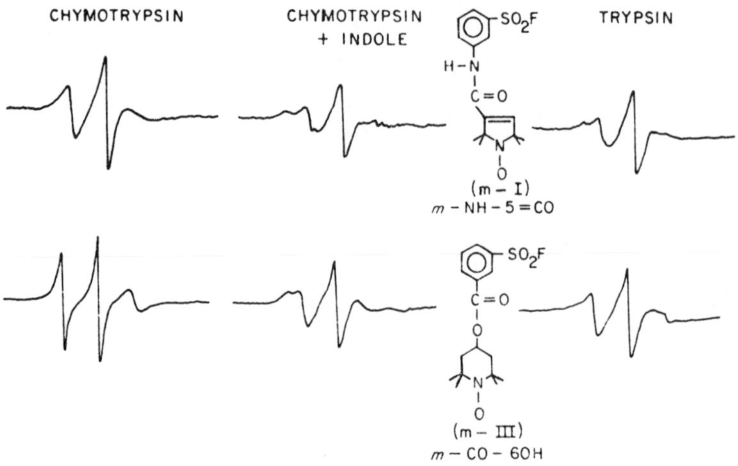

FIGURE 21. ESR spectra of spin labeled α-chymotrypsin at pH 3.5 (left); the latter in the presence of saturated indole (center); and the corresponding spin labeled trypsin, pH 3.5, 0.02 M $CaCl_2$. (From Berliner, L. J., *Methods Enzymol.*, 49G, 418, 1978. With permission.)

b. Conformation Changes — Allostery

As a striking example of a conformation change we describe another active site labeled α-chymotrypsin derivative, which is (reversibly) exposed to increasing concentrations of the denaturant urea. Figure 22 shows spectra for α-chymotrypsin labeled at the active site serine 195 with the phosphorofluoridate nitroxide

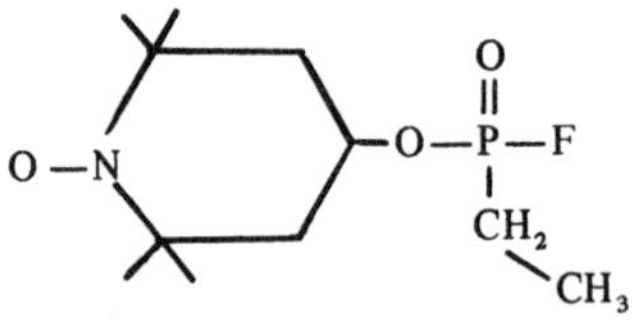

at 0, 4, and 8 M urea concentrations, respectively. This is one example of the detection of rather global conformational changes. On the other hand, an example of a more subtle nature is included in Figure 23 for human hemoglobin spin labeled at the Cys 93 of the two β-chains of the α(CN-met) tetramer with the iodoacetamide label.

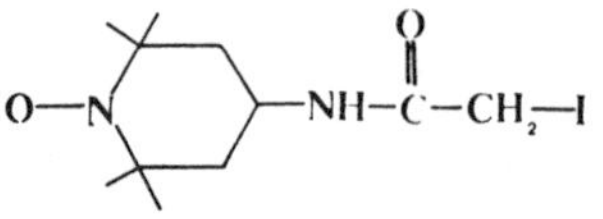

Upon oxygenation of the deoxy form the spectral changes depicted (Figure 23) reflect the changes in conformation mediated through the heme porphyrin to the distant Cys (β-93) via both intra- and intersubunit conformational changes within the hemoglobin tetramer.[31]

E. Orientation

1. General

The asymmetric line broadening and spectral changes that occur when nitroxide tumbling slows down are due to the contributions of the anisotropic electron-nuclear hy-

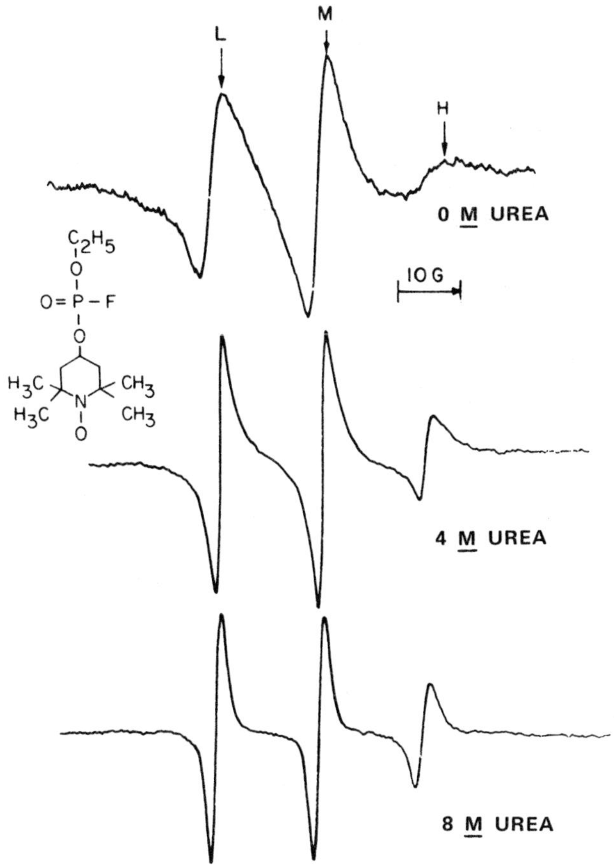

FIGURE 22. ESR spectra of spin labeled α-chymotrypsin at the urea concentrations shown (pH 2.0, 0.1 *M* NaCl). (Reprinted with permission from Berliner, L. J., *Biochemistry,* 11, 2921, 1972. Copyright by the American Chemical Society.)

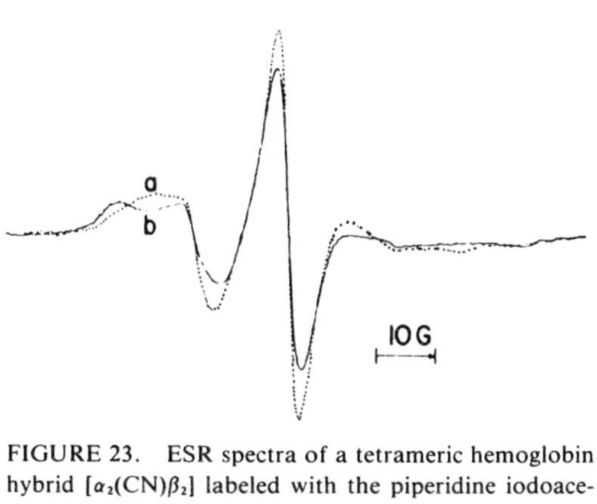

FIGURE 23. ESR spectra of a tetrameric hemoglobin hybrid [α_2(CN)β_2] labeled with the piperidine iodoacetamide spin label and with the α chains in the cyanomet form: (a) Deoxy β (b) Oxy β. Conditions were 0.05 *M* phosphate, pH 7.2, 25°C, 0.4% hemoglobin. (From Ogawa, S., McConnell, H. M., and Horwitz, A., *Proc. Natl. Acad. Sci. U.S.A.*, 61, 401, 1968. With permission.)

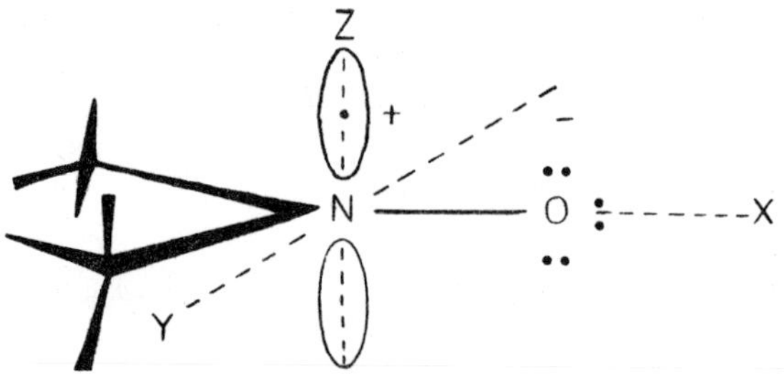

FIGURE 24. The geometry of the nitroxide hyperfine tensor with respect to the N-O moiety. The principal axis system x, y, z, formally describes both the anisotropic hyperfine and g-factor contributions.

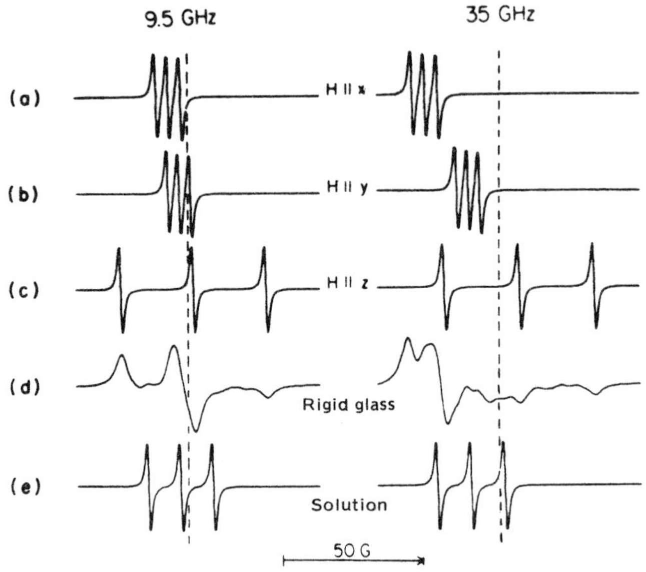

FIGURE 25. Calculated spectra for 2-doxyl propane with field orientation: (a) to (c) along each principal axis of a rigidly oriented nitroxide; (d) a randomly oriented collection of nitroxides — no motion; (e) that in (d) undergoing rapid isotropic motion. Parameters used were T_{xx} = 5.9 G, T_{yy} = 5.4 G, T_{zz} = 32.9 G, g_{xx} = 2.0088, g_{yy} = 2.0058, g_{zz} = 2.0022. (From Griffith, O. H. and Jost, P. C., *Spin Labeling: Theory and Applications*, Berliner, L. J., Ed., Academic Press, New York, 1976, chap. 12. With permission.)

perfine interaction between the free electron and the ^{14}N nucleus, as well as the anisotropy in the **g** factor. Figure 24 reminds us of the relationship between molecular axes in the nitroxide group and the principal axes of the hyperfine **g** tensor (interactions). If we could hold this nitroxide group at a fixed orientation in the magnetic field we would observe a small splitting (5 to 6G) along *x* or *y* while a very large splitting (30 to 34G) occurs along **z**; each direction is also associated with unique g-factors. Figure 25 shows calculated spectra for these special directions along with the rigid glass ("powder") spectrum (d), which is a general composite of all directions, and a rapidly tumbling solution spectrum, (e), for comparison. In any unique fixed orientation a nitroxide spectrum will have a

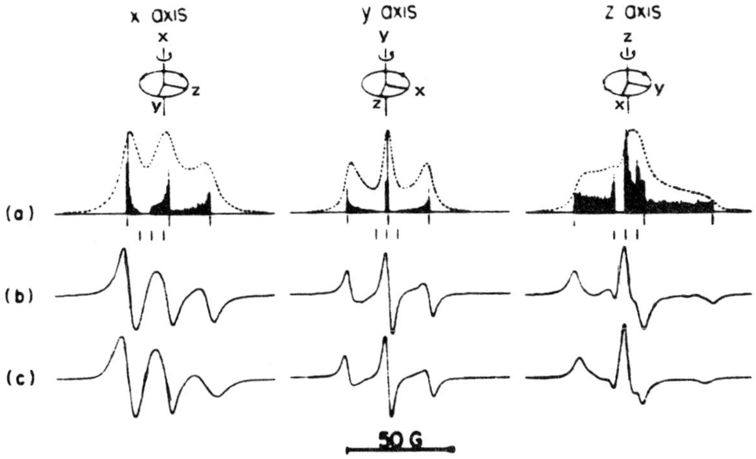

FIGURE 26. Theoretical (simulated) absorption (a) and first derivative (b) spectra for 2-doxyl propane using the parameters in Figure 25 and line widths of 5.0 G, 3.0 G and 3.8 G for x-, y-, and z-axis rotation, respectively. (c) Experimental examples of x-axis rotation (2,2,6,6-tetramethyl-4-piperidinol-l-oxyl in β-cyclodextrin); y-axis rotation (di-*t*-butylnitroxide in thiourea); and z-axis rotation (7-doxylstearate in β-cyclodextrin). (From Griffith, O. H. and Jost, P. C., *Spin Labeling: Theory and Applications*, Berliner, L. J., Ed., Academic Press, New York, 1976, chap. 12. With permission.)

$$T_{obs} = [T_{\parallel}^2 \cos^2\theta + T_{\perp}^2 \sin^2\theta]^{1/2} \qquad (16)$$

splitting where $T_{\parallel} = T_{zz}$, $T_{\perp} = T_{xx} = T_{yy}$, and θ is the angle between the external magnetic field direction and the *z* axis. A similar expression is used for the **g**-factor; however, while the hyperfine values T_{xx} and T_{yy} are usually identical (axial symmetry) this is not as valid an approximation for the **g**-factor ($g_{xx} \neq g_{yy}$),

$$g_{obs} = [g_{zz}^2 \cos^2\theta + g_{xx}^2 \sin^2\theta \cos^2\phi + g_{yy}^2 \sin^2\theta \sin^2\phi]^{1/2} \qquad (17)$$

where ϕ is the other Euler angle (found by projecting the magnetic field direction onto the *xy* plane and measuring from $\phi = 0$ at the *x*-axis).

The orientation in space of a nitroxide is theoretically related to its observed hyperfine splitting and **g**-factor. Thus we may learn about aspects of spin label orientation in a biological system from the hyperfine and **g**- parameters of its ESR spectra. In cases where the nitroxide is undergoing rotational tumbling motion, but about only one axis, the resultant anisotropic motion spectra will have hyperfine splittings that are the properly weighted average of principal hyperfine values along the two other axes. Examples for rapid rotation about the *x*, *y*, and *z* axes are shown in Figure 26.

2. Single Crystal Studies

We can take unique advantage of these anisotropy relations above in determining both precise spin label orientation and symmetry relationships in single crystals of spin labeled proteins. As an example we will take α-chymotrypsin labeled with the *S*(−) enantiomer of the spin labeled (substrate) analog whose solution ESR spectrum is shown first in Figure 27. The ESR spectrum is strongly immobilized implying the possibility (although not conclusively) that the label is rigidly held, perhaps uniquely, by

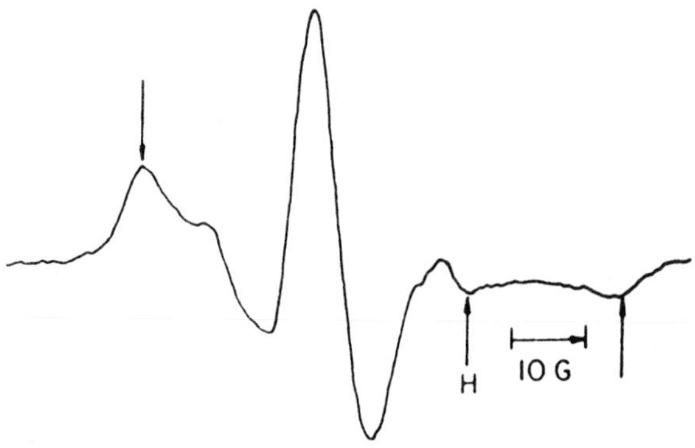

FIGURE 27. X-band ESR spectrum of α-chymotrypsin acylated with the S(−) nitrophenyl ester at pH 2.0, 0.1 *M* NaCl. The arrow (H) designates one of the lines of a very small "free" species spectrum superimposed with the major "strongly immobilized" labeled protein spectrum (designated by the arrows at the high and low field extrema). (From Berliner, L. J., *Methods Enzymol.*, 49G, 418, 1978. With permission.)

the enzyme at its active site.[26] If uniquely oriented, all labeled chymotrypsin molecules in the crystal unit cell will show the same orientation behavior in their ESR spectra. Actually, the asymmetric unit of the chymotrypsin unit cell (space group $P2_1$) is a dimer that is related by a symmetry (twofold rotation) axis coincidentally parallel with the crystal direction **a***. It can be shown that whenever the magnetic field direction is either parallel or perpendicular to a twofold rotation axis that relates two nitroxides, their spectra will be identical (superimposed) while at "general" magnetic field orientations there will be two spectra. Figure 28 shows spectra for a single crystal of α-chymotrypsin, labeled by soaking native crystals in a solution of the nitroxide, at one general and the three special orientations along the crystal axes **a***, **b**, and **c**, respectively. We may conclude from these data that the nitroxide is uniquely oriented in the enzyme active site, since the number of nitroxide orientations (sets of spectra) never exceeds the number of asymmetrically related subunits in the unit cell (two), and that a twofold rotation relation exists along **a***, **b**, or **c** since all of the spectra are superimposed on one spectrum.

From the hyperfine and **g**-values at these special orientations we may write Equation 16 as $T_a^* = [T_{\parallel}^2 \cos^2 \alpha + T_{\perp}^2 \sin^2 \alpha]^{1/2}$ where T_a^* is the observed splitting with the magnetic field along **a*** and α is the angle between the nitroxide *z*-axis and **a***. Similar expressions exist for T_b and T_c. Figure 29 shows the geometric relationships for the nitroxide axis system within the crystal axis system. Here the angle β is identical to the angle θ defined in the last section in Equation 16. We may solve for each angle, for example,

$$\sin^2 \alpha = \frac{T_{a^*}^2 - T_{\parallel}^2}{T_{\perp}^2 - T_{\parallel}^2} \qquad (18)$$

and so on to obtain the direction cosines of the angles α, β, and γ. For the example shown in Figure 28 the values for the angles specifying the directionality of the nitroxide *z*-axis were 60 ± 2°, 38 ± 2°, 69 ± 2° for α, β, and γ, respectively.[33]

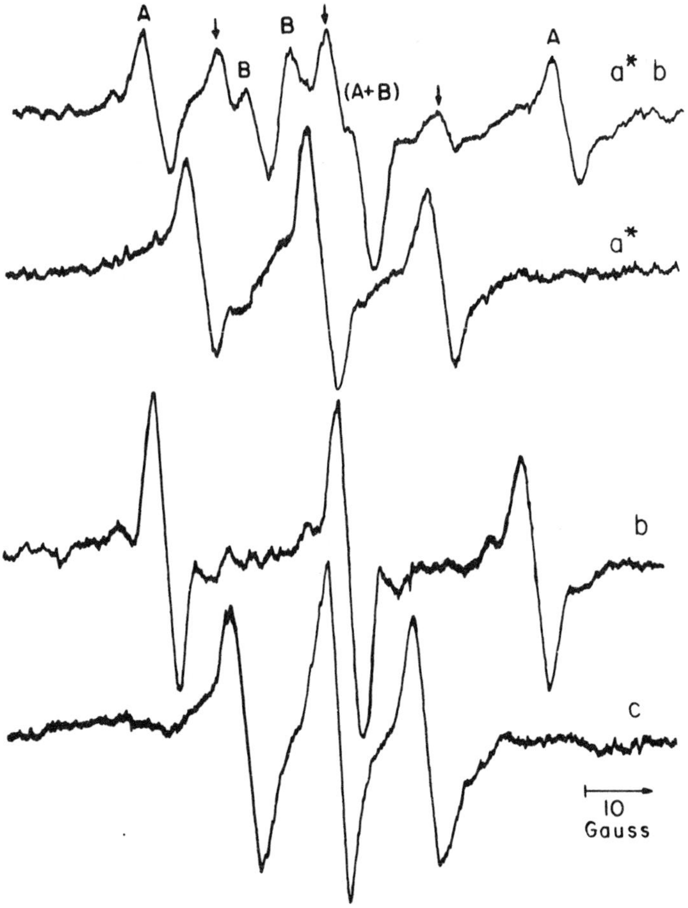

FIGURE 28. Special orientations of α-chymotrypsin single crystals spin labeled with the S(−) enantiomer of the nitrophenyl ester in Figure 27. Conditions were pH 4.2, 2.1 *M* $(NH_4)_2SO_4$. In the first three spectra, the crystal was oriented with the c axis always perpendicular to the external magnetic field. The crystal was rotated about c so that the observed spectra reflected the a*b plane. The top spectrum marked a*b was an orientation with the magnetic field direction in the a*b plane (⊥c) at 40° from a*. Note the two sets of spectra from the two nonequivalent nitroxide orientations marked A and B. The arrows (↓) denote free, unbound nitroxide included for reference. When the magnetic field was parallel to a*, b, or c, these special orientations were reflected in the unique spectra a*, b, c, below. (With permission from Bauer, R. S. and Berliner, L. J., *J. Mol. Biol.*, 128, 1, 1979. Copyright by Academic Press Inc.(London)Ltd.)

In order to define the complete orientation of the nitroxide group, however, the other principal axes, *x* and *y*, must be located from the **g**-factor variation. We know that $g_{xx} > g_{yy} > g_{zz}$ and that the *xy* plane is perpendicular to the *y*-axis. Along the **a*** axis $g_a^* = [g^2 \cos^2 \alpha + g^2 \sin^2 \alpha]^{1/2}$ where $g_{\|} = g_{zz}$ and $g_{\perp} = g_{xy}$ the composite **g**-factor at some projection in the *xy* plane. We may then note that $g_{xy} = [g_{xx}^2 \cos^2 \varepsilon + g_{yy}^2 \sin^2 \varepsilon]^{1/2}$ where ε is the angle between g_{xy} and the *x*-axis.

Thus from knowledge of T_{zz}, $T_{\perp}$, g_{xx}, g_{yy}, and g_{zz} and determination of α, β, and ε, we may derive that nitroxide group orientation with respect to host (protein) crystal structure. In the example used here, since the enzyme structure was known and the

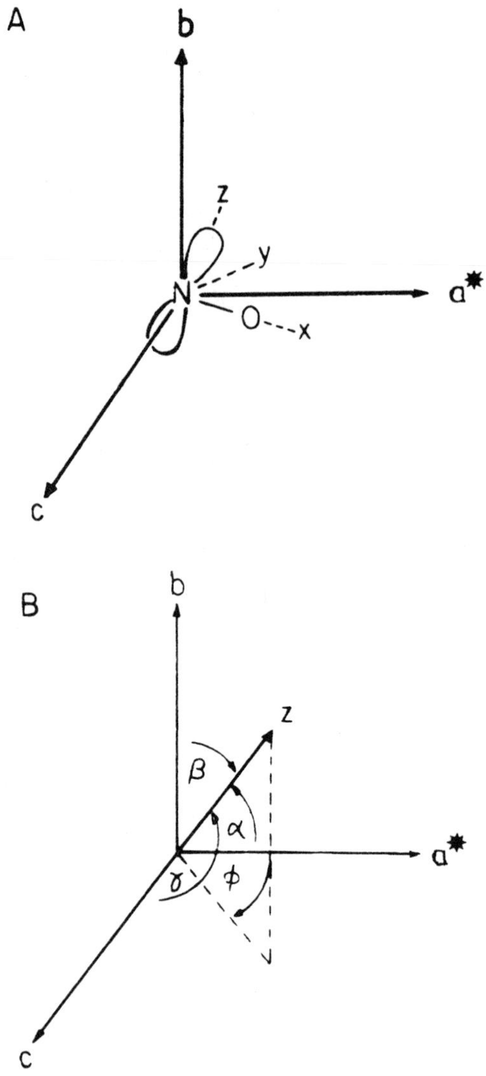

FIGURE 29. Angular relationship for a nitroxide group oriented in a crystallographic axis reference coordinate system.

molecular structure of the spin label was also known, it was possible to exactly place the nitroxide molecule at the α-chymotrypsin active site with the orientation found from the ESR experiments. A structural diagram of the α-chymotrypsin active site region and covalently bound label are shown in Figure 30. The structural information derived from this work also sheds more light on mechanistic hypotheses regarding substrate orientation and the efficiency of its subsequent catalyzed hydrolysis. Thus we see above that the ESR spin label technique accomplished effectively a crystallographic difference mapping of an enzyme-substrate complex!

F. Intramolecular Distances

While the basic ESR observation of the nitroxide label in a biological system yields much information already, there are in fact other interactions that take place over

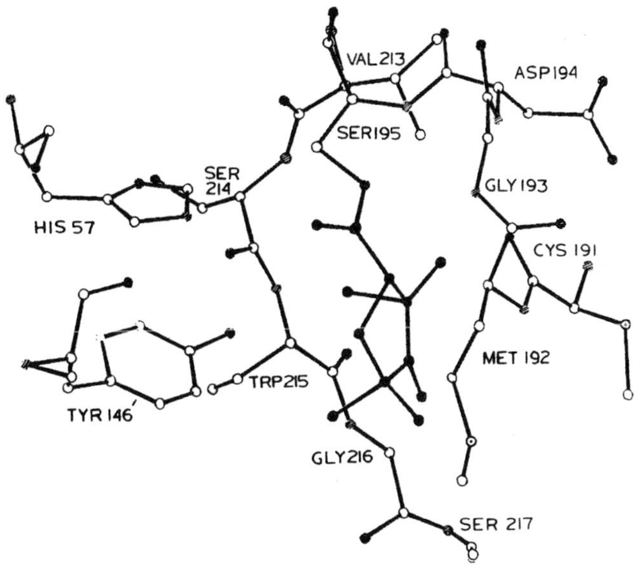

FIGURE 30. ORTEP diagrams of the α-chymotrypsin active site labeled with the S(−) acyl nitroxide in Figures 27 and 28. The nitroxide structure is shown entirely in block except for the N-O nitrogen. Symbols: O, C; ⊖, N; ●, O, ⊙, S. (With permission from Bauer, R. S. and Berliner, L. J., *J. Mol. Biol.*, 128, 1, 1979. Copyright by Academic Press Inc.(London)Ltd.)

intramacromolecular distances (5 to 25 Å) between the free electron and other nuclei and between the nitroxide radical and other paramagnets.

1. Spin-Nuclear Interactions

The spin lattice relaxation time of a nucleus, T_1, is normally dominated by (magnetic) dipole-dipole interactions with surrounding (usually nuclei) magnetic dipoles. Since the electron magnetic moment exceeds that for a proton by 700-fold, a nearby paramagnet will have a profound effect on the nucleus of interest. A dominant factor on the extent of "relaxation enhancement" of the nucleus of interest by the bound paramagnet is a correlation time, τ_c, which is usually determined by the rotational tumbling time, τ_r, of the macromolecular complex. The scope of types of experiments with proteins is depicted in Figure 31.

The appropriate form of the Solomon-Bloembergen equations that describes spin label induced nuclear relaxation is shown below, where $1/T_1$ and $1/T_2$ are the paramagnetic contributions to the longitudinal and transverse nuclear relaxation rates, where γ_I is the magnetogyric ratio, β is the Bohr magneton,

$$\frac{1}{T_{1M}} = \frac{2}{15} S(S+1) \frac{\gamma_I^2 g^2 \beta^2}{r^6} \left[\frac{3\tau_{c_1}}{(1+\omega_I^2 \tau_{c_1}^2)} + \frac{7\tau_{c_2}}{(1+\omega_S^2 \tau_{c_2}^2)} \right] \tag{19}$$

$$\frac{1}{T_{2M}} = \frac{1}{15} S(S+1) \frac{\gamma_I^2 g^2 \beta^2}{r^6} \left[4\tau_{c_1} + \frac{3\tau_{c_1}^1}{(1+\omega_I^2 \tau_{c_1}^2)} + \frac{13\tau_{c_2}}{(1+\omega_S^2 \tau_{c_2}^2)} \right] \tag{20}$$

g is the electron **g** factor, r is the electron-nuclear distance, S is the nuclear spin quantum number, ω_I and ω_S are the nuclear and electron Larmor precession frequencies,

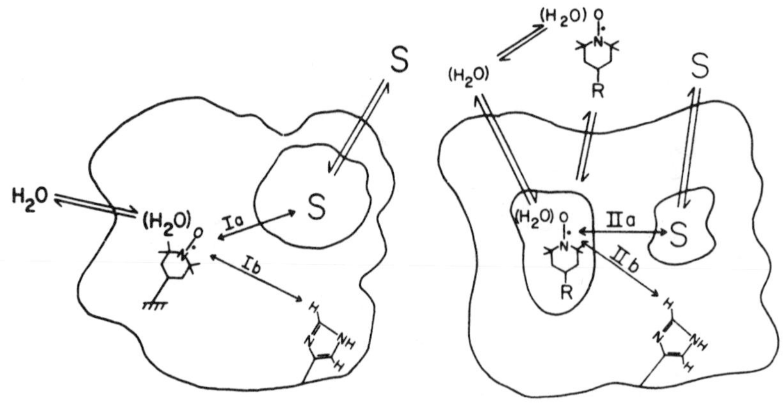

FIGURE 31. Possible NMR experiments using a spin labeled enzyme. In the left example, the covalently labeled piperidine nitroxide induces relaxation in the bound substrate (Ia), the C-2 proton of the histidine (Ib), and the exchanging water molecules. In the example on the right the piperidine nitroxide spin probe reversible binds at one site and induces relaxation similarly to substrate (IIa), histidine (IIb), and water molecules. (From Krugh, T. R., *Spin Labeling: Theory and Applications*, Berliner, L. J., Ed., Academic Press, New York, 1976, chap. 9. With permission.)

Table 4
CONSTANTS TO BE USED WITH EQUATIONS 21 and 22

Nuclei observed	Frequency (MHz)	C_{T1}	C_{T2}	ω_I ($\times 10^{-8}$) (rad sec^{-1})	ω_S($\times 10^{-11}$) (rad sec^{-1})
Protons	220	540	480	13.8	9.12
Protons	100	540	480	6.28	4.14
Protons	60	540	480	3.77	2.49
Fluorine-19	94.08	529	471	5.91	4.14
Phosphorus-31	40.5	400	356	2.54	4.14
Carbon-13	25.15	341	304	1.58	4.14

respectively, and τ_{c1} and τ_{c2} are the respective correlation times for the spin-lattice (longitudinal) and spin-spin (transverse) relaxation processes. For the final distance calculation the solution for r may be simplified to

$$r = C_{T_1} [f(\tau_c) T_{1M}]^{1/6} \tag{21}$$

$$r = C_{T_2} [f'(\tau_c) T_{2M}]^{1/6} \tag{22}$$

when C_{T1} and C_{T2} are collections of constants and where f (τ_c) and f$'(\tau_c)$ are the bracketed expressions comprising τ_{c1} and τ_{c2} in Equations 19 and 20, above.

For a particular nucleus at a specified frequency we already know C_{T1}, C_{T2}, ω_I, and ω_S as tabulated in Table 4; thus knowledge of e.g., $1/T_{1M}$ and τ_c for the system of interest allows calculation of an intramolecular distance in an enzyme in solution. While a sample calculation for r would be instructive, especially since r is much less sensitive to experimental error, as a function to the 1/6 power, r is e.g., only 12% off for a twofold error in τ_c. However the calculation (estimation) of τ_c requires several experimental measurements and crosschecks to ensure its accuracy. Second, it must be absolutely determined that no more than one paramagnetic species could conceivably bind to the protein under the conditions of the experiment, otherwise shorter distances (via additional paramagnetic relaxation) will be calculated. The methodology relating to the NMR measurements of nuclear relaxation rates and the calculation of τ_c are beyond the scope of this chapter, but may be found in other chapters and specific articles relating to these types of measurements.[34]

2. Spin-Spin (Nitroxide-Nitroxide) Interactions

In analogy to the interaction between a paramagnet and a nucleus, similar interactions occur between two paramagnets. Two general cases are important here: [1] paramagnet(metal)-nitroxide interactions, and [2] nitroxide-nitroxide interactions. The first case is particularly applicable to metalloenzymes or enzymes utilizing metal ion cofactors in substrate catalysis, where a paramagnetic ion [e.g., Mn(II)] may be substituted for the natural metal, [e.g., Mg(II) or Zn(II)]. The distance range over which a pronounced effect on the nitroxide ESR spectral line shape occurs is 10 to 25 Å, while nitroxide-nitroxide interactions are detectable up to 10 to 12 Å distances.

Addressing first the case of a paramagnetic metal ion, distances between the metal and nitroxide group are most accurately measured when both species are rigidly bound, that is, where their reorientation time is long compared to the electron relaxation time. As exemplified in Figure 32, the observed spectral change in the nitroxide spectrum upon metal ion binding (or upon substituting a paramagnetic metal for a diamagnetic ion) is an apparent "quenching" of the spectrum brought about by a very large broadening of the spectral line shape. If a line is broadened to infinity, the peak height approaches zero. That fraction of the total spectral intensity that decreases is directly related to those nitroxide spins that are in an orientation for optimal dipole-dipole broadening by the paramagnetic metal. The equation that describes this broadening, δH is expressed as

$$\delta H = (g\beta\mu^2 \tau / r^6 \hbar)(1 - 3\cos^2\theta_{R'})^2 + \delta H_o \qquad (23)$$

where g, β, and $\hbar$ are physical constants; μ is the magnetic moment of the (metal ion) dipole; τ is the correlation time for this interaction; r is the distance between the nitroxide group and the metal; θ_R is the angle between the magnetic field and the electron-metal vector; and δH_o is the residual (intrinsic) nitroxide linewidth. The correlation time τ is that of the electron spin relaxation time of the metal ion, and the rotational correlation time of the macromolecular complex is assumed to be significantly slower (that is the spins are assumed to be relatively rigid). If there is nitroxide motion, both the uncertainty of its position and a reduction of the average dipolar interaction with the metal ion will occur, consequently increasing the value of r, which is calculated. Figure 33 shows a theoretical plot that enabled us to "read off" the distance r for the correlation time and observed signal height reduction for a Mn(II)-nitroxide pair on a macromolecule. Note also the sensitivity to both correlation time and spectral reduction. Note again here that a reasonably accurate estimate of the correlation time is crucial to the accuracy of the calculated distance. Measurement of electron spin relaxation times is more direct than, e.g., the correlation time of an electron-nuclear dipolar interaction, although the most accurate instrumentation is not available in most biochemical laboratories. However, reasonable estimates for Mn(II) spin relaxation times may be available from studies on similar systems.

Nitroxide-nitroxide interactions are likely to arise in proteins where two different spin labels are bound at two specific sites that place the respective N – O groups <10Å. These interactions are manifested in two forms: spin-spin exchange and spin-spin dipolar interactions. In the case of two nonrigid nitroxides, the exchange interaction is immediately apparent by the appearance of two additional lines in the ESR spectra

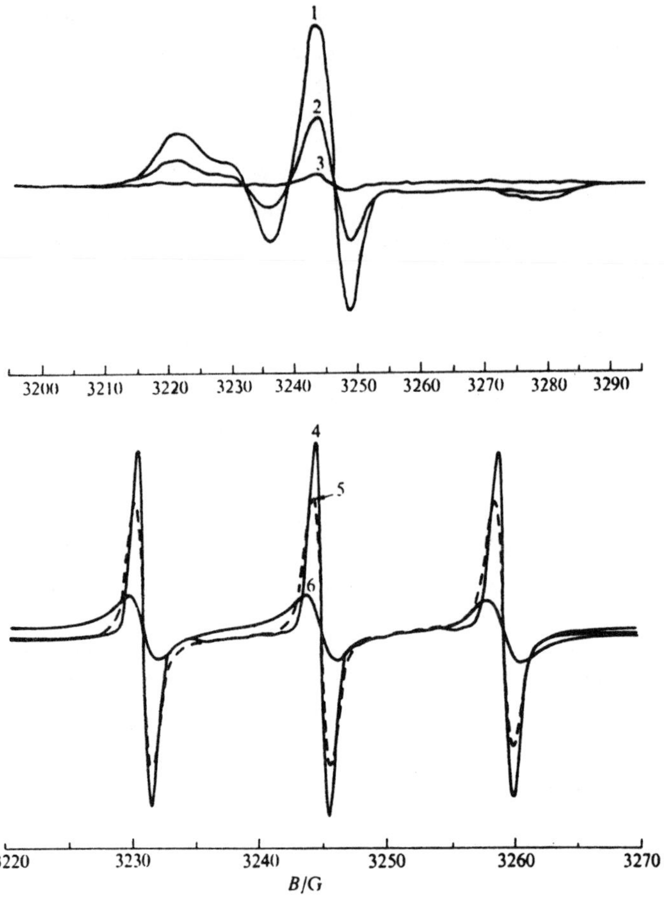

FIGURE 32. Mn·ADP effects on bound and free spin label. (A) Creatine kinase spin labeled with a pyrrolidinyl iodoacetamide spin label at its reactive sulfhydryl group. Effects of Mn·ADP on the ESR difference spectra (the Mn·ADP spectrum alone has been subtracted away) curve (1) no Mn·ADP added; (2) 2.5 m*M*Mn·ADP, (3) 7.7 m*M*Mn·ADP. (B) Spin label (N-l-oxyl-2,2,5,5-tetramethyl-3-pyrrolidinyl iodoacetamide) alone. Curve (4) 0.1 m*M* label; (5) plus 5.5 m*M* Mn·ADP; (6) plus 55.0 m*M* MnADP. Conditions were pH 7.9 20°C. (From Cohn, M., Diefenbach, H., and Taylor, J. S., *J. Biol. Chem.*, 246, 6037, 1971. With permission.)

spaced in between the normal three line spectrum. The resultant five line spectrum is physically the result of electron-electron exchange between one nitroxide and the other. Where the spins are in a rigid orientation, dipolar interactions become visible in a complex, very broad triplet state spectrum which may be analyzed to obtain the parameter 2D from the extreme splitting, where

$$D = \frac{3g^2\beta^2}{2r^3} \qquad (24)$$

(theoretically D is one of the zero field splitting parameters of the electron triplet state). Thus one finds splittings far exceeding the typical 64 to 68 G extreme separation of an isolated nitroxide. Splittings of 56, 109, 258, and 869 G, are related to distances of

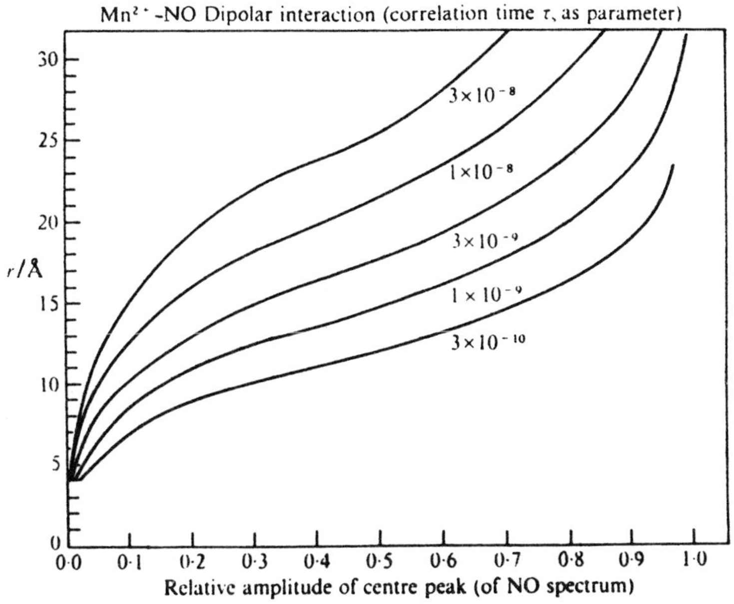

FIGURE 33. The distance between Mn(II) and the nitroxide spin plotted as a function of the amplitude of the center peak of the nitroxide ESR spectrum. The series of curves represents the function (22) at different values of the correlation time, τ_s [the longitudinal electron spin relaxation time of Mn(II) in the enzyme complex]. (From Cohn, M., Diefenbach, H., and Taylor, J. S., *J. Biol. Chem.*, 246, 6037, 1971. With permission.)

10, 8, 6, and 4 Å. Figures 34 and 35 depict examples for rapidly tumbling biradicals undergoing spin exchange and a rigid two nitroxide (biradical), with dipolar interaction, respectively. Again, it is emphasized here that if the two spins undergo free rotational tumbling with respect to one another, the dipolar interactions (splittings) (see Figure 35) will disappear, and only the exchange interaction (Figure 34) may remain. If the spins are significantly greater than 10 Å apart they will be magnetically isolated, resulting in a spectrum showing no interactions.

IV. EXAMPLES

This section covers specific examples of what can be done with spin labeling where we have not already covered example studies in the previous sections. Also included are examples of potential pitfalls of interpretation of data that one might encounter. Collectively, these examples span the major breadth of spin label applications.

A. Measuring Catalytic Rate Constants: α-Chymotrypsin[26,38]

The serine proteases, including α-chymotrypsin, share a common mechanism, which may be described by the following expression

$$E + S \underset{k_{-1}}{\overset{k_1}{\rightleftharpoons}} ES \xrightarrow{k_2} \underset{+P_1}{ES'} \xrightarrow{k_3} E + P_2 \qquad (25)$$

where the rate constant k_2 describes the acylation of Ser 195 by the carboxyl moiety of e.g., an ester substrate, and k_3 the deacylation (or hydrolysis) rate constant of the

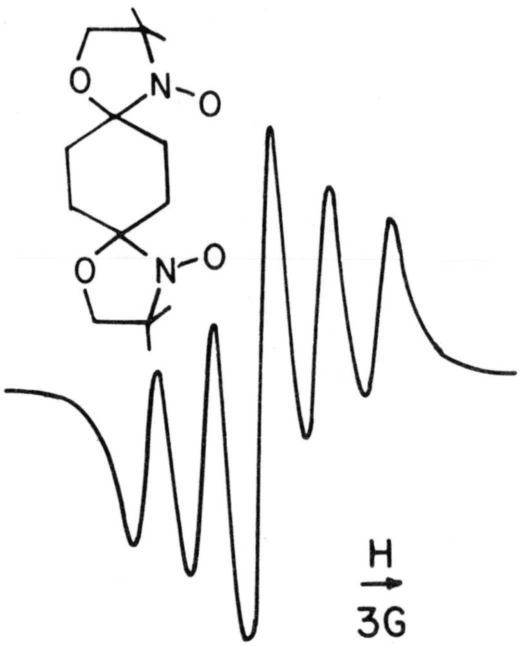

FIGURE 34. ESR spectrum of the "cyclohexane type" biradical in chloroform. (Reprinted with permission from Michon, P. and Rassat, A., *J. Am. Chem. Soc.*, 97, 696, 1975. Copyright by the American Chemical Society.)

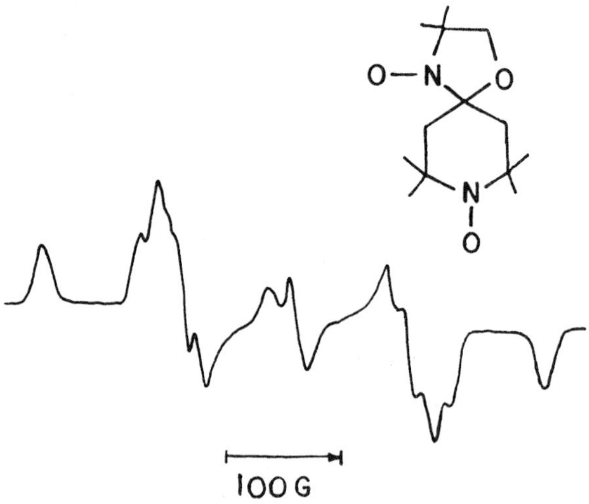

FIGURE 35. ESR spectrum of the dinitroxide in 5:5:2 diethylether/isopentane/ethanol rigid glass at 77°K. The maximum splitting of 438 G indicated that the two N-O electrons must be 5.0 ± 1.0 Å apart. (From Ciecierska-Tworek, Z., Van, S. P., and Griffith, O. H., *J. Mol. Struct.*, 16, 139, 1973. With permission.)

leaving acyl group, P_2, yielding the free enzyme, E, again. In the case of the spin labeled substrate, S,

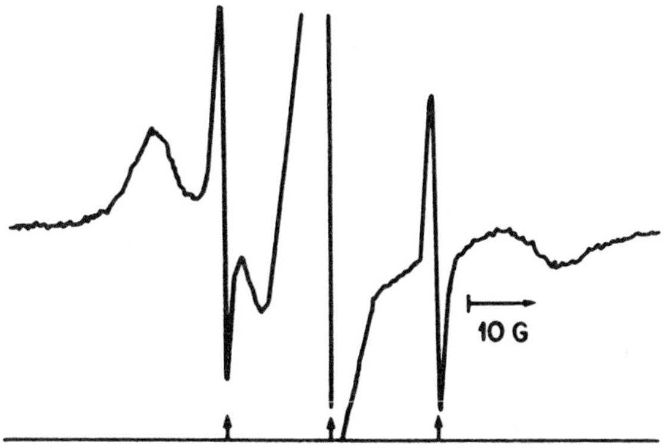

FIGURE 36. X-band ESR spectrum of spin labeled acyl α-chymotrypsin (a short time period after raising the pH to 6.8). Note the growth of the narrow three line spectrum which represents the liberated carboxylate nitroxide resulting from the hydrolytic deacylation reaction. (Reprinted with permission from Flohr, K., Paton, R. M., and Kaiser, E. T., *J. Am. Chem. Soc.*, 97, 1209, 1975. Copyright by the American Chemical Society.)

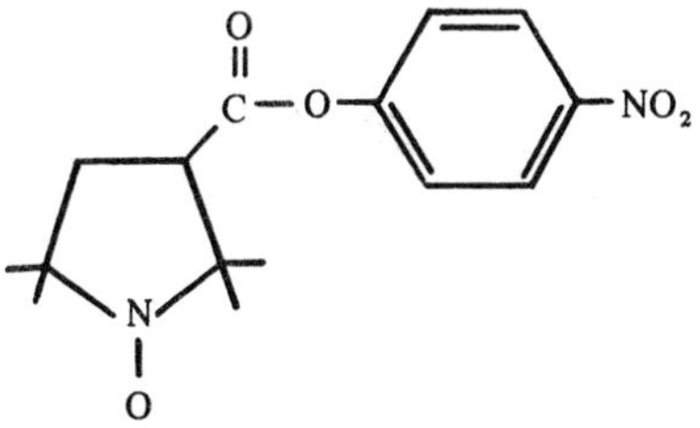

The product P_1 would be *p*-nitrophenolate while P_2 would be the corresponding pyrrolidine nitroxide carboxylate, which results from the hydrolysis of the acyl moiety from Ser 195.

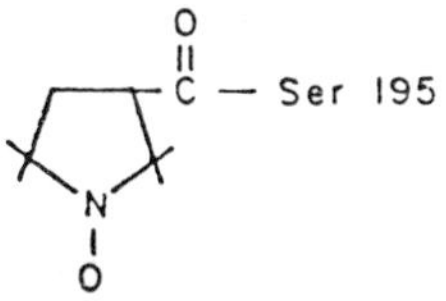

For ester substrates the deacylation step described by k_3 is rate limiting, allowing one to isolate the acyl enzyme species, ES′, at low pH and subsequently study the deacylation rate step independently. Figure 36 contains a multitude of information in that it represents the acyl enzyme isolated at pH 3.5 a very short time after the pH was raised to 6.8, where the k_3 rate step ensues. The broad line shape spectrum was that seen in Figure 27 in the discussion of spin labeled single crystals, which represented the rigid uniquely oriented S(−) acyl nitroxide group at the enzyme active site.[33] The growing narrow line spectrum (arrows) is that of the released free nitroxide carboxylate, P_2, the product of the deacylation step. We may plot the peak height increase with time as shown in Figure 37 to obtain the decay-like curve expected for a first-order rate process. The plot is analyzed by either a computer curve fitting routine or

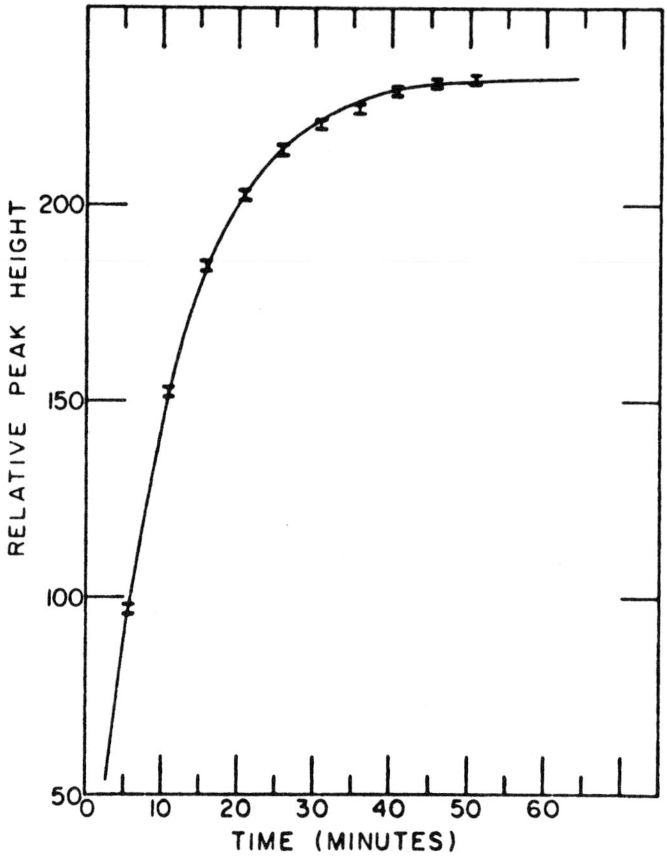

FIGURE 37. Plot of the relative peak height vs. time for the paramagnetic product P_2 arising from deacylation of the spin labeled acyl α-chymotrypsin ES′, (pH 6.8, 0.05 *M* phosphate) in Figure 36. (From Berliner, L. J. and McConnell, H. M., *Proc. Natl. Acad. Sci. U.S.A.*, 55, 708, 1966. With permission.)

by plotting log $[(h_{-1})_\infty - (h_{-1})_t]$ vs. time where $(h_{-1})_\infty$ and $(h_{-1})_t$ are the high field line peak height after complete deacylation has occurred and at time, *t*, respectively. Of course $(h_{-1})_\infty$ can be related to the "spin count" to obtain the concentration of spin labeled acyl enzyme at the beginning of the experiment. Actually, this above example was carried out for both the isolated **R** (+) and **S**(−) enantiomers of this substrate.[38] The isolated acyl-enzyme spectra for each were identical, i.e., both were strongly immobilized, but the deacylation rate constants, k_3, were different for each. The values of k_3 were $(4.5 \pm 1.0) \times 10^{-3}$ sec^{-1} and $(2.3 \pm 0.3) \times 10^{-4}$ sec^{-1} for the **R**(+) and **S**(−) derivatives, respectively, at pH 7.0, 25°C. These rate constants also agreed quite well with steady-state rate constants obtained spectrophotometrically by observing the release of *p*-nitrophenolate at 400 nm.

Summarizing, the spin label technique allowed us here to look individually at one particular rate step in an enzyme catalyzed reaction to obtain both rate and conformation information.

B. Conformational Change or Molecular Artifact?: Trypsin

Some potent irreversible inhibitors for many serine enzymes are the phosphonyl- or phosphoryl-fluorides, discussed earlier, which are depicted below, respectively.

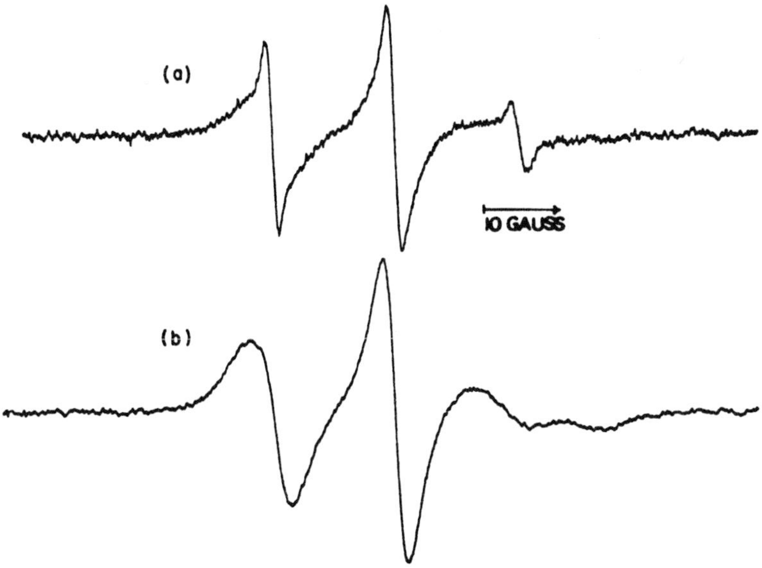

FIGURE 38. ESR spectra of spin labeled trypsin samples at pH 3.5 (0.006 *M* acetic acid, 0.1 *M* NaCl). (a) Treated with either fluorophosphate label at pH 5.5 or 7.5 followed by chromatography on Sephadex G-25 or dialysis at pH 3.5. (b) Labeled as above except that an additional purification step was performed on Sephadex SP-50 at pH 7.1 to remove autolyzed components. This spectrum was also frequently obtainable after concentrating the sample (a) in a collodion bag apparatus (Schleicher and Schuell). (From Berliner, L. J. and Wong, S. S., *J. Biol. Chem.*, 248, 1118, 1973. With permission.)

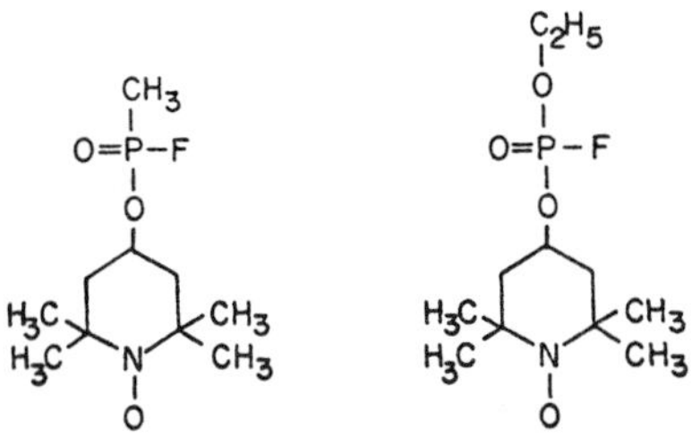

These react specifically and stoichiometrically with Ser 195 of the respective serine enzyme (acetylcholinesterase, α-chymotrypsin, trypsin, elastase, thrombin, etc.). Figure 38 shows two spectra obtained for trypsin labeled with either of the two labels shown above and isolated at pH 3.5, 0.1 *M* NaCl. Spectrum 38a (in Figure 38a) contains two motional components, i.e., is comprised of two spectra, one of high mobility, the other (less obvious) of low mobility. When this sample Figure 38a was carefully chromatographed on Sephadex G-25, the spectrum in Figure 39 was obtained, which is again two components, but emphasizes the broader immobilized component more than the narrow line (arrows) component. What were the origins and significance of each of these spectral components? Actually these components may have been misinterpreted in earlier work by authors studying the biradical phosphorylating agent

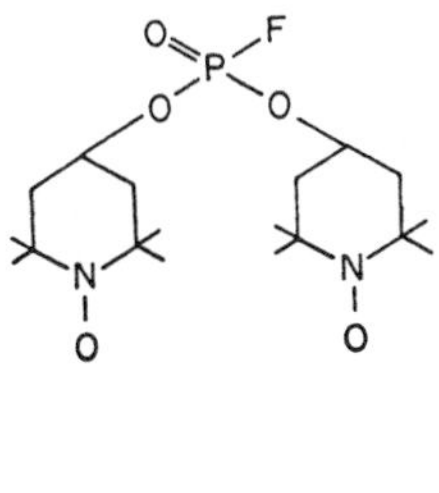

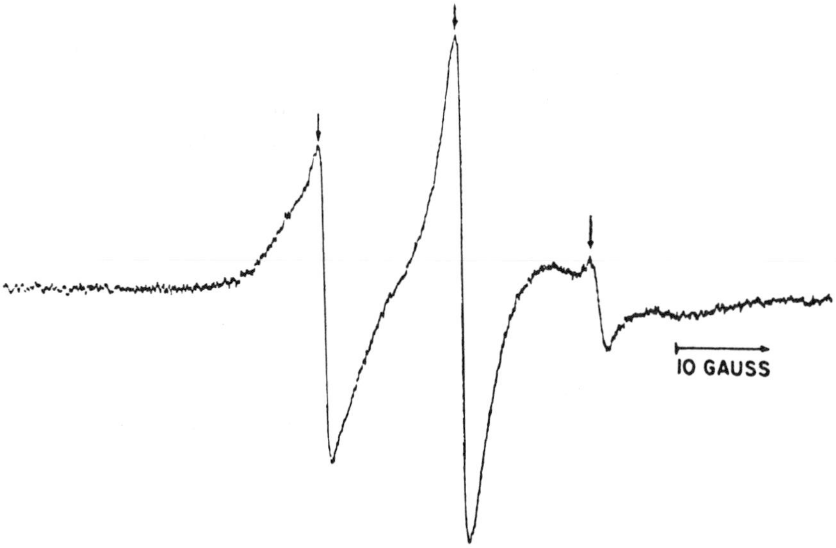

FIGURE 39. ESR spectrum of the sample in Figure 38(a) after rechromatography (small fractions) on Sephadex G-25 at pH 3.5. This spectrum was the first protein fraction (4.5 mg/mℓ) off the column. (From Berliner, L. J. and Wong, S. S., *J. Biol. Chem.*, 248, 1118, 1973. With permission.)

when they observed an irreversible change (15 to 30 min) from a broad line to narrow line spectrum upon changing the pH from 3.0 to 7.5.[40] If one chromatographs sample 38a or 39 on Sephadex SP-50, pH 7.1, where trypsin separates into its inert (autolyzed or denatured), and active α- and β-forms respectively, one obtains Spectrum 38b for α- or β-trypsin, where the narrow line component is completely lost. On the other hand the "inert" (autolyzed) peak yields the spectrum in Figure 40a. If one raises sample 39a from pH 3.5 to pH 7.7 or also adds a small fraction of free unlabeled trypsin to the spin labeled sample, Spectrum 39a irreversibly converts to spectrum 40b within 15 to 30 min. Furthermore, Figures 40a and 40b chromatographically represent the same trypsin species — inert (or autolyzed). What does this mean? When the trypsin was labeled with the spin labeled inhibitor, the very small fraction of unlabeled, active trypsin was free to proteolyze (autoproteolysis) some labeled trypsin in the sample, yielding a damaged form of the labeled enzyme. While trypsin is optimally active at pH 7 to 8, upon raising the pH of a labeled sample, the small amount of unlabeled, active enzyme may catalyze this autoproteolysis further. What is striking here is that the narrow line component spectrum of Figure 39 (arrows), when isolated from Figure 40a and quantitated by the "spin count" technique, represented only 1% of the total spin labeled enzyme in Figure 39 (also quantitated with the spin count technique). What should be emphasized is that the visual differences between Spectra 38a and 38b are pronounced to one without a very experienced eye for nitroxide ESR spectra (or computerized data handling equipment). The presence of a very small percentage of a molecular artifact — autolyzed trypsin — serves to change the ESR spectrum from that of the pure isolated labeled enzyme (38b) to a quite different appearing composite spectrum (38a), which has almost lost all semblance to Figure 38b. Thus one should always be suspicious of "additional" narrow line components that are not reversibly related to the major labeled enzyme species in the spectrum (Figure 38b).

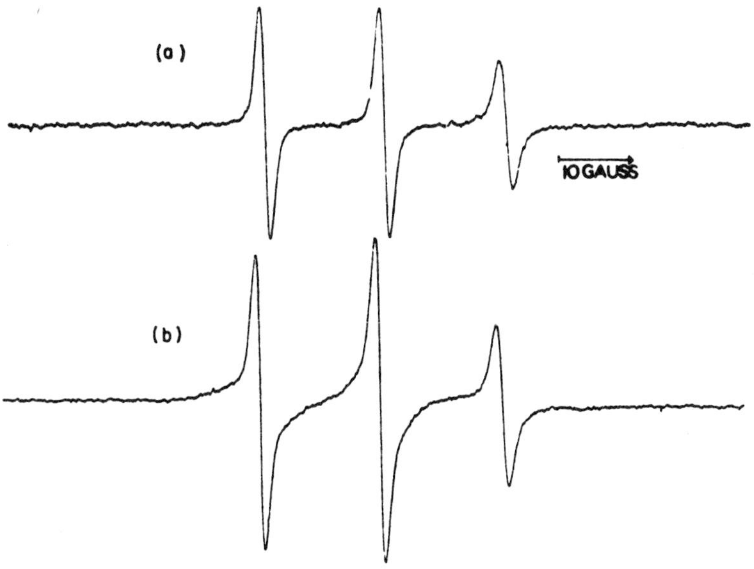

FIGURE 40. ESR spectra of (a) "inert" (autolyzed) trypsin fraction from Sephadex SP-50 chromatography of the spin labeled enzyme in Figure 39. ESR spectra were identical whether measured at pH 7.1 or 3.5 (b) Labeled trypsin (∿2-9 mg/mℓ) which was then treated with a few mg of active trypsin for 30 min at pH 7.7, room temperature. (From Berliner, L. J. and Wong, S. S., *J. Biol. Chem.*, 248, 1118, 1973. With permission.)

C. The Spin Label Reports Local, Not Global Events: Denaturation of α-Chymotrypsin

A change in the spin label mobility may reflect changes in the local protein environment. A spin label, at the active site of an enzyme that is being denatured in urea, should reflect the unfolding of the enzyme by a sharp change in the mobility of the nitroxide moiety with increasing urea concentration. In studies with α-chymotrypsin labeled at Ser 195 with the fluorophosphonate spin label

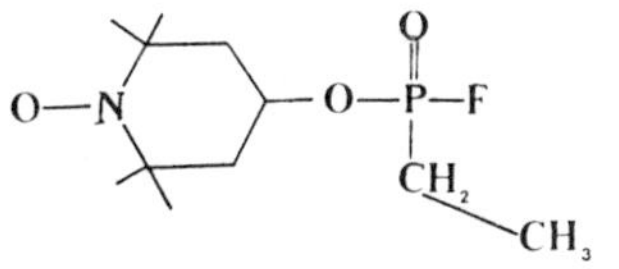

Berliner[30] and Grunwald[41] followed the unfolding by plotting the peak height ratios of the high and mid-field lines, a principal parameter in the rotational correlation time expressions (see Section III A.1). An example of the spectra obtained in none, intermediate, and fully unfolded urea concentrations was shown in Figure 22 (p.27). Now, if one compared the peak height ratios H/M with difference UV measurements at 293 nm or with fluorescence quantum yield data for the active site (Ser 195) labeled anthraniloyl moiety

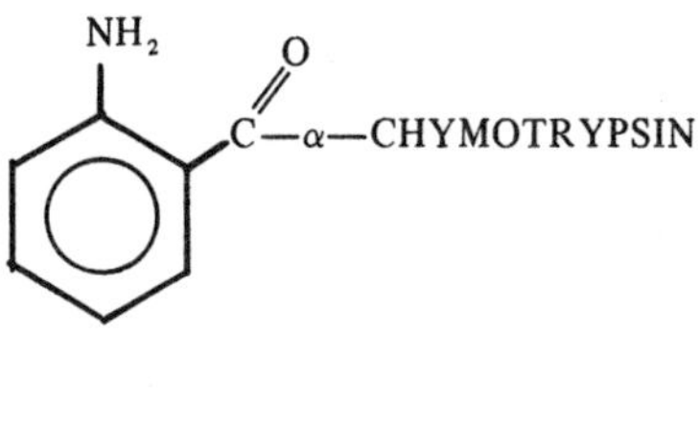

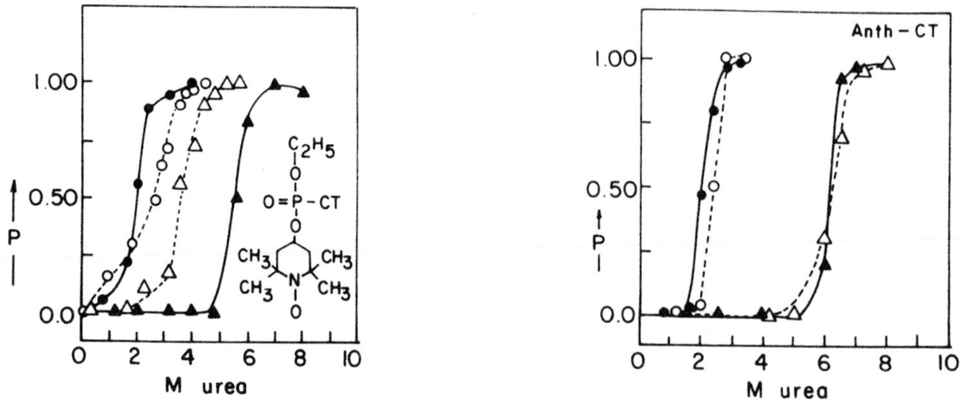

FIGURE 41. Denaturation profiles of (left) spin labeled α-chymotrypsin in urea at pH 2.8 (0.5 *M* citrate) and pH 7.9 (0.1 *M* phosphate). In each case denaturation was measured by two methods: O, ΔA_{293nm}, pH 2.8; Δ , ΔA_{293nm}, pH 7.9; O ESR, pH 2.8; -Δ-, ESR, pH 7.9. The parameter, P, is the fraction change in ΔA_{293nm}, or the high to low field peak height ratio (H/M) in the ESR spectrum. Note how the pH 2.9 transition is higher by ESR and the contrary result at pH 7.9. (right): Anthraniloyl -α-chymotrypsin in the same buffers as above by two methods: ●, ΔA_{293nm}, pH 2.9; Δ, ΔA_{293nm}, pH 7.9; O, (pH 2.9); Δ , (pH 7.9) fractional change in fluorescence intensity of the bound anthraniloyl group. Note here the excellent agreement between both methods and their agreement with the ΔA_{293nm} data for the spin labeled enzyme in the left example. Thus, the ESR experiment monitored local changes while the two optical methods responded only to global changes. (From Grunwald, J., Ph.D. thesis, Technion, Haifa, 1975. With permission.)

all under similar pH (2.8 and 7.9) and temperature conditions, the composite plot of unfolding vs. urea concentration as measured by each technique is shown in Figure 41. We immediately note that the unfolding transition for the spin labeled enzyme occurred at different urea concentrations, 2.8 and 3.7 *M* than that found by the other techniques 2.0 and 5.5 to 6.0 *M*. Why was the spin labeled enzyme apparently less (or more) stable to urea than the native or anthraniloyl derivative? First, it seems unlikely that the Ser 195 blocked anthraniloyl derivative should be identical with the native enzyme, while the (Ser 195 blocked) spin labeled enzyme should be intrinsically less stable. If we consider the fact that while the anthraniloyl moiety is somewhat smaller and is most likely bound in the aromatic substrate specificity pocket, while the larger spin label nitroxide moiety cannot physically fit into this pocket, the latter is sensing conformation changes farther out from the catalytic apparatus (8 to 10 Å) than would the fluorescent label. On the other hand, while the difference UV measurement monitors the denaturation of aromatic moieties (Trp, Tyr) over all of the molecule ("global changes"), the ESR experiment "sees", to the first order, only its immediate environment. Consequently, the changes detected in Figure 41 by the spin label could be either a local denaturation, which occurs before the global denaturation (or before the active site unfolds), not a destabilization of the enzyme structure in the case of the spin labeled enzyme vs. the other derivatives as postulated earlier.

D. Measuring Molecular Dimensions: Carbonic Anhydrase

A simple yet clever approach to measuring the depth or dimensions of a protein combining site is to follow the nitroxide mobility with an increase in the distance between the nitroxide moiety and the combining site. This is particularly suited to proteins such as antibodies, which have very strong affinities for the antigenic hapten moiety. In the case exemplified here, however, an enzyme, carbonic anhydrase, was studied with a series of sulfonamide spin probes that were potent druglike inhibitors

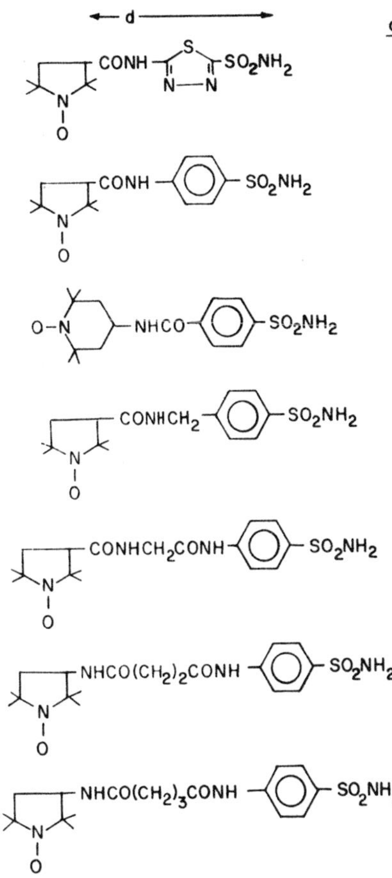

	d^* (Å)
XCIII	7.7
XCIV	8.2
XCV	7.8
XCVI	8.9
XCVII	11.3
XCVIII	12.7
XCIX	14.7

FIGURE 42. Sulfonamide spin probes for carbonic anhydrase. The distances d were calculated from fully extended CPK models. (From Chignell, C. F., *Spin Labeling II: Theory and Applications,* Berliner, L. J., Ed., Academic Press, New York, 1979, chap. 5. With permission.)

of the action of this enzyme in hydrating CO_2 in the blood. A series of these spin probes was synthesized with the basic structure

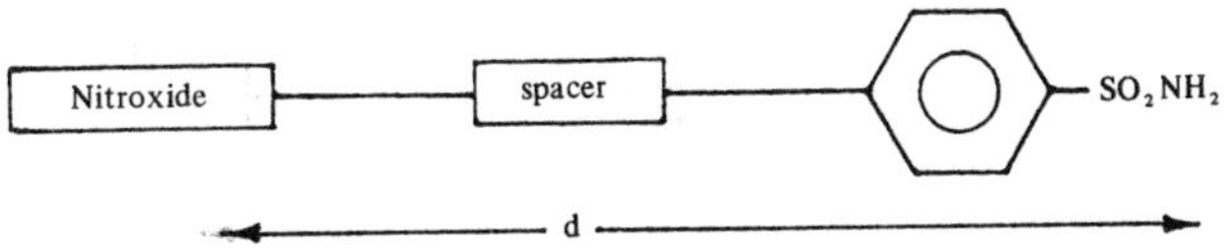

Their structure and (fully extended) dimensions are shown in Figure 42. These spin probes bind very strongly to any of the human or bovine carbonic anhydrase isozymes. An example spectrum is shown in Figure 43 where essentially no narrow line component exists for this noncovalently bound spin probe. The dissociation constants for these derivatives are well below 10^{-6} *M,* such that under the conditions shown (Figure 43) the unbound spin probe concentration is negligible. In this work Chignell and coworkers plotted nitroxide mobility vs. the distance *d,* from the sulfonamide binding site[43] (see Figure 44). Thus for bovine carbonic anhydrase *B,* a sharp change in the ESR spectral mobility occurred between labels XCVII and XCVIII, or at 13 Å. The

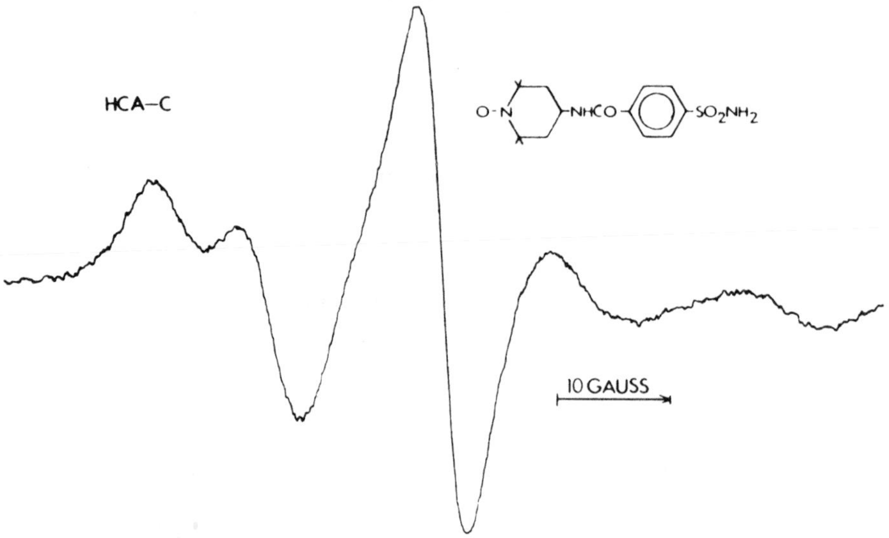

FIGURE 43. ESR spectrum of sulfonamide spin probe XCV (2×10^{-5} *M*) and human carbonic anhydrase C (2.62×10^{-4}*M*) at pH 7.4, 0.1 *M* phosphate, 25°. (From Berliner, L. J., *Methods Enzymol.*, 49G, 418, 1978. With permission.)

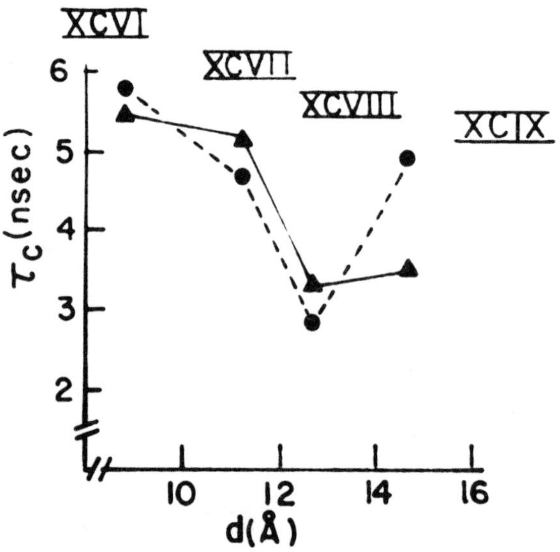

FIGURE 44. Rotational correlation time, τ_c of spin labels XCVI to XCIX bound to bovine carbonic anhydrase B (•) and human carbonic anhydrase B (▲). (From Chignell, C. F., *Spin Labeling II: Theory and Applications*, Berliner, L. J., Ed., Academic Press, New York, 1979, chap. 5. With permission.)

elegance of the technique was corroborated by the dimensions learned from the subsequent X-ray structure determination.

There are two potential pitfalls, however, that one most consider when attempting such an approach on a new protein. If the spin probes were bound in e.g., a curved, rather than extended conformation, the estimated dimension *d* would be longer than actual. Of more concern would be the case where a spin probe, of length *d* greater than the depth of the combining site, bent back and bound rigidly to some nearby hydrophobic site (secondary binding). Thus instead of the nitroxide rotating freely and unhindered, since it was free of the sulfonamide combining cavity and expected to be out in the external aqueous environment, the results would have suggested that the extent of the sulfonamide site had not been exceeded. This pitfall is overcome by increasing the distance *d* in very small increments (i.e., by utilizing several spin probes such as shown in Figure 42). In fact, Erlich and co-workers found that with human carbonic anhydrase *C* label XCIX was more immobilized than was the shorter XCVII.[44] Since it was well established from the entire series in Figure 42 that the value of *d* for this enzyme species was approximately 13 Å, some type of secondary binding had occurred between the nitroxide moiety of XCIX and the enzyme. Therefore, had not the site of probes used increased in small increments as shown in Figure 42, a gross overestimation of the binding site depth might have occurred.

E. Detecting Subtle Differences in Macromolecular Structure: Thrombin

By using spin labels and comparing different species, forms, or conformations of an enzyme or family of enzymes we may learn about subtle differences in structure and conformation that may be complementary to other techniques in solution.

The human blood clotting enzyme, α-thrombin, has been discovered to exist in a degraded yet catalytically active, three noncovalently associated chain species called γ-thrombin. Basically, the principal, yet critical difference between the two forms is the high coagulant ability of α- vs. loss of coagulation ability of γ-thrombin. Both forms are almost identical in other functions that thrombin catalyzes. Of interest to biochemists are the structural changes that occurred when the totally covalent structure of α-thrombin was converted by two proteolytic cleavages to the three (noncovalently associated) chain γ-form.

The series of spin labeled active site serine (205) directed inhibitors shown in Figure 45 was examined with both α- and γ-thrombin under exactly the same conditions. Seven of these spin labels, m-I, m-III, m-IV, m-V, m-VI, m-VII, and m-VIII displayed greater immobilization with γ-thrombin compared with α-thrombin, while the remainder gave identical spectra for both forms. An example is shown in Figure 46. At first glance we might ask why all labels did not show a difference since the two forms are in some aspects, different. The answer came from an analysis of structural (space filling) models of each of these labels, which clearly showed that the seven labels noted above occupied the same volume in space, which was distinct from that volume in space occupied by the other nitroxides. Structurally, the space filling models showed that all nitroxide moieties that occupied the position roughly *meta* to the sulfonyl group were sensitive to the conformational difference between α- and γ-thrombin.

Thus as a more detailed model of thrombin molecular structure evolves, we can pinpoint in the three dimensional model where the α-γ differences arise. The spin label technique here displayed a unique sensitivity to quite subtle structural differences.

F. ESR is Very Sensitive to "Impurities" (Spectral Artifacts): Label Hydrolysis

In the previous section a series of "irreversible" serine protease inhibitors of the fluorosulfonyl type was introduced. Actually the sulfonylated enzyme derivative is a sulfonyl ester intermediate, analogous to the acyl intermediate described in Section IV.A for an ester substrate; that is, while this spin labeled inhibitor is, in practice, irreversible, it can undergo desulfonylation — hydrolysis off of the enzyme. Figure 47 depicts the fate of a sulfonylated enzyme derivative. In either side reaction, desulfon-

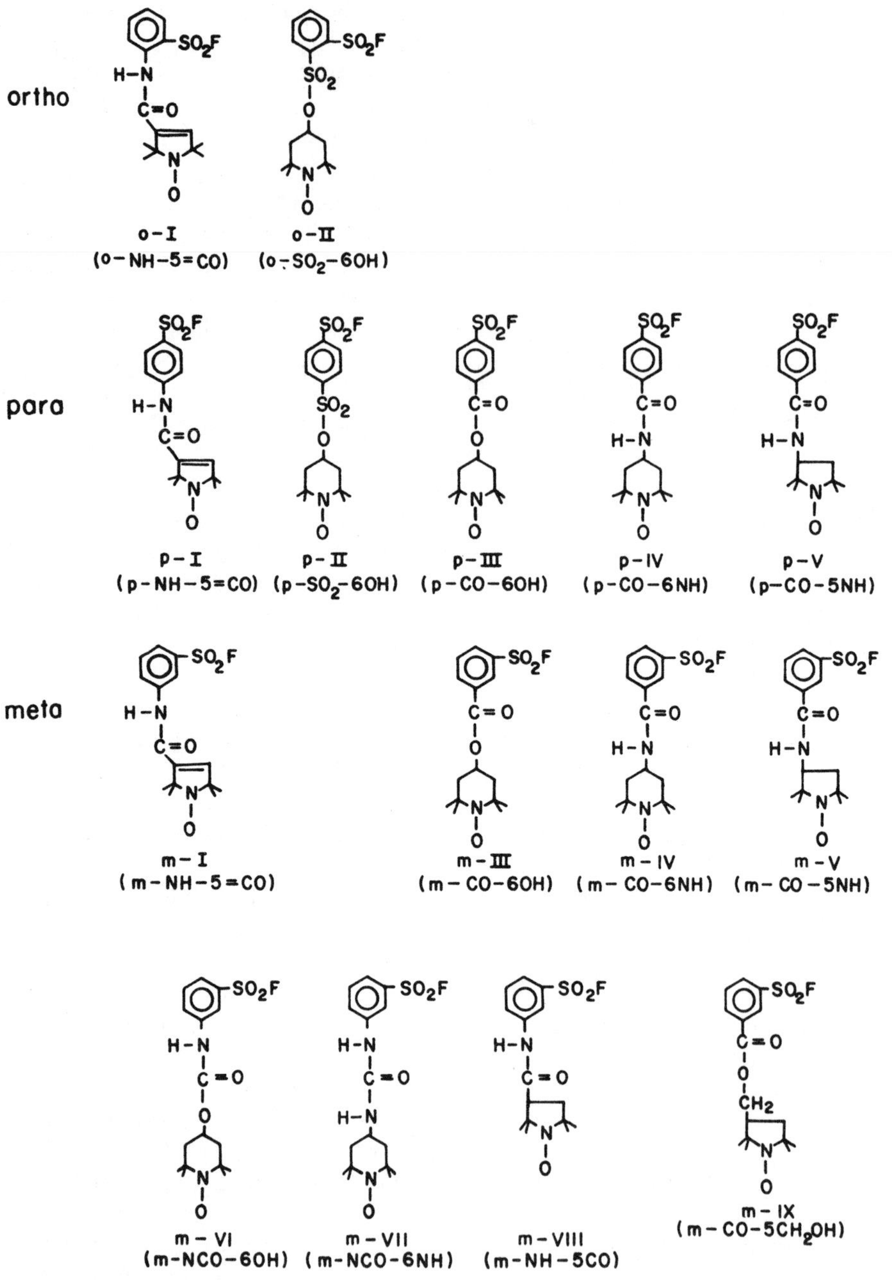

FIGURE 45. Structures of fluorosulfonyl spin labels. (From Berliner, L. J., in *Progress in Bioorganic Chemistry,* Kezdy, F. and Kaiser, E. T., Eds., Wiley Interscience, New York, chap. 1. Reprinted by permission of John Wiley & Sons, Inc.)

ylation to yield native enzyme and the freely tumbling spin label sulfonate, or intralabel hydrolysis to yield a diamagnetic sulfonylated enzyme and a freely tumbling nitroxide amine, the resultant ESR spectrum will be a super position of the bound (spin labeled) enzyme derivative and a sharp narrow line spectrum, the latter of which will be orders of magnitude greater in peak height than that for the bound spectrum due to the inverse relationship between linewidth and the peak height.

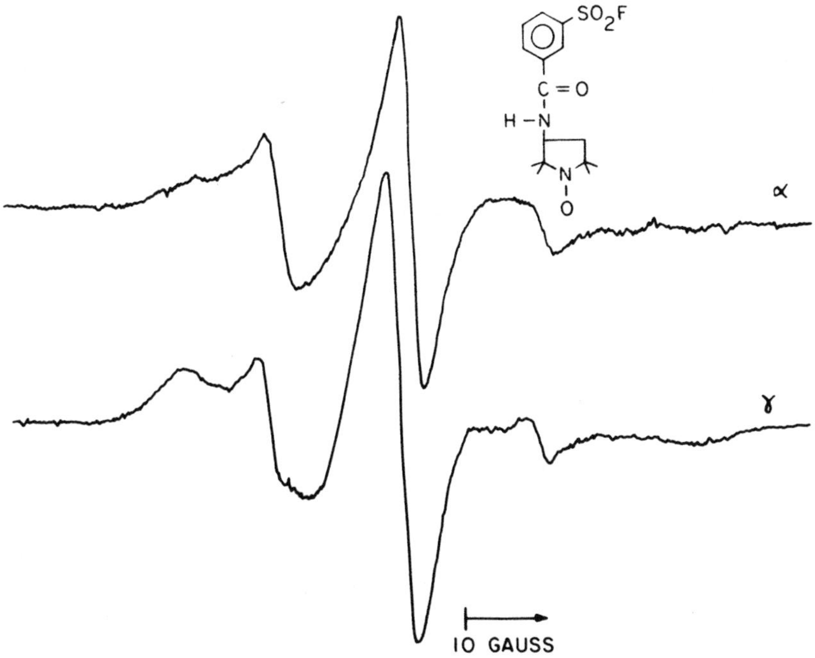

FIGURE 46. Spin labeled α- and γ-thrombins. pH 6.5 (0.05 *M* phosphate buffer, 0.75 *M* NaCl, 26°C).

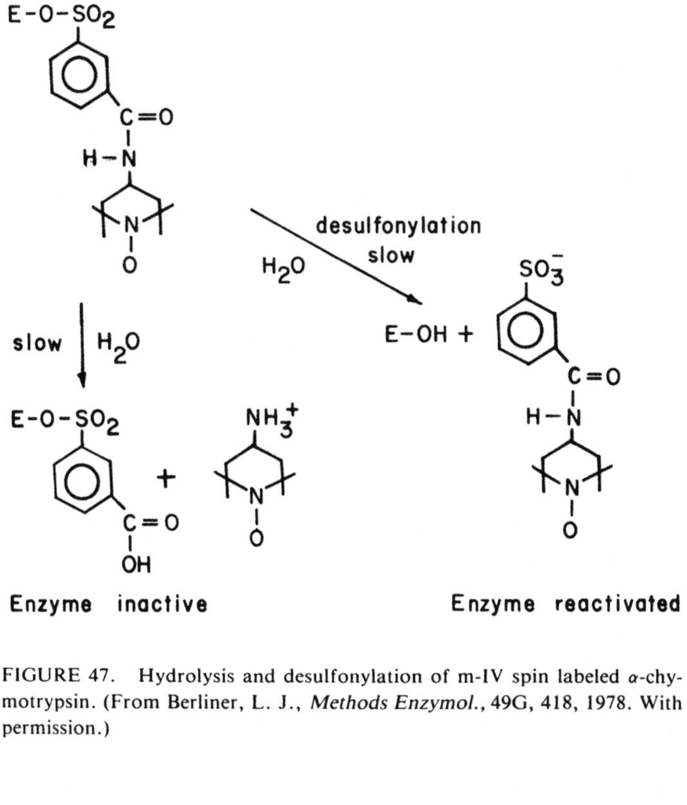

FIGURE 47. Hydrolysis and desulfonylation of m-IV spin labeled α-chymotrypsin. (From Berliner, L. J., *Methods Enzymol.*, 49G, 418, 1978. With permission.)

What is pertinent here is that this sharp-line spectral component will become prominant after only a few tenths of a percent of the spin label is hydrolyzed and is even more predominant at higher pH. But an experienced observer might construe this sharp line component as a pH-dependent conformational change in the enzyme, or, the broader line spectral component might go completely undetected in the first place. Such a conformational change would seem quite consistent with e.g., a relatively rapid appearance of a narrow line spectrum at pH 9, which either remained when the pH was lowered (an irreversible change) or disappeared if the enzyme were dialyzed vs. a lower pH (a reversible change). While such very narrow line spectra are occasionally observed as real conformational forms of a labeled enzyme, the linewidth should still be greater than that of a free nitroxide. One must always be careful to rule out potential spectral artifacts due to what we term as impurities (due to secondary or chemical side reactions, which are not of significance to the biological system under study). The enormous sensitivity of the ESR experiment to free nitroxide impurities can occasionally mislead our interpretations if we are not aware of the possible origins of such narrow line (yet substantial signal/noise) spectra.

G. Assessment of the Effects of Subsequent Chemical Modification of a Protein

Many protein modification studies leave uncertain interpretation where the modification does not alter, e.g., the catalytic activity or Michaelis constant. The spin label technique offers a powerful method of assessing the structural (conformational) effects of such a modification on specific regions of the enzyme.

1. Immobilized Trypsin[45]

An industrially and biochemically useful "chemical modification" of many proteins involves their immobilization or covalent attachment to an insoluble support or chromatographic resin. Once attached to such a support, e.g., Sepharose or glass beads, the study of its dynamic physical properties are impossible by most spectroscopic methods since the sample is no longer optically clear. In order to answer whether any major conformational changes have occurred to the enzyme as a result of immobilization, a spin label approach is quite valuable.

Figure 48a depicts the same trypsin sample discussed in Section IV.B, which has been labeled specifically at Ser 195 with the phosphonyl spin label.

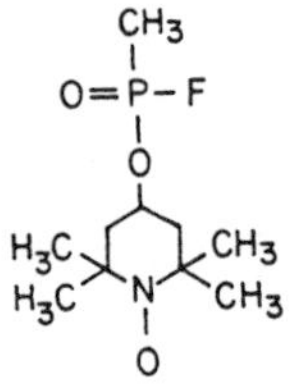

The effects of immobilization were examined in two ways. In the first case, spectrum 48b, the trypsin was first attached to (aryl diazonium activated) glass beads and subsequently spin labeled. Thus the spectrum shown represents those enzyme molecules that retained enzyme activity after immobilization, and which also were physically accessible to the spin label. In the second instance, Figure 48c represents enzyme that was first labeled in solution to give Spectrum 48a, and then subsequently attached to the glass beads. Lastly, as a control, the labeled native enzyme was examined in saturated sucrose solution to mimic tumbling conditions that approached total immobilization.

Note that the immobilized enzyme derivatives 48b and c reflected the now totally

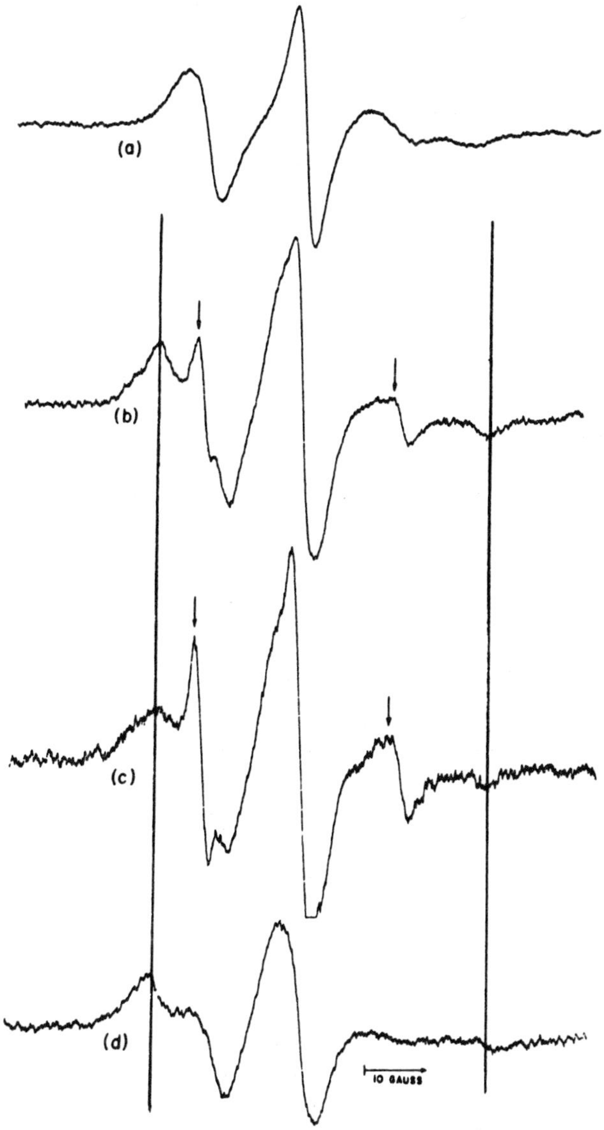

FIGURE 48. ESR spectra of phosphonate spin labeled trypsins (pH 3.5, 0.05 *M* acetic acid, 0.02 *M* $CaCl_2$, 26°C). (a) In solution; (b) labeled then immobilized on glass beads; (c) immobilized, then labeled; and (d) sample (a) in saturated sucrose. The arrows denote small amounts of autolyzed trypsin that could not be removed from these preparations. (From Berliner, L. J., Miller, S. T., Uy, R., and Royer, G. P., *Biochim. Biophys. Acta*, 315, 195, 1973. With permission.)

reduced enzyme tumbling as compared to 48a, and the similarity in line shape and outer hyperfine splittings, which strongly suggests that the active site conformations were identical regardless of whether the enzyme was first labeled or first immobilized. The narrow line spectral components (arrows) were due to a very small percentage of autolyzed trypsin (see Section IV.B), which could not be completely removed from these preparations.

That a change in active site conformation did not occur upon immobilization could not be ruled out, since the spectrum for the "control" Figure 48d, while of similar line shape, was of 1 to 2 G greater separation in its outer hyperfine extrema than for the immobilized derivatives in Figure 48d and c. However, if such a change occurred, it was not of any severity to the catalytic apparatus as judged by the reactivity and similarity of either enzyme derivative in 48b or c. Thus the spin label method here found the active site conformation to be quite intact despite the covalent attachment of some enzyme surface residues to a glass support.

2. Crosslinked α-Chymotrypsin Crystals

If one were to study chemical modification reactions on a single crystal of an enzyme, the symmetry and packing properties of the crystalline state would dictate that the modifications would be on the average uniform throughout the crystal at specific residues, and probably smaller in number than would be possible in the solution state where isolated protein monomers are present. In the example presented here, the spin labeled chymotrypsin single crystal discussed earlier in Section III.E.2 was crosslinked with glutaraldehyde, a reagent that forms either Schiff's base or perhaps Michael addition products with lysine residues. While in solution an inordinate number of crosslinked dimers, trimers, and so on are possible, the crystalline case would be expected to be quite simpler by virtue of the fixed orientation (and accessibility) of surface lysine groups as dictated by the crystal packing environment. Figure 49 shows spectra of labeled α-chymotrypsin after exposures of 0, 9, and 23 hr in 0.5% (v/v) glutaraldehyde at pH 4.2, in 2.2 *M* $MgSO_4$. The criterion for crosslinking was the insolubility of the crystal in buffer alone without the high sulfate salts normally required to keep it from redissolving. Note that as the crosslinking modification progressed, a new set of spectral lines appeared that showed the same symmetry properties of the crystal as a whole discussed earlier (Section III.E.2). The crosslinking reaction created a new set of chymotrypsin subunits in the crystal that were identifiable by the distinct orientation of the nitroxide group at their active sites. The orientation of the *z*-axis in the crosslinked subunits was 43° away from that in the "native" or unaltered subunits. While it was unlikely that the entire subunit rotated 43° upon crosslinking, the active site binding region for the nitroxide acyl group was altered sufficiently to cause the spin label to assume an orientation shift 43° from that in the "native" subunits. Again here, the modification did not alter the catalytic properties of the enzyme as the crystal (which was further labeled after crosslinking) showed increased label intensity for the crosslinked as well as native orientation. Thus in this second example the spin label method offered an even more detailed description of the structural consequences of chemical modification.

H. What Motion is the Spin Label Reporting — The Macromolecule, the Local Site, or the Nitroxide Moiety?: Galactosyl Transferase

As we discussed in Section III.A, the ESR spectrum of a nitroxide bound to a protein may be described in terms of a rotational correlation time or tumbling rate. For a spin labeled protein we must consider whether this motion is due principally to relatively unrestricted nitroxide motion or to the protein (environment). If the spin label is rigid with respect to the protein, then the usual strongly immobilized spectrum will have a nitroxide correlation time which is that of the macromolecule. The criteria for assessing if the nitroxide is rigid with respect to the protein are both of a qualitative and semiquantitative nature. If the protein mol. weight falls within the slow motion range (*ca* > 50,000 daltons), the observed spectrum should be a strongly immobilized "powder" spectrum, i.e., the label is rigid. For proteins of mol. weight <50,000 the techniques

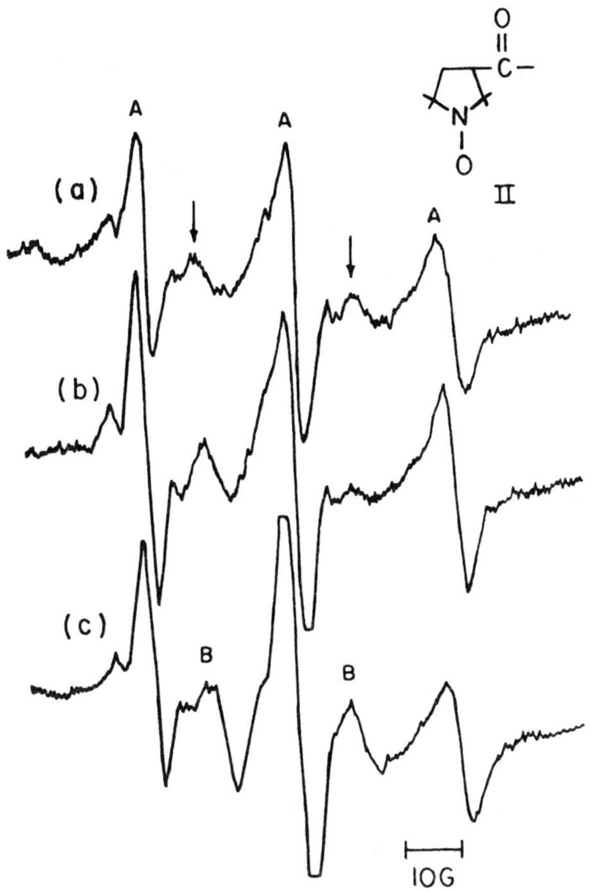

FIGURE 49. The effect of glutaraldehyde crosslinking on spin labeled α-chymotrypsin crystals. All spectra were recorded with the magnetic field along the b-axis. Arrows denote the position of lines due to free label. The original set of lines are denoted as A, the set appearing after crosslinking. (B). (a) No glutaraldehyde treatment; (b) 9 hr in 0.5% glutaraldehyde; (c) 23 hr in 0.5% glutaraldehyde. (With permission from Berliner, L. J. and Bauer, R. S., *J. Mol. Biol.*, 129, 165, 1979. Copyright by Academic Press Inc.(London)Ltd.)

described in Section III.A.2 in the intermediate to slow tumbling range may be employed. If the spin label tumbles faster than the macromolecule, an intermediate or weakly immobilized spectrum (τ_R = 10^{-10} – 10^{-12} sec) is observed. The protein itself may be immobilized (see Section IV.G.1), or slowed down considerably in high viscosity media such as sucrose, leaving the nitroxide motion and local macromolecular segmental motion as the dominant contributions to the resultant nitroxide motion.

The use of sucrose has been quite beneficial as a method of slowing the protein by increasing macroviscosity; however one cannot rule out partial microviscosity effects on the nitroxide motion itself. The experiment in Figure 50 depicts such an example. Bovine galactosyl transferase was spin labeled at a reactive thiol with the maleimide spin label

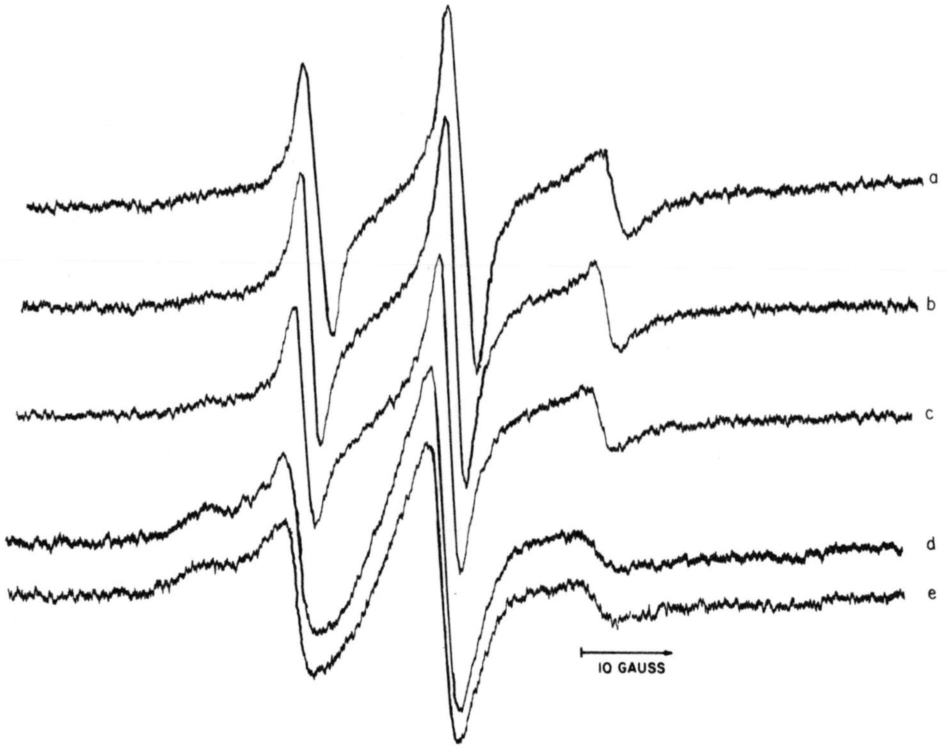

FIGURE 50. ESR spectra of maleimide spin labeled bovine galactosyl transferase, pH 8.0, 0.06 *M* N-methyl-morpholine buffer, 5% w/v) $(NH_4)_2SO_4$, 26°C. Enzyme concentration was $\sim 10^{-5}$ *M*. (a) No additions; (b) with 120 m*M* UDPgalactose; (c) with saturated *N*-acetylglucosamine; (d) with saturated glucose; and (e) with saturated sucrose.

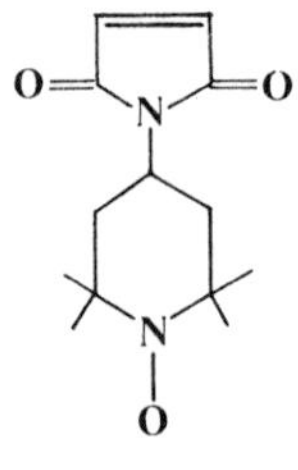

to give the relatively mobile spectrum shown in Figure 50a. Since this thiol group was partially protected by the binding of UDP galactose, the galactosyl donor substrate for this enzyme, the labeled enzyme was exposed to 120 m*M* UDP galactose (Figure 50b); however, no effect on the nitroxide motion was observed. Spectrum 50c shows the apparent effects of one of the substrates for this enzyme, *N*-acetyl glucosamine (NAG), while 50d is in the presence of saturated glucose, the acceptor substrate for lactose biosynthesis. While both Spectra 50c and 50d showed the presence of a new broad line spectral component, which is quite apparent at the low field (left) side of the spectrum, it was important to note that these saccharides form relatively viscous solutions at saturating concentrations. In order to determine whether these ESR spectral changes were due to the specific binding of glucose or simply to microviscosity effects, the spin labeled enzyme was exposed to sucrose solution of the same viscosity as the glucose solution in Spectrum 50d. The sucrose-treated sample in Figure 50e is identical to that in 50d, yet sucrose does not bind to enzyme. The observed immobilizations in Spectra 50c and 50d were due to changes in the (solvent) microviscosity in the label environment. Therefore, if the nitroxide is relatively mobile and exposed in part to the external solvent, its motion can be influenced by macroviscosity changes. While the employment of high viscosity solutes such as sucrose is by all means a quite valuable technique for "separating" out the label motion from that of the protein, the resultant ESR spectrum is then a reflection of the nitroxide motion in the high viscosity (saturated sucrose) solvent and not necessarily that of the label in the normal buffer.

REFERENCES

1. **Reed, G. H. and Ray, W. J., Jr.,** Electron paramagnetic resonance studies of manganese(II) coordination in the phosphoglucomutase system, *Biochemistry,* 10, 3190, 1971.
2. **Cohn, M., Leigh, J. S., Jr., and Reed, G. H.,** Mapping active sites of phosphoryl-transferring enzymes by magnetic resonance methods, *Cold Spring Harbor Symp. Quant. Biol.,* 36, 533, 1972.
3. **Andree, P. J. and Berliner, L. J.,** Metal ion and substrate binding to bovine galactosyltransferase, *Biochemistry,* 19, 929, 1980.
4. **Uchida, H., Berliner, L. J., and Klapper, M. H.,** Three new iron-ligand complexes of human methemoglobin A, *J. Biol. Chem.,* 245, 4606, 1970.
5. **Ohnishi, S. I. and McConnell, H. M.,** Interaction of the radical ion of chlorpromazine with deoxyribonucleic acid, *J. Am. Chem. Soc.,* 87, 2293, 1965.
6. **Francis, F.,** Preparation of triacetonamine hydrate, *J. Chem. Soc.,* 1927, 2897, 1927.
7. **Sosnovsky, G. and Konieczny, M.,** Preparation of triacetonamine II, *Z. Naturforsch. Teil B,* 32, 338, 1977.
8. **Jost, P. and Griffith, O. H.,** Electron spin resonance and the spin labeling method, in *Methods in Pharmacology,* Vol. 2, Chignell, C. F., Ed., Appleton-Century-Crofts, New York, 1972, 223.
9. **Rozantsev, E. G.,** *Free Nitroxyl Radicals,* Plenum Press, New York, 1970.
10. **Gaffney, B. J.,** The chemistry of spin labels, in *Spin Labeling: Theory and Applications,* Berliner, L. J., Ed., Academic Press, New York, 1976, chap. 5.
11. **Keana, J. F. W.,** New aspects of nitroxide chemistry, in *Spin Labeling II: Theory and Applications,* Berliner, L. J., Ed., Academic Press, New York, 1979, chap. 3.
12. **Schwartz, M. A. and McConnell, H. M.,** Surface areas of lipid membranes, *J. Am. Chem. Soc.,* 17, 837, 1978.
13. **Keana, J. F. W., Dinerstein, R. J., and Baitis, F.,** Photolytic studies on 4-hydroxy-2,2,6,6-tetramethylpiperidine-1-oxyl, a stable nitroxide free radical, *J. Org. Chem.,* 36, 209, 1971.
14. **Lai, C. S. and Piette, L. H.,** Hydroxyl radical production involved in lipid peroxidation of rat liver microsomes, *Biochem. Biophys. Res. Commun.,* 78, 51, 1977.
15. **Berliner, L. J. and Shen, Y. Y.,** Probing active site structure by spin label (ESR) and fluorescence methods, in *Chemistry and Biology of Thrombin,* Lundblad, R. L., Fenton, J. W., II, and Mann, K. G., Eds., Ann Arbor Science, Ann Arbor, Michigan, 1977, 197.
16. **Stone, T. J., Buckman, T., Nordio, P. L., and McConnell, H. M.,** Spin labeled biomolecules, *Proc. Natl. Acad. Sci. U.S.A.,* 54, 1010, 1965.
17. **Hoffman, B. M., Schofield, P., and Rich, A.,** Spin-labeled transfer-RNA, *Proc. Natl. Acad. Sci. U.S.A.,* 62, 1195, 1969.
18. **Freed, J. H.,** Theory of slow tumbling ESR spectra for nitroxides, in *Spin Labeling: Theory and Applications,* Berliner, L. J., Ed., Academic Press, New York, 1976, chap 2.
19. **Berliner, L. J., Ed.,** *Spin Labeling: Theory and Applications,* Academic Press, New York, 1976, Appendix I.
20. **Shimshick, E. J. and McConnell, H. M.,** Rotational correlation time of spin-labeled α-chymotrypsin, *Biochem. Biophys. Res. Commun.,* 46, 1, 1972.
21. **Hyde, J. S. and Dalton, L. R.,** Saturation transfer of spectroscopy in *Spin Labeling II: Theory and Applications,* Berliner, L. J., Ed., Academic Press, New York, 1979, chap. 1.

22. **Kusumi, A. S., Ohnishi, T., Yoshizawa, T., and Ito, K.**, Rotational motion of rhodopsin in the visual receptor membrane as studied by saturation transfer spectroscopy, *Biochim. Biophys. Acta*, 507, 539, 1978.
23. **Griffith, O. H., Dehlinger, P. J., and Van, S. P.**, Shape of the hydrophobic barrier of phospholipid bilayers. Evidence for water penetration in biological membranes, *J. Membr. Biol.*, 15, 159, 1974.
24. **Lassman, G., Ebert, B., Kuznetsov, A. N., and Damerau, W.**, Characterization of hydrophobic regions in proteins by spin labeling technique, *Biochim. Biophys. Acta*, 310, 298, 1973.
25. **Humphries, G. M. K. and McConnell, H. M.**, Antibodies against nitroxide spin labels, *Biophys. J.*, 16, 275, 1976.
26. **Berliner, L. J. and McConnell, H. M.**, A spin-labeled substrate for α-chymotrypsin, *Proc. Natl. Acad. Sci. U.S.A.*, 55, 708, 1966.
27. **Berliner, L. J. and Wong, S. S.**, Manganese(II) and spin-labeled uridine 5′-diphosphate binding to bovine galactosyltransferase, *Biochemistry*, 14, 4977, 1975.
28. **Berliner, L. J.**, Spin labeling in enzymology. Spin labeled enzymes and proteins, *Methods Enzymol.*, 49G, 418, 1978.
29. **Berliner, L. J. and Wong, S. S.**, Spin labeled sulfonyl fluorides as active site probes of protease structure. I. Comparison of the active site environments in α-chymotrypsin and trypsin, *J. Biol. Chem.*, 249, 1668, 1974.
30. **Berliner, L. J.**, Urea denaturation of active-site spin-labeled α-chymotrypsin, *Biochemistry*, 11, 2921, 1972.
31. **Ogawa, S., McConnell, H. M., and Horwitz, A.**, Overlapping conformation changes in spin-labeled hemoglobin, *Proc. Natl. Acad. Sci. U.S.A.*, 61, 401, 1968.
32. **Griffith, O. H. and Jost, P. C.**, Lipid spin labels in biological membranes, in *Spin Labeling: Theory and Applications*, Berliner, L. J., Ed., Academic Press, New York, 1976, chap. 12.
33. **Bauer, R. S. and Berliner, L. J.**, Spin label investigations of α-chymotrypsin active site structure in single crystals, *J. Mol. Biol.*, 128, 1, 1979.
34. **Krugh, T. R.**, Spin-label-induced nuclear magnetic resonance studies of enzymes, in *Spin Labeling: Theory and Applications*, Berliner, L. J., Ed., Academic Press, New York, 1976, chap. 9.
35. **Cohn, M., Diefenbach, H., and Taylor, J. S.**, Magnetic resonance studies of the interaction of spin-labeled creatine kinase with paramagnetic manganese-substrate complexes, *J. Biol. Chem.*, 246, 6037, 1971.
36. **Michon, P. and Rassat, A.**, Nitroxides. LXIX. 1,4-bis(4′,4′-dimethyloxazolidine-3′-oxyl)cyclohexane structure determination by electron spin resonance and nuclear magnetic resonance, *J. Am. Chem. Soc.*, 97, 696, 1975.
37. **Ciecierska-Tworek, Z., Van, S. P., and Griffith, O. H.**, Electron-electron dipolar splitting anisotropy of a dinitroxide oriented in a crystalline matrix, *J. Mol. Struct.*, 16, 139, 1973.
38. **Flohr, K., Paton, R. M., and Kaiser, E. T.**, Studies on the interactions of spin-labeled substrates with chymotrypsin and with cycloamyloses, *J. Am. Chem. Soc.*, 97, 1209, 1975.
39. **Berliner, L. J. and Wong, S. S.**, Evidence against two ′pH locked conformations of phosphorylated trypsin, *J. Biol. Chem.*, 248, 1118, 1973.
40. **Hsia, J. C., Kosman, D. J., and Piette, L. H.**, ESR probing of macromolecules: spin-labeling of the active sites of the proteolytic serine enzymes, *Arch. Biochem. Biophys.*, 149, 441, 1972.
41. **Grunwald, J.**, Studies on the Denaturation of Proteolytic Enzymes: α-Chymotrypsin and Papain, Ph.D. thesis, Technion, Haifa, 1975.
42. **Berliner, L. J.**, Applications of spin labeling to structure-conformation studies of enzymes, in *Progress in Bioorganic Chemistry*, Kezdy, F. and Kaiser, E. T., Eds., John Wiley & Sons, New York, 1974. chap. 1.
43. **Chignell, C. F.**, Spin labeling in pharmacology, in *Spin Labeling II: Theory and Applications*, Berliner, L. J., Ed., Academic Press, New York, 1979, chap. 5.
44. **Erlich, R. H., Starkweather, D. K., and Chignell, C. F.**, A spin label study of human carbonic anhydrases B and C, *Mol. Pharmacol.*, 9, 61, 1973.
45. **Berliner, L. J., Miller, S. T., Uy, R., and Royer, G. P.**, An ESR study of the active-site conformations of free and immobilized trypsin, *Biochim. Biophys. Acta*, 315, 195, 1973.
46. **Berliner, L. J. and Bauer, R. S.**, Structural changes within unit cell subunits in crosslinked α-chymotrypsin crystals, *J. Mol. Biol.*, 128, 165, 1979.

Chapter 2

BIOCHEMICAL APPLICATIONS OF ELECTRON SPIN RESONANCE SPECTROSCOPY

Daniel J. Kosman and Robert D. Bereman

TABLE OF CONTENTS

I. INTRODUCTION

There are numerous review articles on electron spin resonance spectroscopy (ESR), as well as on the application of this technique to biological systems.[1-10] For the research worker already involved in the area of paramagnetic resonance, these excellent reviews offer a great deal of insight into and discuss in depth the application of paramagnetic resonance spectroscopy. Yet, we are not aware of any work whose primary goal has been to introduce researchers to the practical considerations of paramagnetic resonance. Since we are both "chemists" who employ ESR to study biological problems and have had to go through some trying experiences as we learned about this technique and struggled to introduce our students to it, we feel we are in a position to save the newcomer some time by considering a more practical approach to paramagnetic resonance.

It may be, therefore, that some readers will find this chapter lacking in rigor. To those, we recommend the more advanced reviews. Since this work has resulted from our practical work, you may find references to subjects of your particular interest lacking. For this we apologize and encourage you to follow up these areas on your own.

II. BASIC PRINCIPLES OF ESR

A. Introduction

Magnetic resonance relies on the behavior of particles that possess both a magnetic moment and angular momentum. Some nuclei and all unpaired electrons possess these characteristics. The angular momentum of an electron is represented by a spin vector, S, which can have the values of $\pm\hbar/2$; this is also called the spin of the electron and in units of $\hbar$ can take only the values $\pm$ ½ in any direction specified by the experiment (these are the spin quantum numbers, since $S_z = m_s\hbar$). The magnetic moment vector of the electron μ is related to the angular momentum or spin by Equation 1.

$$\mu = \frac{-g\beta \mathbf{S}}{\hbar} \text{ or if } \mathbf{S} \text{ is expressed in units of } \hbar, \qquad (1)$$

$$\mu = \pm \frac{g\beta}{2}$$

β is the Bohr magneton (μ_o is sometimes used) and has the value 0.92732×10^{-20} erg/gauss in the cgs system. The value g is a dimensionless factor that is simply the proportionality constant between the angular momentum or spin and magnetic moment of an electron. For a "free" electron, it has the value 2.002319; however, the g value of an electron in an atomic, or more commonly, molecular orbital will be different from this. That is, the characteristic angular momentum or spin of an electron, S, will give rise to different magnetic moments depending upon the atomic or molecular orbital to which the electron is constrained. Although this aspect will be discussed more fully in Sections II.D and F, one should appreciate the information obtained about the electronic system being investigated by the determination of the system's g value(s).

The minus sign in Equation 1 shows that the magnetic moment and angular momentum vectors point in opposite directions. This can be rationalized if one uses the "left hand rule" that predicts the relationship between the electric current in a wirecoil and the direction of the lines of force of the resultant magnetic field. Since by definition the direction of the magnetic moment is the same as the direction of the positive charge of the system (the right hand rule), for a spinning electron, the opposite is true. Thus, if the electron is placed in a magnetic field, the electron magnetic moment will align itself so that the angular momentum (spin) will be opposed to the field. If the energy of such a moment in a field is

$$\mathcal{H} = -\mu \cdot \mathbf{H} \tag{2}$$

or if we define the direction of the magnetic field as H_z,

$$E \text{ or } \mathcal{H} = -\mu H_z \text{ (or } H_o) = \frac{g\beta}{\hbar} \mathbf{S}\, H_o \tag{3}$$

then the Zeeman energy of the electron can take the values $-g\beta H_o/2$ and $+ g\beta H_o/2$. As noted above, the minimum energy state is that in which the spin angular momentum vector $(-\hbar/2)$ is aligned opposite to the magnetic moment vector $(g\beta/2)$. The ground spin state of an electron in a magnetic field is $-\frac{1}{2}$; the "excited" spin state is $+\frac{1}{2}$. The electron paramagnetic (or spin) resonance experiment involves the excitation of an electron from the ground to the excited state, in essence, the realignment of both the angular momentum and magnetic moment vectors in the applied magnetic field, H_o. The difference between these states is

$$\Delta E = h\nu = g\beta H_o \tag{4}$$

Equation 4 defines the "resonance" condition — that combination of electromagnetic radiation ($h\nu$), magnetic field (H_z defined as H_o), and orbital behavior of the electron (the g value), which satisfies the equation. Obviously, in the experiment only g is invariant; it is a fundamental property of the electron. Thus, resonance can be obtained at a variety of values of ν and H_o, the only requirement being that $h\nu/H_o = g\beta$. Put another way, the energy difference between the spin states depends upon the value of H_o. This behavior is commonly represented by a splitting diagram (Figure 1). What distinguishes one unpaired electron from another are the slopes of energy dependencies illustrated in Figure 1, which are equal to $\pm\, g\beta/2$; that is, as emphasized above, on the g value of the electron. The larger the g value, the larger the dependence of ΔEon H_o (see later, Figure 2).

This type of behavior distinguishes magnetic resonance from other spectroscopic techniques. The experimentalist can "select" the energy difference between the ground and excited state(s). The choice is dictated by two considerations: (1) the sensitivity and resolving power of the experiment, and (2) the availability of sources of electromagnetic radiation ($h\nu$) and corresponding magnetic fields (H_o). The sensitivity of the spectrometer does, in part, depend upon the magnetic field used. Clearly, the absorption of energy by the sample depends upon the number of electrons in the ground, $m_s = -\frac{1}{2}$ spin state (N^-). However, there is a concurrent stimulation of spins dropping into the ground from the excited state whose population is N^+. Thus, the net change, n, is given by

$$n = N^- - N^+ \tag{5}$$

The difference, n, is determined by a Boltzman distribution, of course, $N^+/N^- = e^{-\Delta E/kT}$; with $\Delta E = g\beta H_o$, this becomes $e^{-g\beta H_o/kT}$. Thus, as H_o becomes larger, so does n, and for a given number of unpaired spins, the greater the absorption of energy.

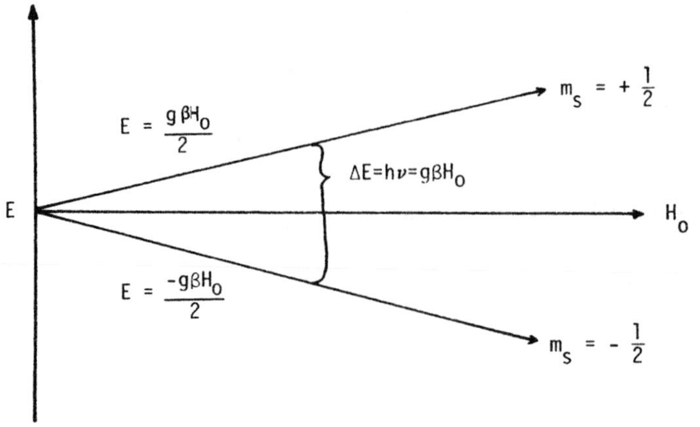

FIGURE 1. Simple splitting diagram indicating the relationship between the energy of the two electron spin states and the magnitude of H_o.

To appreciate the importance of this relationship between field strength and sensitivity, consider a quantitative example. At a field of 3000 G (an X-band spectrometer operating with $\nu \approx 9$ GHz as discussed later), $e^{-g\beta H_o/kT} = N^+/N^- = 0.999$ for T = 298°K and g = 2. That is, the difference between the ground and excited states (n) is approximately 2 parts in one thousand. To compare this value with that obtained at 12,000 G (a Q-band instrument operating with $\nu \approx 35$ GHz), the use of Equation 6 is convenient.

$$n = \frac{Ng\beta H_0}{2kT} \tag{6}$$

N is the concentration of spins (total) in the sample and k is the Boltzmann constant, 1.38×10^{-16} erg/°K. Equation 6 predicts that the population difference is linearly dependent on the magnetic field strength. Assuming other factors to be the same, a Q-band spectrometer should be four times more sensitive than an X-band one. Not surprisingly, that is not true, quantitatively, although an increase is generally obtained.

Equation 6 also predicts that the sensitivity is inversely proportional to the temperature. This means that absorption observed at the temperature of liquid nitrogen (77°K) should be, and generally is, four times as intense as the same resonance observed at room temperature. A further increase in sensitivity is possible by approaching the temperature of liquid helium (4.2°K), although other factors, particularly saturation, then become important as will be discussed in a later section (II.C).

The third "variable" in Equation 6 is the g value. The equation indicates that at a given field and temperature, an electron spin system with a larger g value will exhibit a more intense absorption. This is reasonable since the term $\Delta E = g\beta H_o$ is larger, $N^+/N^- = e^{-g\beta H_o/kT}$ will be smaller, and thus $n = N^- - N^+$ will be larger. However, in order to stimulate excitation at this given value of H_o (fixed field), for the larger g value a larger value of $h\nu$ is required. That this is so is illustrated in Figure 2. Conversely, at a fixed value of $h\nu$ (fixed frequency), less H_o is required to bring the electron with the higher g value into resonance. This is indicated in Figure 2 as well. In fact, ESR spectrometers are fixed frequency, variable field instruments. Thus, as we will develop further below, the spectrum generated by going from a low to high (weak to strong) magnetic field is, in part, a "g-value" spectrum, with systems with larger g values at

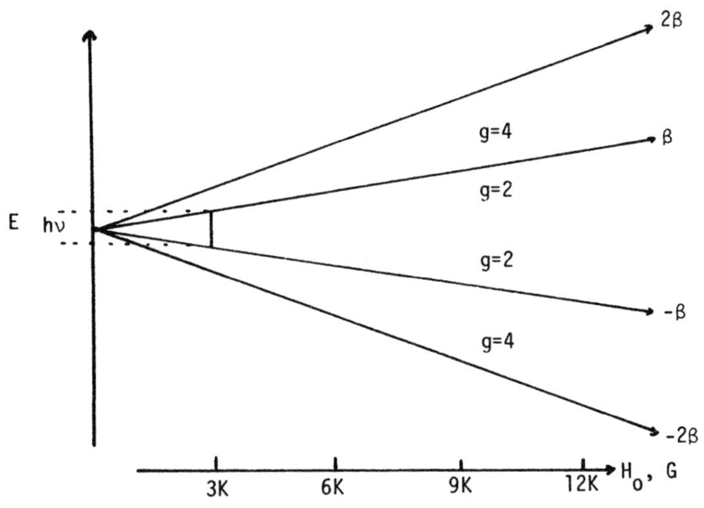

FIGURE 2. Splitting diagram for two spins which have g values of 2 and 4, respectively, as noted in the figure. The values of βH_o are the slopes of the four states. The energy of the exciting electromagnetic radiation is indicated as hν. Of sufficient energy to cause resonance (excitation) at 3000 G for g = 2, this radiation is only ½ of that needed to cause resonance of a g = 4 electron at this field strength. A g = 4 electron is in resonance, therefore, at lower field (1.5 kG in figure).

resonance in weaker fields than those with smaller g values. The result is, of course, that signal intensity is unrelated to g value, since ΔE is a constant of the experiment.

This discussion relates directly to consideration of the resolving power of the instrument noted above. Again, Figure 2 is useful. For convenience, let us assume that ν as indicated in the figure is ∿ 9 GHz. The g = 2 electron comes into resonance at about 3 kG, while the g = 4 electron is in resonance at half that value, 1.5 kG. There is a 1.5 kG separation. Compare this separation to that obtained using a Q-band instrument, one that employs a frequency of ∿ 35 GHz. The two g values will be separated by ∿ 6 kG, with the resonances appearing at 6 (g = 4) and 12 kG (g = 2). Thus, by employing a higher frequency (energy) of exciting electromagnetic radiation, one spreads out the g-value spectrum in the varying magnetic field, improving the potential spectral resolution. The usefulness of employing varying values of ν (and H_0) to spread out the g value spectrum will be noted later in the section on nuclear hyperfine couplings; these latter interactions are field-independent and are thus distinguishable from g value differences.

Although sensitivity and resolution improve with increases in frequency and field, the choice of these dependent experimental variables is limited. Only two frequencies are commonly available, 9 and 35 GHz, which as noted above, require fields of ∿ 3 kG and ∿ 12 kG for resonance at g = 2. The klystrons (radar "tubes") that generate the microwave radiation (3.2 cm and 8 mm, respectively) and crystals that act as detectors of the energy absorbed by the sample are very sensitive and are tunable only over a very limited range of frequencies. Thus, a fixed frequency instrument is necessary. On the other hand, the magnetic field required for a 4 mm spectrometer, for example, cannot be produced by the simple electromagnets commonly used in ESR spectrometers; a superconducting magnet is necessary. So, while higher frequencies and fields do yield improvements in resolution and sensitivity, practical considerations limit the choice. In fact, the g-value resolution at 35 GHz is as good as is often necessary. At 12 kG (g = 2) and an inherant linewidth of 1 G, g value differences of ∿ 0.02%

should be detectable. Since the linewidths of biological samples are commonly greater than 5 to 10 G, the resolution obtainable using a Q-band instrument is more than sufficient.

In summary, electrons possess spin and thus angular momentum. Since they are also charged, this angular momentum produces a magnetic moment. The magnetic moment and the spin of an unpaired electron in a metal ion or organic radical will be oriented in an applied magnetic field H_o. Since the spin is quantized ($m_s = \pm \frac{1}{2}$) two energy states will be available, $\pm g\beta H_o/2$. Excitation of an electron from the ground spin state ($m_s = -\frac{1}{2}$, $E = -g\beta H_o/2$) to the excited spin state ($m_s = +\frac{1}{2}$, $E = +g\beta H_o/2$) requires the presence of incident electromagnetic (microwave) radiation, ν, such that $h\nu = g\beta H_o$, the energy difference between the two spin states. In practice, ν is fixed, and H_o is varied; "resonance" occurs when $H_o = h\nu/g\beta$, i.e., the sample absorbs energy. g is a fundamental property, a "fingerprint", of the electron. Sensitivity and resolution increase with increasing values of $h\nu$ and H_o; sensitivity also increases with decreasing temperature.

One final point — not all systems possessing unpaired electrons exhibit ESR spectra. In particular spin even systems (those that have integer spins, i.e., 1,2, etc.) do not have allowed transitions between the ± spin states. Thus, the Mo (IV) oxidation state — a $4d^2$, s = 1 system — is not detectable by ESR. Similarly, another d^4 system, the Mn (III) ion with s = 2, is also ESR silent. This characteristic of the allowedness of spin transitions must be kept in mind when interpreting the meaning of the absence (or presence) of an ESR spectrum attributable to a specific oxidation state of a metal ion.*

B. Generating an ESR Spectrum: Field Modulation and the First Derivative Mode

As discussed above, absorption of microwave energy, $h\nu$, by the sample (the electron) occurs when H_o, the magnetic field experienced by the electron reaches a value that satisfies the relationship, $h\nu = g\beta H_o$. Since the energy interval available from the incident electromagnetic radiation is fixed at $h\nu(=\Delta E)$, a spectrum is generated by steadily changing the differences between ground and excited spin states of the electron spin systems by changing H_o, making the differences equal to $h\nu$ for each system, in turn. In going from lower to higher values of H_o, one brings systems into resonance that have steadily decreasing g values; i.e., as H_o increases, that value of g necessary to satisfy $h\nu = g\beta H_o$ decreases.

In principle, the absorption of the microwave energy could be monitored as in any other absorption spectrophotometer, by sampling directly the difference in power between the incident and transmitted or reflected radiation. However, as outlined above, the net change in the populations of the ground and excited states during resonance is very small due to the very small energy difference between these two states. Consequently, the net absorption is small and typically less than the random noise of an electronic circuit.

To allow for the necessary signal amplification, the d.c. absorption is converted to an a.c. one by modulating the external magnetic field, H_o, at a fixed frequency with a secondary field H_m. The detector is driven at the same frequency and is therefore sensitive only to (selects out) input which varies with the frequency of modulation; essentially, the d.c. noise is ignored. How this modulation works is outlined briefly below.

As H_o is swept through the g-spectrum (at, for example, 125 G/min), the small modulating field H_m oscillates at a frequency, typically 100 kHz, with an amplitude of 0.1 to 10 G. (These are limit values common to biological ESR and do not represent the

* A special, but relatively common case is the triplet state that exhibits a weakly allowed $\Delta m_s = 2$ transition at "half-field" (0.5 H_o of g = 2); this generally intense line is a strong indication of a triplet state.

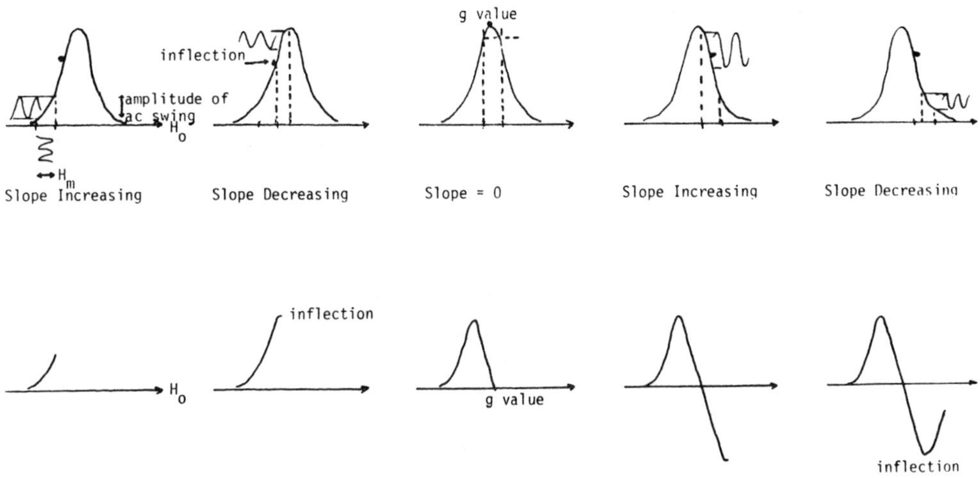

FIGURE 3. Relationships between absorption and first derivative curves as encountered in ESR.

actual limits of most instruments.) The detector circuit is operating at 100 kHz, also, in such an instrument. Thus, at a given value of H_o, the absorption will vary sinosoidally with a frequency of 100 kHz and an amplitude determined by the absorption difference between the limits of the modulation field, $H_o \pm H_m$. The detector circuit will respond only to this difference, not to the absorption itself. Amplification of this difference allows for better signal-to-noise, since only the 100 kHz noise in the circuit will be amplified, too; all the low frequency noise characteristic of the d.c. detector is lost.

The result of this modulation is that the "absorption" is displayed in the first derivative, or dispersion, mode. That is, the absorption difference between $H_o \pm H_m$ represents the instantaneous slope of the absorption curve at H_o. As H_o is swept through the absorption envelope, the change in slope per increment in H_o is measured and thus a first derivative is generated. This is illustrated in Figure 3, which focuses on the three unique regions in the absorption, and corresponding first derivative curves. These are the two inflection points in the absorption curve, each of which is associated with a maximum (or minimum) in the derivative spectrum (the peak and trough), and the absorption maximum that corresponds to zero slope or the "crossover" point on the derivative curve. This latter point defines the "g value" for the system, or more precisely, the position of the transition in the magnetic field.

The important variable in this method of signal detection is the magnitude of H_m. Based on the discussion above and Figure 3 one can recognize that the larger H_m the larger the potential difference between the absorption measured at the associated limits of the modulation field. Consequently, a larger a.c. swing would result and yield more signal. However, one should also appreciate the limits to this conclusion, namely, at values of H_m comparable to the inherent absorption bandwidth (or greater), real differences might actually be obscured, and the amplitude of the a.c. deflection reduced. Such "over-modulation" reduces the amplitude of the detected signal and, perhaps more importantly, distorts the first derivative envelope and causes a loss of definition of the true slope of the absorption curve. Over-modulation causes a loss of both sensitivity and resolution. The limits of H_m common to biological ESR as noted above are determined by the range of linewidths typical of paramagnetic, biological samples. In general, for the highest spectral resolution, a modulation amplitude of no more than 0.1 of the linewidth of the narrowest line is recommended. Larger H_m values, up to about $\sqrt{3}$ of the linewidth, are sometimes used when maximization of instrument sensitivity is sought.

C. Relaxation and Saturation

A variety of ways to increase an ESR signal have been presented. These methods include those that increase the difference in the population of ground and excited spin states, n, and thus the net flux of electron spins from the former to the latter state. However, the intensity of the signal, I, is not only dependent upon the value of n, but also on the energy density or power of the incident microwave radiation. As the power increases the transition probability increases, i.e., the signal becomes more intense. However, such simple behavior is not observed experimentally. While the signal intensity does in general respond as some function of the microwave power, at a certain level, the signal actually starts to decrease, to broaden, and to eventually disappear altogether. This saturation broadening or power saturation is an important property of paramagnetic centers, since different types of systems, e.g., organic free radicals, metal ions, etc., saturate at different and characteristic levels of microwave power.

The saturation phenomenon is due simply to the equalization of the ground and excited state spin populations since as $n \to 0$, I also $\to 0$. Thermal processes operate to maintain the equilibrium value n determined by the field strength, the g value, and the temperature. These processes "relax" the system, return it to equilibrium. The time constant for this thermal equilibration is the spin-lattice relaxation time, T_1. This process allows for the dissipation of the energy absorbed by the electron spins as heat transferred to the medium surrounding these spins, reestablishing the normal Boltzmann distribution. This energy is that associated with the magnetization of the electron spins along the direction of the external magnetic field, that is, those spins aligned with and against this field, and thus this time constant, T_1, is also called the longitudinal relaxation time.

In order for the electron spin system to relax efficiently, (for T_1 to be short), the system must be coupled to the lattice; that is, there must be available in the lattice thermal (vibrational) energy levels exhibiting energy differences (quantized as phonons) similar to $g\beta H_o$. Since the thermal energy levels are temperature-dependent, the marked temperature dependence noted for power saturation is not urprising. Thus, while very low temperatures increase n, they also tend to lengthen T_1. Under such conditions, $n \to 0$ at even very low microwave power and consequently the anticipated increase in sensitivity due to decreasing temperature (Equation 6) cannot be realized completely.

Although the spin-lattice relaxation process is what dissipates the energy absorbed by the electron spin in going from the ground to the excited state and thus is the mechanism that reestablishes the Boltzmann distribution and determines the saturation behavior of the spin system, another relaxation mechanism is also of importance. This is the spin-spin relaxation process. The time constant for this process, T_2, is also called the transverse relaxation time since $1/T_2$ is the rate at which the net magnetization of the xy component of the spin system at resonance changes. This process will be considered in more detail in a later section; thus suffice to say here that those spins that absorb microwave energy possess a net magnetization aligned with the field, H_o (along the z axis), as well as $\perp$ to it, in the xy plane. (Strictly speaking, the magnetization is a vector; the two components of the magnetization referred to here are the projections of this vector.) While $1/T_1$ is the rate of change of the z component of the magnetization (and the return to the equilibrium value of n), $1/T_2$ is the rate of change of the xy component. As this latter component breaks up so that the individual spin vectors, while still in a "excited" T_1 state, exhibit a variety of xy orientations, the laboratory field, H_o, will be modified slightly by these randomly oriented magnetic dipoles. Each electron will be experiencing a slightly different resultant field and will thus be in resonance at a slightly different value of H_o, i.e., the resonance line will broaden. As these local fields become more random, the line broadens still further. Thus, the relationships between linewidth and the rate of the relaxation processes associated with T_1 and T_2 are inverted. While short T_1 values contribute to narrow lines (not easily saturated), small values of T_2 cause line broadening. The linewidth is, in fact, given by

$$\frac{1}{T_2} = \frac{\pi g \beta \Delta H_{1/2}}{h} \tag{7}$$

where $\Delta H_{1/2}$ is the width of the resonance (absorption) line at half-height given in gauss. In most cases, the value of $\Delta H_{1/2}$ can be taken as the peak-to-trough displacement in the first derivative spectrum (cf. Fig. 3). The effect of microwave power and temperature are discussed further in Appendix B.

D. Anisotropy and the g Value

The anisotropy or directionality of the interactions between the various magnetic moments in a molecular system containing an unpaired electron, and an external magnetic field is a dominant characteristic of biological ESR. This is so because in most cases, the molecular system is fixed (as in a frozen solution) or is rotating more slowly than the frequency of exciting microwave radiation. Therefore, an unpaired electron confined to a molecular orbital whose rotation correlates with that of a biological macromolecule will itself exhibit a rate of orbital reorientation relative to the external magnetic field so slow as to appear fixed. In most cases, the nature of the ESR spectra derived from paramagnetic biological systems is very dependent on the relative orientations of the field and orbital motion of the electron.

It is the g value that exhibits this directionality or anisotropy. In Section II.A, the g value was introduced as the proportionality constant that related the spin (angular momentum) of an electron to the magnetic moment. However, the spin of an electron is not the only motion that an electron exhibits; it also has orbital motion and can generate a magnetic moment associated with its path in an atomic or molecular orbital. Whether or not it does so depends on the angular momentum associated with the orbital. An s orbital, for example, has no angular momentum associated with it ($L = 0$), while p and d orbitals do. Also true is that s orbitals are isotropic or are spherically symmetric; they have no "directionality". Consequently, an unpaired electron confined to an s orbital would have a g value very close to the "free-spin" value of 2.0023. In contrast, an electron in an orbital that possessed a substantial degree of p or d character would be expected to exhibit a g value quite distinct from 2.0023, and, furthermore, since such orbitals are not spherically symmetric, the g "value" measured would very likely depend on the relative orientation of the electron's orbital motion (the magnetic moment associated with its orbital angular momentum) and the external magnetic field. There are two very important concepts here: first, how the magnitude of a g value reflects the orbital angular momentum, and thus the orbit (or wave function) of the electron; and secondly, since orbits that possess angular momentum are inherently anisotropic, so should the g value of an electron in such an orbit.

In the simplest way, the g value indicates that for a given spin angular momentum one gets a certain magnetic moment. Larger values of g suggest that for the same amount of spin angular momentum, a "stronger" moment is generated; one can reasonably picture this as being due to the contribution to the total moment of the electron by the electron's orbital angular momentum. There is a "coupling" between the spin and orbital angular momenta — the term spin-orbit coupling is often used. Indeed,

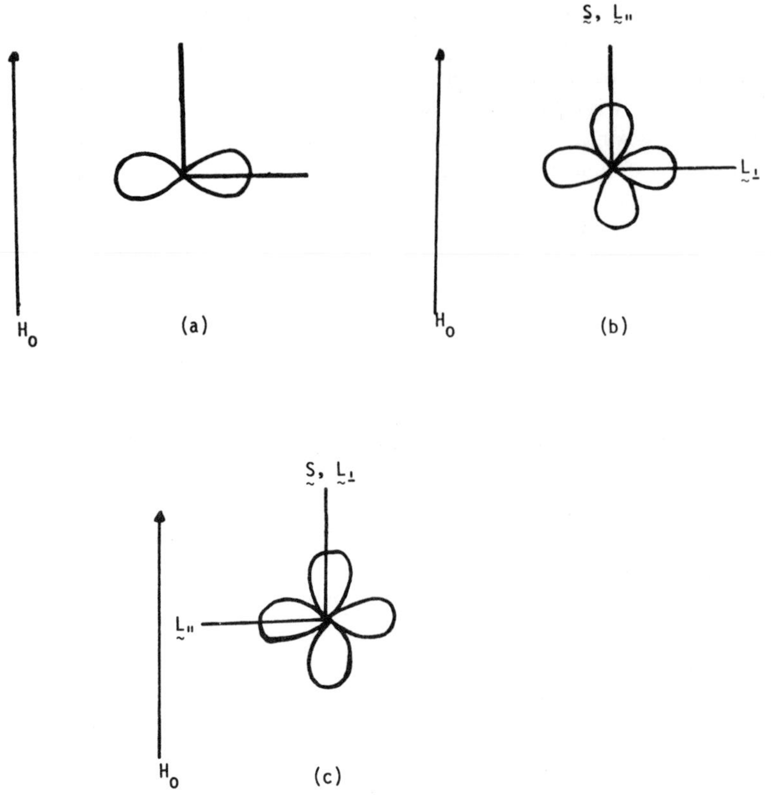

FIGURE 4. Relationship between the orientations of the external magnetic field, H_o, the spin moment, S, and orbital moment, L. In (a) H_o (and thus S) is orthogonal to L; there is no coupling between S and L. In (b) H_o and S are again orthogonal to $L_\perp$, but are aligned (parallel to) $L_\|$. Thus S and $L_\|$ can couple and the orbital motion associated with $L_\|$ does contribute to the g value measured ($g_\|$) with the system in the orientation indicated. In (c) the system has been rotated 90°; H_o and $L_\perp$ are now aligned. S and $L_\perp$ can couple, and $l_\perp$ contributes to the g value measured ($g^\perp$).

the efficiency (measured as energy in cm^{-1} or, preferably, the Kaiser, K) of this coupling varies from system to system, thus spin-orbit coupling is an important component of the g value that makes characterization of the paramagnetic center possible.

At the same time this coupling must have directionality; it must be anisotropic. For example, if the spin and orbital magnetic moments are orthogonal (Figure 4a), there can hardly be coupling between them. Consequently, the orbital motion associated with the orbital shown, makes no contribution to the g value for the system in this orientation. It will be the motion of the electron in orbitals that have significant probability in the same direction as the external magnetic field (and thus have angular momenta that are aligned with the spin moment of the electron), which will couple to the spin. This is illustrated in Figures 4b and c in which axes of the system are defined arbitrarily relative to the external magnetic field (parallel, ‖, and perpendicular, ⊥).

Figures 4b and c illustrate the extreme orientations of the system; clearly, intermediate orientations are also possible in which H_o is aligned between $L_\|$ and $L_\perp$. Associated with these orientations will be g values to which <u>both</u> L and $L_\perp$ contribute as indicated by Equation 8.

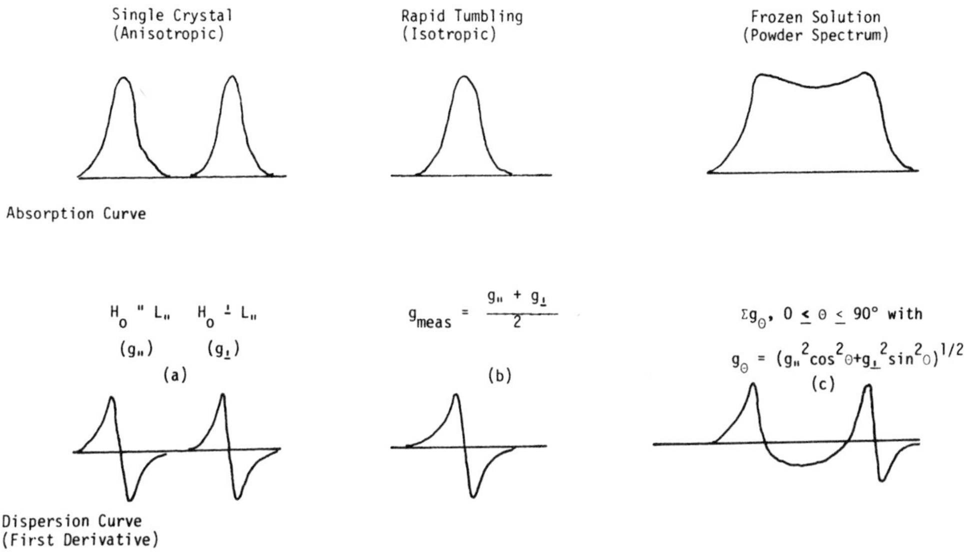

FIGURE 5. Absorption and first derivative curves for the resonance behavior of an electron in an orbital pair as in Figure 4. In (a), two distinct orientations of the crystal yield the limiting values, $g_\parallel$ and $g_\perp$. In (b), rapid tumbling of the molecular framework averages out the orientations and thus the g values. In (c), the freezing out of all possible orientations generates an envelope of g_θ values. This same envelope can be generated in (a) by rotating $g_\parallel$ into $g_\perp$.

$$g_{\mathrm{meas}} = (g_\parallel{}^2 \cos^2\theta + g_\perp{}^2 \sin^2\theta)^{1/2} \tag{8}$$

θ is defined as the angle between $L_\parallel$ and H_o.

The preceding discussion and Equation 8 focused on a system held in a particular orientation relative to H_o as one might encounter in a study of a single crystal. There are two other situations of interest. One is that associated with a system as in Figures 4b and c, which has a rotational lifetime similar to or less than a spin state lifteime; that is, one in which $L_\parallel$ and $L_\perp$ are realigned relative to H_o during the resonance process. The resulting resonance will be isotropic or be independent of the physical orientation of the sample in H_o, and the measured g value will be the average of $g_\parallel$ and $g_\perp$, i.e., $g_{meas} = g_\parallel + g_\perp/2$. The other situation is that in which this "solution" of randomly oriented, fast tumbling molecules is frozen. The resulting glass or (for aqueous solutions in particular) "powder" contains all orientations and, for each, a g value as described by Equation 8. Thus, the "powder" spectrum obtained will be a composite of all of these values, with the limits defined by $g_\perp$ ($\theta = 90°$) and $g_\parallel$ ($\theta = 0°$). The spectra illustrated for these three cases are presented in Figures 5a, b, and c.

This discussion can be extended to a more realistic three-dimensional model, one in which $g_\perp$ is composed of two g tensors, g_{xx} and g_{yy}, which may or may not be equivalent. If $g_{xx} = g_{yy} = g_\perp$, the only addition to what has been presented above is the numerical weighting of $g_\perp$ necessary since two principle orientations contribute to this value, while only one contributes to g_{zz} ($g_\parallel$). Thus, when H_o is $\parallel$ to $L_\parallel$ only that orientation is contributing to the resonance. When H_o is $\perp$ to $L_\parallel$ ($\parallel$ to $L_\perp$) the g_{meas} reflects both g_{xx} and g_{yy} and thus is sampling twice the number of orientations. This is readily apparent in the powder spectrum of such a system, commonly termed an axial one ($g_{xx} = g_{yy} \neq g_{zz}$) as in Figure 6a. Compare this spectrum to that in Figure 5c.

If $g_{zx} \neq g_{yy}$ ($\neq g_{zz}$), a rhombic system is indicated (Figure 6b). The transitions now

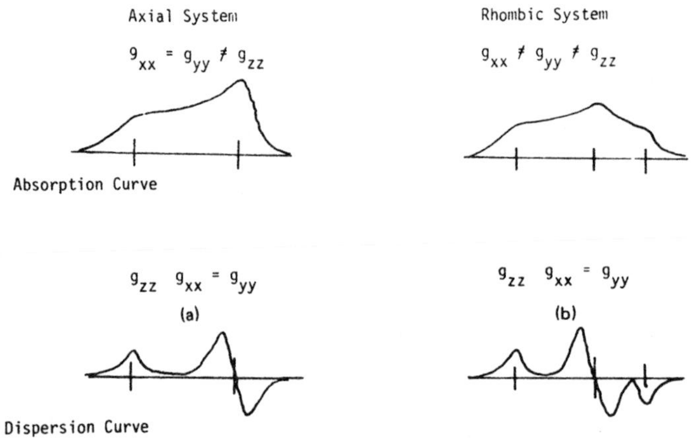

FIGURE 6. Powder spectrum of the two most common symmetry systems encountered in biological ESR: (a) axially symmetric orbital environment with two characteristic g values; (b) the rhombic system in which $g_{xx} \neq g_{yy}$ and the resulting three, unique principal orientations of the molecular system.

are spread out in the g-spectrum. As discussed in Section II.F, these g values can be associated with specific atomic or molecular orbitals, and can be used in a semiempirical way to describe the characteristics of such orbitals.

E. The Hyperfine Interaction

As noted above, the spin Hamiltonian, or energy function of an unpaired electron is given simply as $\mathcal{H} = g\beta/\hbar\ SH_o$. Also it was stated that the field, H_o, could be modified by the orbital motion of the electron. That is, the orbital motion contributed its own magnetic moment or field, which then altered the field actually experienced by the electron itself. This "spin-orbit coupling" is represented by the deviation of the g value from 2.0023. The field experienced by the electron can also be changed by yet another internal magnetic moment, that associated with nearby nuclei that have non-zero nuclear spins. This coupling between the electron and nuclear moments is what is known as the hyperfine interaction.

The spin Hamiltonian can be written to include the contributions due to the magnetic properties of coupled nuclei as given in Equation 9.

$$\mathcal{H} = \frac{g\beta}{\hbar} SH_o - g\beta g_N\beta_N \left[\frac{\mathbf{S}\cdot\mathbf{I}}{r^3} - \frac{3(\mathbf{S}\cdot\mathbf{I})(\mathbf{I}^r\cdot\mathbf{r})}{r^5}\right] + \frac{8\pi}{3} g\beta g_N\beta_N |\psi(0)|^2\, \mathbf{S}\cdot\mathbf{I} + \text{nuclear Zeeman term} \qquad (9)$$

where g_N is the magnetogyric ratio or nuclear "g" factor of the coupled nucleus, β_N is the nuclear magneton, r and **r** are the distance between and vector joining the electron and nucleus, respectively, and $|(\psi(0)|^2$ is the unpaired electron spin density at the nucleus. The first term is, of course, the electron Zeeman interaction energy. The second describes the through space, dipolar magnetic interaction between the magnetic moments associated with the electron, **S**, and nuclear, **I**, spin angular momenta. The last term, the Fermi contact term, describes the effect of the field produced at the nucleus when **S** is at the nucleus. These latter two interactions are distinctive in two ways; first, the dipolar and not the Fermi contact interaction is distance-dependent. Secondly, and importantly, the dipolar and not the Fermi contact interaction is anisotropic as represented by the radius vector, r. For this reason, the Fermi contact term is often called the isotropic hyperfine interaction. As detailed later, it is only unpaired s electron density that can give rise to a contact interaction, while within the limitations set by the r^3 dependence, unpaired spin density in any orbital type could potentially contribute a dipolar interaction.

How these interactions alter the resonance spectra associated with an unpaired electron is a most significant aspect of ESR. An example is the best way to quickly introduce this. Copper(II) is a d^9 system ($S = 1/2$). The nuclear spin of either copper-63 or copper-65 is 3/2, and associated with this are four nuclear spin states, $+3/2$, $+1/2$, $-1/2$, and $-3/2$, listed in order of their energies in an applied magnetic field (the right hand rule). Thus, the ground (A) and excited (B) electron spin states, $-1/2$ and $+1/2$, respectively, are both changed by the presence of the four, different nuclear moments associated with the four nuclear spin states (see below).

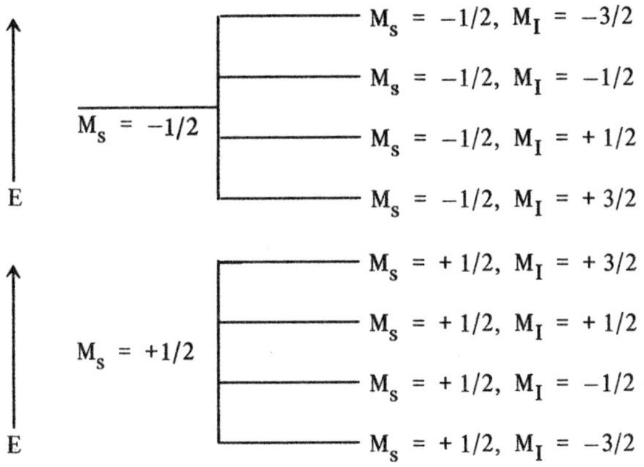

Effect of metal-based nuclear spin on M_S splitting.

The "ground state" spin system in (B) is 1/2, $-3/2$; this is because this "orientation" of M_s, the excited state, becomes more stable as the field it experiences decreases. In the $-3/2$ "orientation" of the nucleus, the nuclear moment is aligned against H_o, and thus subtracts from it. Conversely, the $+3/2$ state adds to H_o and makes the $M_s = +1/2$ state less stable, of higher energy.

Allowed electron spin transitions are those that do not involve changes in nuclear spin state, and for copper(II), then, there are four: the $+3/2$, $+1/2$, $-1/2$, and $-3/2$ transitions as they are commonly called. Their position in the magnetic field, H_o, can be understood by inspection of a splitting diagram as shown below.

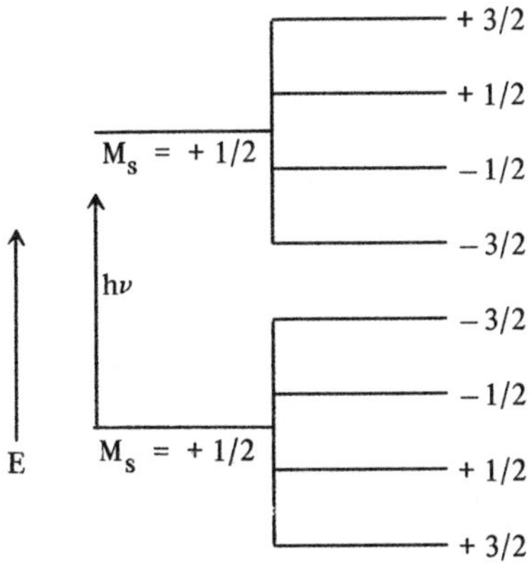

Transition between $M_S \pm 1/2$ states.

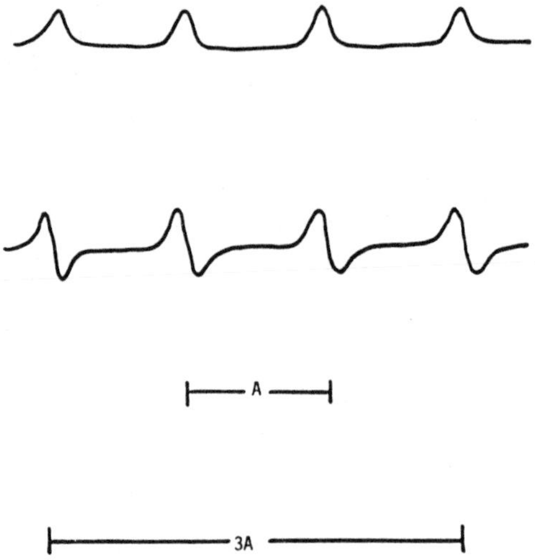

FIGURE 7. The A value for a nuclear spin 3/2 case.

Clearly, if the energy available for the transition in the absence of the nuclear hyperfine splitting is $h\nu$, the energies of these spin states must be equalized by varying H_o. The $+3/2$ transition requires less external field to reach resonance, since the nuclear moment is aligned with this field and adding to it. The $-3/2$ transition requires a higher field to overcome the dipole contribution opposed to H_o associated with the $-3/2$ nuclear spin state. Thus, the A value, or hyperfine spectrum, of this system could be represented by Figure 7.

The separation of the transitions represents the hyperfine coupling, and can result from any combination of contact and dipolar interactions. The g value is now near the center of the hyperfine pattern, and not represented by an absorbance, per se. The refined calculation of A and g values is covered in Appendix C.

F. Hyperfine Tensor

In most, but not all transition metal systems, the unpaired electron density that originates from the d orbitals of the uncomplexed ion is localized in molecular orbitals in the complex, which are localized on the metal. The effect of an unpaired electron in an orbital near the core electrons (primarily the 1s and 2s orbital electrons) is to polarize these electron pairs; that is, to attract s electron density of an opposite spin to that of the unpaired "d" electron and to repel electron density of a similar spin. This results in an effective unpaired s electron, which then yields an isotropic coupling between the nuclear spin and the electron spin. This is the isotropic term that can be obtained from a solution ESR experiment and is denoted as A_{iso} or <a>. It is difficult to give a quantitative interpretation to the value of the isotropic coupling term, but it is normally assumed that the value is roughly related to overall covalency and decreases as the covalent nature of the molecular orbitals containing the unpaired electron increases.

The principal value of the A_{aniso} term ($A_{\|}$-A_{iso} or 2/3 ($A_{\|}$ - $A_{\perp}$) is however, directly proportional to the unpaired electron density in the d portion of the molecular orbital of interest. ($A_{\|}$ and $A_{\perp}$ refer to splittings of the $g_{\|}$ and $g_{\perp}$ values.) Values for A_{aniso} for those cases in which the molecular orbital is totally ionic (100% d character) have been tabulated for various metal ions in various oxidation states.[11]

The same mechanism that results in an A_{iso} term for a metal also produces the superhyperfine splitting due to ligand atom nuclei. The term superhyperfine (shf) is applied to the additional influence of a ligand nuclear spin on the unpaired electron. The absolute magnitude of A_{ligand} is often less than 20% of the A_{metal} and thus is detected as fine structure on the individual peaks resulting from the metal hyperfine interaction.

Appendix C gives a detailed example of the determination of the g and A values for a metal complex including the application of second order corrections to the g value calculations. An accurate set of ESR parameters, the g tensors and metal or ligand nuclei hyperfine values, allows one to determine various details of the molecular orbitals in a metal complex. Two very important points need to be considered, however, if one decides to carry out these semiempirical molecular orbital calculations. Perhaps most important is that these calculations are primarily concerned with the covalency of antibonding molecular orbitals (nonbonding in some cases). The ESR experiment does not relate in a direct way to the filled bonding molecular orbitals that are primarily ligand in nature. What results is an idea about the nature of the bonding situation, which to a first order approximation is reflected by the antibonding orbitals. Secondly, no single calculation of a bonding parameter is relevant by itself. That is, one should never determine the nature of a covalent orbital in one complex and then compare this to a calculation employing ESR parameters that someone else has determined. This results from the various approaches and their approximations that have led to the many independent (and often useful) sets of equations relating bonding parameters and ESR parameters. It is also acknowledged that any approach that neglects the simultaneous use of optical data is somewhat weaker than those approaches that take the molecular orbital energy differences into account. If the ligand hyperfine tensors have been determined, then these allow a potentially more useful and accurate set of molecular orbital coefficients to be determined (see below).

The usefulness of ESR for the study of metal-containing biological systems obviously depends on the ability to detect and accurately determine the g and A tensor values. Of the many systems that have been characterized, those containing copper(II) have been the most useful, primarily because so many small molecule systems have been characterized in some detail. It is therefore often desirable to carry out metal replacement reactions to yield copper(II) substituted systems. For this reason, it is informative to confine this section to the specific case of a copper(II) complex.

Copper exists as both ^{63}Cu and ^{65}Cu and even though each nucleus has the same spin, their magnetic moments are slightly different. This produces two nearly overlapping sets of hyperfine lines in the ESR experiment. Since linewidths of biologically ESR spectra are rather broad, it is difficult to detect each set of lines and one commonly observes only the average of the two isotopes. The magnitude of the $A_{\parallel}$ (A_{zz}) and $A_{\perp}$ (A_{xx} and A_{yy}) coupling constants can be represented by the following expressions.

$$A_{\parallel} = P\left[-k\frac{4\alpha^2}{7} + (g_{\parallel} - 2.0023) + \frac{3(g_{\perp} - 2.0023)}{7}\right] \quad (10)$$

$$A_{\perp} = P\left[-k\frac{2\alpha^2}{7} + \frac{11}{14}(g_{\perp} - 2.0023)\right] \quad (11)$$

In these equations, α represents the coefficient of the d portion of the molecular orbital containing the unpaired electron (normally the $d_{x^2-y^2}$ orbital, which by symmetry consideration must be σ in nature). P and K are normally taken as constants, 35 mK (millikaiser = 10^{-3} cm^{-1}) and 0.35, respectively.

The additional equations relating the g tensor terms to the similar molecular orbital coefficients are:

$$g_{\|} = 2.0023 - \left[\frac{8\lambda}{(E_{x^2-y^2} - E_{xy})}\right] \alpha^2 \beta^2 \qquad (12)$$

$$g_{\perp} = 2.0023 - \left[\frac{2\lambda}{(E_{x^2-y^2} - E_{xz})}\right] \alpha^2 \sigma^2 \qquad (13)$$

where the $E_{x2-y2} - E_{xy}$ corresponds to the optical transition between the molecular orbitals composed primarily of the d_{x2-y2} and d_{xy} orbitals, respectively. λ is the spin orbit coupling constant for the free ion and is taken as -828 cm^{-1}, β and δ are molecular orbital coefficients for in plane and out of plane π bonding molecular orbitals, respectively.

$$\psi B_{1g} = \alpha d_{x^2-y^2} - \tfrac{1}{2}\alpha\,[-\sigma_x^{(1)} + \sigma_y^{(2)} + \sigma_x^{(3)} - \sigma_y^{(4)}] \qquad (14)$$

$$\psi B_{2g} = \beta d_{xy} - \tfrac{1}{2}(1-\beta^2)^{1/2}\,[p_y^{(1)} + p_x^{(2)} - p_y^{(3)} - p_x^{(4)}] \qquad (15)$$

$$\psi A_{1g} = \gamma d_z{}^2 - \tfrac{1}{2}(1-\gamma^2)^{1/2}\,[\sigma_x^{(1)} + \sigma_y^{(2)} - \sigma_x^{(3)} - \sigma_y^{(4)}] \qquad (16)$$

$$\psi E_g = \sigma d_{xy} - (1-\sigma^2)^{1/2}\,[p_z^{(1)} - p_z^{(3)}]/\sqrt{2} \qquad (17)$$

$$= \sigma d_{yz} - (1-\sigma^2)^{1/2}\,[p_z^{(4)}]/\sqrt{2}$$

The normal technique in evaluating α, β, and δ is to solve Equations 10, 12, and 13 with an iterative approach and then calculate $A_\perp$ using Equation 11. If the agreement is good between the calculated and observed value of $A_\perp$, then the α, β, and δ values are acceptable. If they are not, then most likely, the wrong optical transition energies were used in Equations 16 and 17 and the process must be repeated.

While a number of coordination ligand atom environments with copper(II) are known, the most useful to biochemists are the nitrogen coordination systems because in favorable cases, superhyperfine splittings (shfs) may be observed. The number of superhyperfine lines depends on the number (n) and spin quantum number (I) of the nuclei that are coupled to the unpaired electrons. The number of lines is given by 2nI + 1.

If we refer back to the discussion of the metal hyperfine interaction, we need only consider this superhyperfine interaction as a perturbation of that representation. For example, a new "stick" diagram can be imagined (Figure 8). Thus, the observed transitions are not now a set of four lines, but a set of four lines each one of which is split into a triplet (I = 1). The separation between the individual triplet peaks would be A_N or the superhyperfine coupling constant.

As noted, ligand nitrogen atoms are perhaps the most common in biological systems. The number of lines produced by extra coupling to one nitrogen could be 3, to two nitrogens could be 5, to three nitrogens could be 7, and so on. The relative intensities of these lines for equivalent nuclei will be 1:1:1 for one nitrogen, 1:2:3:2:1 for two nitrogens, 1:3:6:7:6:3:1 for three nitrogens, and 1:4:10:16:19:16:10:4:1 for four nitrogens. Thus with more interacting nitrogens atoms, it becomes increasingly difficult to detect all the superhyperfine lines.

Hyperfine splittings due to nitrogen normally range from 10 to 20G. Superhyperfine lines whose splittings are less than 10G can in some cases be determined by the ENDOR technique as discussed below.

The values for $A_\|$ (^{14}N) and $A_\perp$ (^{14}N) can be related to bonding parameters discussed above by the following equations.

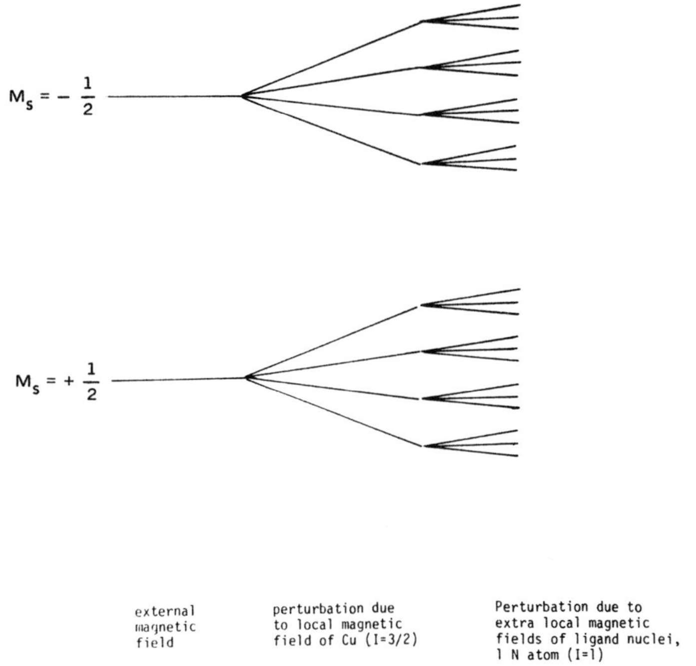

FIGURE 8. "Extra" effect of ligand nuclear spin on Figure 7.

$$A_{\parallel}(^{14}N) = \frac{\alpha^2}{4} g_N\beta_N \left[\frac{8\pi}{9} \mid \psi\, {}^2_0 \mid 2S + \frac{8}{15} <r^{-3}>2p\right] \qquad (18)$$

$$A_{\perp}(^{14}N) = \frac{\alpha^2}{4} g_N\beta_N \left[\frac{8\pi}{9} \mid \psi\, {}^2_0 \mid 2S + \frac{4}{15} <r^{-3}>2p\right] \qquad (19)$$

It is generally assumed that the squared coefficient for the d portion of the molecular orbital represents the percentage spin density at the metal center; the square of the ligand atom represents the percentage spin density at the ligand nuclei, and any remaining "percentage" in spin density is in the overlap region.

III. RESOLUTION OF INHOMOGENEOUSLY BROADENED LINES*

A. Introduction

Resonance lines that arise from identical paramagnetic centers whose magnetic environment fluctuates with time are homogeneously broadened. Saturation of such centers causes line broadening and a decrease in amplitude. Inhomogeneously broadened lines, however, arise from a set of spin packets, each of which is experiencing a different, but constant, local magnetic field. Consequently, each of these spin packets saturates independently of one another, and the resonance line is not broadened by excess microwave power. The effects and uses of power saturation are covered in greater detail in Section II.C.

Clearly, inhomogeneously broadened lines contain significant information in the form of the energies (frequencies) of the local fields modulating the behavior of each

* Section III deals with more specific applications of paramagnetic resonance spectroscopy. Thus we have attempted to supplement this section with a few references to the recent literature.

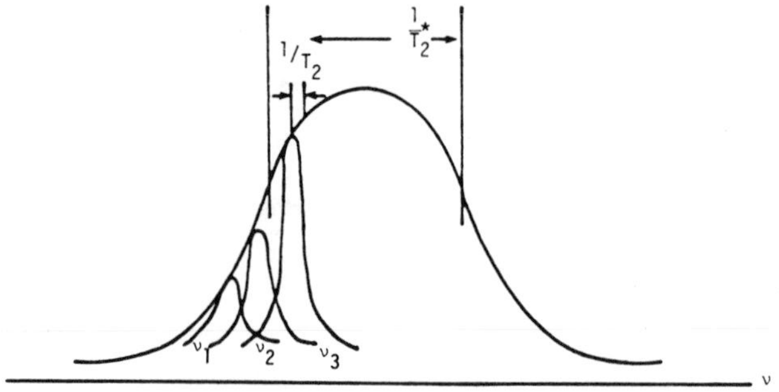

FIGURE 9. An inhomogeneously broadened resonance line expressed as the precessional frequencies (ν_i) of the individual spin packets which contribute to the resonance. The linewidths of the whole line ($1/T_2^*$) and spin packets ($1/T_2$) are indicated.

spin packet. For a given electron spin transition, this information corresponds to the nuclear spins coupling to the electron spin states; if these hyperfine interactions are similar in magnitude to the inherent linewidth (caused by "homogeneous" broadening), then they will go unresolved and, as noted, contribute still further to the width of the resonance line. This is illustrated in Figure 9. Such inhomogeneously broadened lines generally have a Gaussian shape described by the function

$$e^{-0.69}\left(\frac{T_2^*}{T_2}\right)^2$$

where $1/T_2^*$ and $1/T_2$ are the linewidths at half-height (see Section II.C) for the whole (inhomogeneously broadened) line and the individual spin packets, respectively (Figure 9).

Two resonance techniques exist for the resolution of these small nuclear hyperfine interactions: electron nuclear double resonance (ENDOR)[12-14] and the electron spin echo.[15-17] They yield the same information, namely the nuclear frequencies coupled to the electron spin. Since the magnetogyric ratios are known for all stable nuclei,[18] the determination of the nuclear frequencies at a particular magnetic field serves to identify the nuclei types that are coupled to the unpaired electron(s). Both techniques are based on the same phenomenon, as well; that is, the mixing of nuclear spin and electron spin transitions in an electron spin system at resonance.

B. ENDOR

This technique uses partial power saturation and an applied radio frequency (rf) field to stimulate electron spin transitions. The experiment can best be described by reference to a simple example (Figure 10). In this system a nuclear spin, $I = 1/2$, is coupled to an electron spin doublet ($S = 1/2$). The resultant energy levels are indicated. In addition, note also the effects on the level energies caused by the interaction between the externally applied field and the nuclear moment (the nuclear Zeeman term in Equation 9). This results in $E_{a \to b} \neq E_{c \to d}$. This difference can be calculated and is $2\, g_{Ni}\, \beta_N H_o$. Obviously, to be able to measure this energy difference means one can determine the nuclear species, N_i, as was mentioned above. The ENDOR experiment (as does the electron spin echo one) yields the values for these differences.

Consider the $E_{a \to d}$ arrow in Figure 10a. This represents the excitation of a single spin packet in a homogeneously broadened line, chosen by fixing he field and micro-

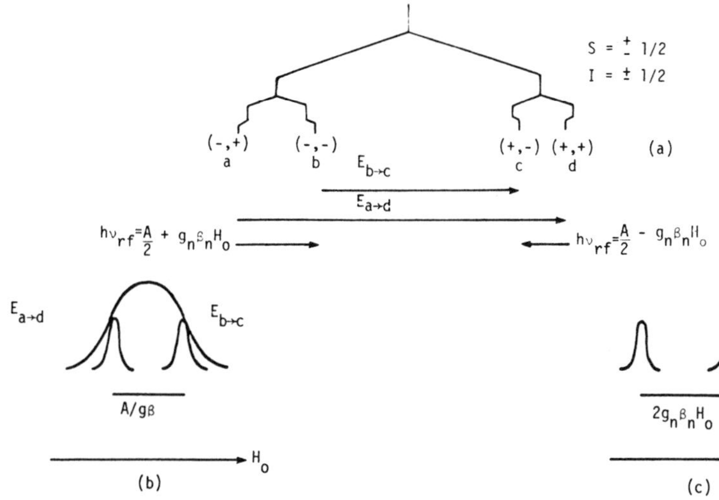

FIGURE 10. (a) Energy diagram for an S = ± ½, I = ± ½ system including nuclear Zeeman splitting ($-g_N\beta_N I\cdot H_o$). The electron spin transitions, $E_{a\to d}$ and $E_{b\to c}$ are indicated. In (b), the inhomogeneously broadened resonance line of which these two transitions are a part is shown as generated by the typical variable field experiment. In (c), the ENDOR spectrum is shown, as generated by monitoring the partially saturated a→d transition while sweeping through an rf field around $h\nu_{rf} = A/2$. The "+" and "−" ENDOR frequencies are characteristic of individual nuclei; g_N can be readily calculated from the value of the ENDOR splitting.

wave frequency of the experiment. The pair of transitions associated with this I = ½ system is indicated as well, in the variable field experiment typical of the electron resonance measurement; characteristically, the hyperfine splitting is $A/g\beta$ (Figure 10b).*

The ENDOR experiment proceeds as follows. The transition a→d is at resonance and is partially saturated by appropriate choice of microwave power. This involves, of course, a decrease in the number of spins in state a relative to d. If an rf field at right angles to H_o is now swept through a frequency range centered around A/2h, consider what will happen. At $h\nu_{rf} = A/2 - g_N\beta_N H_o$ (the energy difference between states d and c) a nuclear spin transition will be stimulated, which will reduce the population of state d (I = +½ → I = −½). This has the effect of reducing the saturation of the electron spin transition a→d, causing a stimulation of absorption of microwave energy by the electron spin packet. It is this ESR absorption that is recorded in the ENDOR experiment and is pictured in Figure 10c.

At an $\nu_{rf} > A/2h$ at which $h\nu_{rf} = A/2 + g_N\beta_N H_o$ (the energy difference between states a and b), another absorption of microwave energy will be observed. This is the result again of a "relief" in the saturation of the electron spin system, this time due to the nuclear spin transition a→b. Stated in a somewhat different way, this relief is due to the connecting link between states c and d (S = +½) that is provided by the a→b nuclear excitation and relaxation processes. Thus, the ENDOR experiment involves detection of electron spin transitions as in a simple ESR measurement, but transitions that are generated by varying an rf field, and that is how the ENDOR spectrum is presented as in Figure 10c.

For the example given, the two absorption lines are separated in rf units by $2g_N\beta_N H_o$; from this can be calculated the nuclear moment of the modulating nuclei, and thus

* Note that the pair of resonances indicated (for I = ½) are not necessarily symmetrical with respect to the spin envelope. They could be associated with a transition in any part of this envelope.

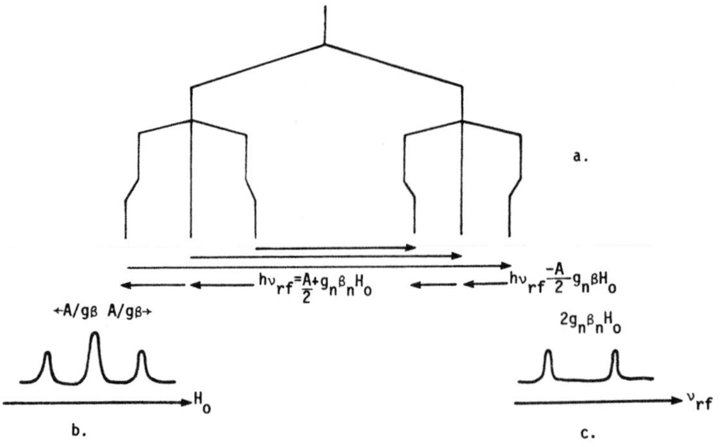

FIGURE 11. a) Energy diagram for S = ±½ coupled to two identical I = ±½ nuclei. The three, different electron spin transitions are noted (see 11b) as are the two nuclear spin transitions (see Fig. 11c). These two spin transition-types are illustrated in b and c as detected in the direct ESR experiment (variable H_o constant microwave frequency (Fig. 11b) and in the ENDOR experiment (variable ν_{rf}, constant H_o and microwave frequency).

identify it. Conversely, at a given experimental field, the frequencies of likely nuclei can be calculated before the measurement is made and the ENDOR spectrum obtained subsequently interpreted by comparison.

Two features of ENDOR not made readily apparent by the discussion thus far are how the experiment can simplify complex ESR spectra and how complete ENDOR spectra are interpreted. The first of these aspects is again best understood by way of example. Consider a system that contains two nuclei, I = ½, which are identical in that they are characterized by the same hyperfine coupling value. This is illustrated in Figure 11. As pictured, in the ESR spectrum itself these two nuclei give rise to three lines. However, in the ENDOR spectrum, only two transitions are observed (stimulated) associated with the modulation of the electron's local field caused by the two orientations of the nuclear moments precessing around the nuclear Zeeman field. In general, there will be 2N ENDOR lines, where N is the number of different coupled nuclei types and/or different hyperfine couplings. For example, the ESR spectrum of a system containing 1 ^{14}N (I = 1), 1 ^{19}F (I = ½), and 3 inequivalent $'$H (I = ½) would be expected to have 48 lines (3 × 2 × 2 × 2 × 2). The ENDOR spectrum, however, would have only 10 lines, 2 pairs separated (in units of $\beta_N H_o$) by g_{14N} and g_{19F}, respectively, and 3 pairs (at different rf values) all separated by g_{1H}.

That these latter 3 pairs of ENDOR lines appear at different values of incident rf energy illustrates the second experimental feature of this resonance technique. The energies of the individual transitions are determined not only by the nuclear angular momentum, but also by the hyperfine coupling, A. Thus, all coupled protons, for example, will be represented by pairs of ENDOR lines. There will be in the ENDOR spectrum as many such pairs as there are different A_{1H} values and they will be displaced in the rf spectrum by the value of this coupling. A complete interpretation of an ENDOR spectrum thus involves the pairing of resonance lines by the a priori tabulation of expected nuclear frequencies (for the applied H_o), followed by the calculation of the hyperfine coupling, A, for each pair. Since the ENDOR lines are inherently narrower than the inhomogeneously broadened ESR transitions, much more precise A values are obtainable in this fashion.

In summary, ENDOR is a highly valuable resonance technique. However, the instrumentation is more sophisticated than that employed in direct ESR experiments. The coupling of the rf and microwave radiation to the sample necessitates a special cavity design, although even the conventional rectangular cavity can be readily modified for ENDOR by the insertion of an rf coil. However, there is also the experimental difficulty of being able to selectively saturate one part of an inhomogeneously broadened line. "Spin diffusion", which in effect causes saturation of other spin packets, broadens the ENDOR line. This can usually be overcome by correct selection of temperature since the diffusion process is temperature independent. T_1 processes, which are temperature dependent, can bring neighboring spin packets into thermal equilibrium at rates fast enough to obviate the unwanted contributions of spin diffusion. Of course, as T_1 becomes very short, the microwave power required to effect saturation may become unobtainable (or undesirable). This obviously represents an experimental difficulty, which for some systems is not surmountable.

Another aspect of ENDOR applied to biological samples is that associated with the experimental conditions often required to detect the inherent paramagnetism of such samples. Because of the relaxation properties of, in particular, biologically relevant transition metal ions, most experiments are conducted on frozen glasses in which the paramagnetic centers generally exhibit a (large) degree of anisotropy. Note, however, that neither the broad lines nor anisotropy characteristic of such systems prohibits a successful ENDOR experiment. At a given microwave frequency and Zeeman field, only one spin packet is stimulated by the incident rf field. Clearly, this anisotropy can be actually of some use in that the nuclei and their associated hyperfine couplings that modulate each principle g tensor can be assigned and thus, theoretically, a quite detailed picture of the metal complex may be obtainable.

C. Electron Spin Echo[15-17]

As in the ENDOR experiment, the electron spin echo begins by the excitation of a single spin packet within an inhomogeneously broadened ESR resonance line. The electron spin echo differs, however, in that it is, in effect, a relaxation measurement; that is, it represents the modulation of the decay of the xy component of the electron spin magnetization following the spin packet's absorption of microwave energy. The information derived from the experiment is contained within the pattern(s) of modulation of this spin-spin relaxation process. As will be outlined here, this pattern of modulation is attributable to weakly coupled nuclei ($I \neq 0$) whose hyperfine couplings are too small to be resolved by direct ("continuous wave") ESR. Again, as in the ENDOR experiment, it is the frequency of the modulation that serves to identify these weakly coupled nuclei.

Consider the behavior of a spin packet, precessing at a frequency ν_o in a magnetic field H_o.[19] All other spin packets, in this and all other resonance lines, will be precessing with frequences greater or less than ν_o. Thus a pulse of microwave energy, $h\nu_o$, will be absorbed by this spin packet only. If such a pulse is of sufficient power and duration (typically 200 W and 20 nsec), the spins in the packet will be tipped into the xy plane; that is, the spin magnetization will rotate around the microwave field (H_1) by 90°.

The experiment now becomes one of monitoring what happens to this collection of coherent spins as a function of time following this 90° pulse. Inasmuch as the energy bandwidth of the exciting radiation was not infinitely narrow, the packet consists of spins experiencing a variety of local magnetic fields. This will be apparent after the pulse is terminated, because spins at higher field will be precessing (faster) at slightly higher frequencies than those at slightly lower field. Thus, the spin packet will "fan out" in the xy plane; the magnetic moment vector of the spin assembly, **M**, will break

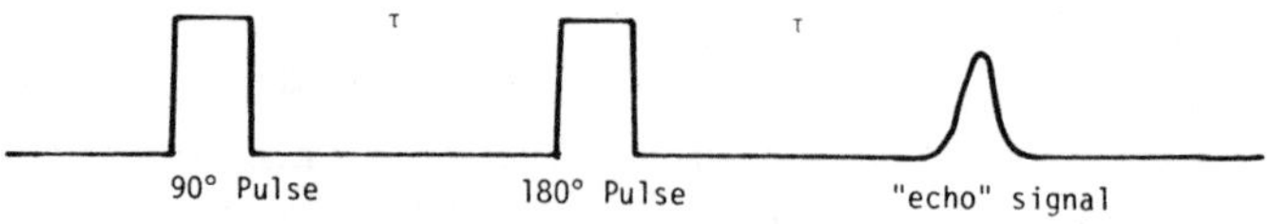

FIGURE 12. Generating spin echo where τ is the time (generally in μsec) between the two pulses, and between the second pulse and the "emission" of the echo signal.

up. The time constant for this process is the transverse relaxation time, and in the spin echo experiment, the length of time for which **M** has a nonzero value (until the spin moments are completely randomized) is termed the phase memory. The loss of precessional coherence within the spin packet is also associated with time-dependent processes, i.e., those factors that contribute to homogeneous line broadening (such as fluctuations in the local, nuclear, fields). Put another way, the "faster" spins are not always faster.

This becomes clearer by considering the actual generation of the spin echo. Consider for the moment the idealized situation in which the precessional frequency of one half the spins is $2\nu_o$ and that of the other half is $\nu_o/2$. Thus, the magnetization vector would fluctuate with a frequency ν, that is, attain its initial amplitude every $1/\nu$ sec. This would persist as long as these two frequencies were not altered by fluctuations in local magnetic fields. This peak amplitude magnetization is an "echo" of the initial state; however, this is not the way in which the echo is actually measured. What, in fact, is done is to rotate the precessing moments around H_1, again, employing a second microwave pulse of sufficient power and duration to effect an 180° "flip". In effect, this "slows" down the more rapidly precessing spins and "speeds" up the slow ones. This shortens the time required to generate the "echo" and the echo produced will be rotated by 180°. At the moment when the "slow" spins catch up to the "fast" ones (in our simple example, when they both have frequency ν_o), current will be generated in the microwave detector circuit (tuned to ν_o) by the dynamo principle; in essence, the microwave bridge is in "resonance" with the sample. The timing of this echo response is pictured in Figure 12.[15,19]

As noted, in real systems, the precessional frequencies of the individual moments are not constant. Random fluctuations in the local fields experienced by each spin cause the spin packet to lose the coherence it possessed at the time of the initial pulse. Thus, as τ increases, the echo signal generated decreases. In biological samples containing paramagnetic metal ion centers, this "phase memory" is typically 3 μsec measured from the first pulse. Although of practical importance (short phase memories limit the amount of usable data derivable from the experiment), this "unmodulated" decay is devoid of information. Rather it is the pattern of echo amplitudes as a function of τ which, when observed, characterizes the electronic environment of the spin packet; specifically, the oscillations in the echo amplitude, caused by regular fluctuations in the local fields, will exhibit the frequencies associated with the precessional frequencies of weakly coupled nuclei. Thus, what is termed the envelope modulation function of the spin echo decay is generated by combinations of nuclear frequencies; as in the ENDOR experiment, the identification of the nuclei types contributing to this modulation process simply involves the matching of the frequencies observed in the modulation envelope to the frequencies expected of various nuclei in the magnetic field of the experiment. In order to discuss this further, one must consider the way in which the echo envelope is actually generated.

Consider first that the periods of nuclear precession obtaining at the magnetic fields of the experiment ($\sim$ 3 kG) are of magnitude 50 to 750 nsec. The shortest time of τ

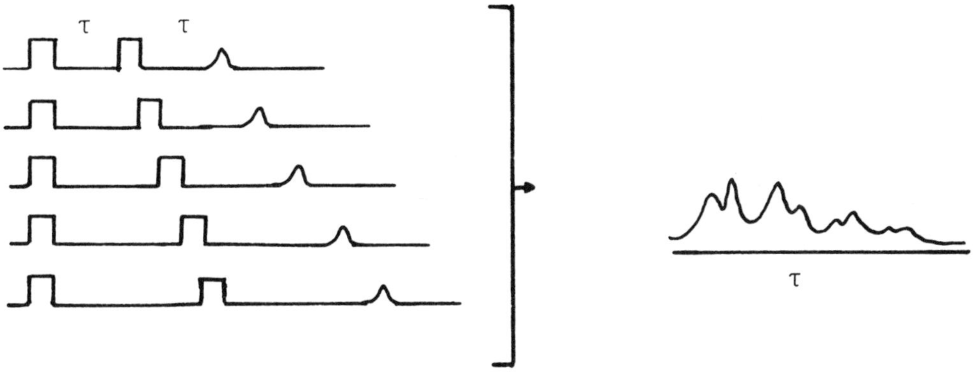

FIGURE 13. Diagrammatic representation of generation, collection, and display of electron spin echoes. The time, τ, between the first and second (90° and 180°) microwave pulses is made progressively longer (from ∼ 50 nsec to ∼ 3 μsec) by ∼ 20 nsec increments. The "echo" generated at each time is detected by the crystal detector and the detector output is stored. At the end of this series of microwave pulses, the stored detector output is fed into a recorder (or oscilloscope) and displayed as a function of the time, τ.

usually employed is 20 nsec, and the value is increased incrementally in a series of pulses until the echo signal is no longer observable (see Figure 13); as noted, this occurs commonly after ∼ 3 μsec. Thus, although the electron spins are going through hundreds of precessional cycles between each pulse, the nuclei with even the shortest periods are sampled more than once within one of their precessional cycles. Thus, this sampling of the echo during one nuclear precessional period will reflect the difference between the nuclear moment adding to and subtracting from the Zeeman field. This fluctuation is what causes the precessing electron to speed up and slow down, respectively, which changes the degree of spin coherence at any time, τ, as indicated in the modulation of the spin echo decay envelope. To emphasize this further, the envelope modulation function is, therefore, a curve representing the fluctuating local magnetic fields experienced by an electron spin. In essence, the technique is sampling a number of S,I states much as is done in the ENDOR experiment. In fact, the frequencies resolvable from modulation functions are just those determined in ENDOR experiments. A quantum mechanical analysis of the mixing of the electron and nuclear states makes this apparent.[15,16] What such an analysis also indicates is that for the modulation to be observed, the electron-nuclear hyperfine interaction (S·**A**·I) must be similar in magnitude to the nuclear Zeeman and quadrupole terms. These latter values for ^{14}N, for example, are ∼ 1 MHz and ∼ 2MHz, respectively (at 3 kG) corresponding to a hyperfine coupling of ∼ 0.5G.

A summary of this discussion is best presented by two examples. First, note the diagrammatic representation of the generation of the echo envelope shown in Figure 13.

In Figure 14 are presented two envelope modulation patterns obtained for two Cu(II)-complexes.[20] In the first, Cu(II)-diethylene triamine, only ^1H modulation is observed, with a period of ∼ 70 nsec. The coordinating ^{14}N do not contribute to the envelope since the **S·A·I** term (∼ 40 MHz) far exceeds the nuclear terms in the spin Hamiltonian. The other is of the same complex in which an imidazole has replaced H_2O as an equatorial ligand. The high frequency ^1H precession is still evident as is a lower frequency modulation attributed to ^{14}N. Again, the coordinating pyridine nitrogen in the imidazole ring is too strongly coupled to modulate the spin echo decay. However, the distal, non-coordinating, pyrrole nitrogen apparently does contribute to the envelope modulation function. The period at ∼ 3 kG is ∼0.7 μsec. Thus, among

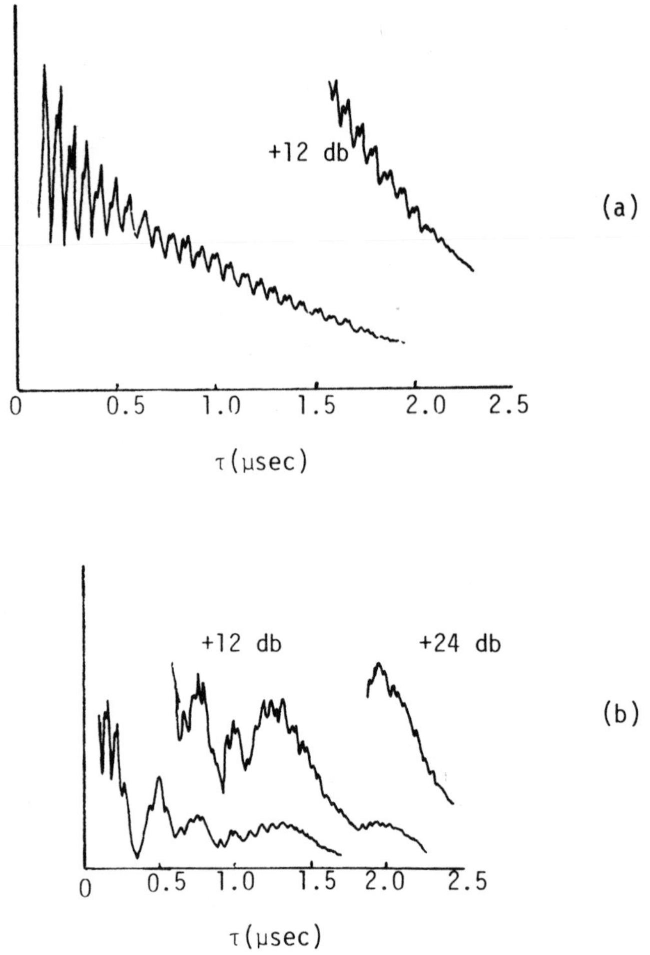

FIGURE 14.[20] Envelope modulation patterns for (a) Cu(II)-diethylenetriamine and (b) Cu(II)-diethylenetriamine imidazole.

amino acid ligands of biological significance, the imidazole ring of the histidine side chain is unique in its pattern of modulation.[21]

D. The Linear Electric Field Effect on the Electron Spin Echo[22]

The Stark effect is associated normally with changes of optical spectra as a result of the interaction of the absorbing molecule with an electric field.[23] The effect is ascribed to the displacement of the electron distribution in the molecule caused by the external field. Since it is the molecular electric field(s) experienced by an electron that gives the unpaired electron its characteristic g value(s), an external electric field would be expected to cause shifts in g values, as well. However, in order for shifts in g to be observable, the paramagnetic center must be noncentrosymmetric. Rather than this being a handicap, this limitation itself provides a first level of interpretation of any observable g shift. Nonetheless, the experimental detection of such shifts is difficult since linewidths (particularly of biological samples) are so large; the imprecision in measuring g values themselves obscures their shifts. However, the "resolution" of an inhomogeneously broadened line into spin packets by the electron spin echo technique makes measurements of the electron Stark effect possible. This is outlined briefly below.

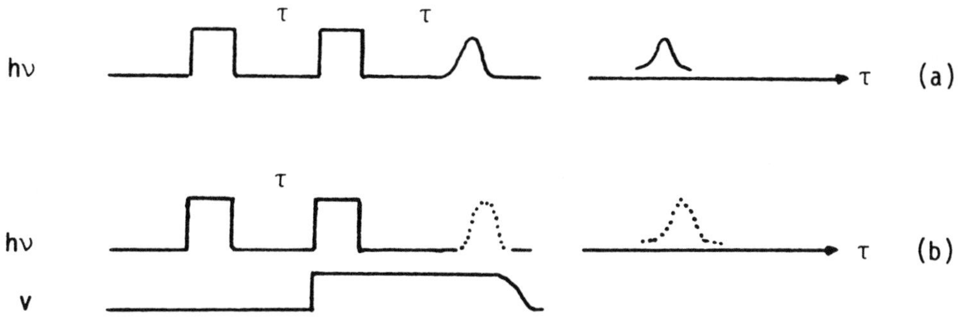

FIGURE 15. In (a) the two-pulse microwave sequence is shown and the echo signal generated at one time, τ. In (b) the same time, τ, between the first and second pulses is employed, but an electric field is turned on at the end of the second pulse. It remains on until the echo signal is detected, but is turned off before the next pulse cycle. In (b), the electric field is shown as having "delayed" the echo, thus at time, τ, there is no echo signal. The g-value shift (and thus the change in electron precessional frequency) induced by the field has caused this alteration in the echo amplitudes at specific times, τ_1.

Consider the sequence of events diagrammed in Figure 15. The spin echo cycle is initiated as described above. Simultaneously with the second (180°) pulse, an electric field is switched on and remains on until after the appearance of the echo signal. This field is typically 100 kV/cm. The presence of the electric field causes a change in the local (orbital) field experienced by the precessing electron spin and thus the electron's precessional frequency. Consequently, the process, which after a time, τ, brings the spin packet into convergence (the echo) is perturbed. For example, suppose of two spin packets, the faster was speeded up and the slower slowed still further. At certain times, τ, a null point in the echo amplitude would be expected because the two spin packets would be out of phase by exactly π. The shift in frequency associated with these times, τ_n, is given by[22]

$$\Delta\omega = (2n + 1)\,\pi\,/\,2\tau_n \qquad (20)$$

The shift in frequency is related to the shift in g by the basic equation for the Zeeman of the electron. Rather than attempting to find these specific null times, one can simply define some finite value of the echo amplitude (e.g., $A_\tau^E/A_\tau = 0.5$, the "half-fall" value) and determine the "half-fall" time, $\tau_{1/2}$, at which this value occurs. The frequency shift at $\tau_{1/2}$ is $\pi/6\tau_{1/2}$.[22] Δω has also been represented by f or Δf, and, as such, is used to define a shift parameter, σ (or S).[22,24] This is defined as $\sigma = d/6f(\tau V)_{1/2}$ where d is the thickness of the sample and $(\tau V)_{1/2}$ is the product of time and voltage that yields the "half-fall" of the echo amplitude. This indicates that (not surprisingly) varying the voltage, at a fixed τ, also will cause a variable shift and thus a varying echo amplitude attenuation. As indicated, however, the product $(\tau V)_{1/2}$ is determined at a fixed voltage.

The detection of a g-value shift, although indicative of noncentrosymmetry, is the minimum information derivable from a LEFE ("linear" electric field effect) experiment.[20,25,26] It should be appreciated that in a metal center that exhibits g-value anisotropy the potential for shifts will be determined by two factors: (1) the orientation of the magnetic and electric fields relative to one another and to the principle g tensor(s), and (2) the symmetry of each g tensor, that is, whether or not an axis contains an odd component (is not, itself, centrosymmetric). This latter feature is characteristic of tetrahedral geometry, for example, or D_{2h} or D_{4h} systems in which a significant ligand difference existed on any particular axis in the two spatial directions (± displacement). Thus, by physically changing the alignment of the electric and magnetic fields relative to one another (generally ∥ and⊥) and, at a fixed microwave frequency, by appropriate choice of magnetic field (to select spin packets of specific g value) one can locate in a ligand field the major "odd component(s)", that ligand element or geometrical distortion, which in greatest measure contributes to the noncentrosymmetry of the paramagnetic center. Thus, coordination differences between the "blue" and "non-blue" copper(II) centers in a variety of proteins can be readily detected,[20] the coordination geometry in cytochromes has been studied,[24,25] and the structure of iron-sulfur centers has been probed.[26]

IV. PARAMAGNETIC KINETIC TRANSIENTS[27]

A. Introduction

Many interesting paramagnetic species are relatively short-lived and have long since decayed within the time required to manually fill a quartz tube or flat cell, position it in the resonance cavity, tune the instrument, and record the spectrum. Furthermore, the rates of formation and decay of such species provide information in addition to the structural features implied by the spectrum itself. Thus, the ability to carry out kinetic experiments on unstable paramagnetic species is of special usefulness.

There are two basic ways by which short-lived species can be "trapped" (observed) and their kinetic behavior assessed. These are continuous flow and stopped-flow; both have been applied with some success in the area of biological ESR as with other spectroscopic techniques. Each has its own advantages and limitations.

B. Continuous Flow[28,29]

The advantage of this technique is also its disadvantage; it uses liquid samples at any temperature, but it uses them in, often prohibitively large quantities. In this method, the components of the system of interest are stored in separate reservoirs, and which are subsequently mixed in a continuous stream at a precise distance, x, from the center of the resonance cavity. By varying x, or the flow rate, one can vary the time between mixing and observation. Thus, one can generate a profile of the concentration of the paramagnetic transient(s) as a function of time after mixing. A unique feature of this technique is that since the concentration of species is held constant by the flow, one can record a complete spectrum of the radical, not simply measure the intensity of absorption at a single field setting. The flow can be driven by syringe or maintained by a pump, but in either case, the volume of the system is typically quite large (1 to 2 mℓ aside from the reservoirs themselves) and, more importantly, the flow rate necessary (∿10 mℓ/sec) to trap very short lived species ($t_{1/2} < 100$ μsec) results in the consumption of large quantities of material (> 100 mℓ/flow). Add to this the fact that to generate the temporal profile of the radical(s) concentration requires several independent flows and the disadvantage of this method is made clear. Continuous flow kinetics are not now the method of choice.

C. Stopped-Flow Methods[30-35]

Obviously, the stopped-flow is a modification of the continuous flow methodology; one simply stops the flow and, at a given distance, x, from the mixing chamber, watches the corresponding radical (its concentration) decay. While one could change both x and the flow rate to vary the time, t (relative to mixing) at which the flow is stopped, it is usually just as easy mechanically and clearly more economical of time and material, to make t as short as possible so that in one "flow" the whole temporal profile of the radical(s) can be established. This is the essence of stopped-flow. Advances in mixing design have brought this mixing down to ∿ 5 msec, a "dead time" only slightly longer than that encountered in other stopped-flow systems.

There are two stopped-flow methods. The first is like the typical system, using liquid samples, and is the direct opposite of continuous flow. To shorten the distance between mixing and observation, an aqueous quartz cell with built-in mixing jets is used. The dead volume of such a cell is ∿ 500 $\mu\ell$; that is the volume needed per flow to renew the solution in the flat cell. However, because of the nature of the cell, very fast flow rates are not possible; the dead-time is limited to at best ∿ 100 μsec. The spectrometer is set at a field position chosen to be a maximum of signal intensity (for sensitivity) or extent of signal amplitude change. For transients, the position often must be determined by trial and error, or, more precisely, though usually more expensive in terms of material, by obtaining the complete spectrum by continuous flow as outlined above. One specific area of biological ESR in which this type of stopped-flow has proven useful is that of spin labeling. Spin labels are not transients, per se, but their degrees of immobilization do change in response to chemical and/or conformational events and thus the rate constants of these processes can be determined. That the labels are not chemical transients makes such experiments inherently easier, of course.

The second method is the one of choice since it is the most versatile. This is the method of rapid quenching by freezing the sample at precisely determined times after mixing and subsequently recording the spectrum of the frozen sample. This method combines the mixing concepts of the typical stopped-flow instrument — rapid flow and efficient mixing, e.g., short dead times — with the low temperatures often needed anyway to observe certain biological paramagnets, particularly the functional transition metals encountered in many macromolecules. This technique was first developed and used successfully by Bray in the early 1960s,[30-32] and was further refined by Beinert[33] and subsequently by Ballou and Palmer.[34-35] The elements of the systems used are essentially the same as in both the continuous and stopped-flow methods — sample reservoirs driven by piston or pump, and a mixing chamber at a given distance from an injection port or nozzle. The reaction is "stopped" by spraying the reaction mixture into a cryogenic liquid, which is commonly isopentane precooled to ∿ 130°K. The frozen pellets are packed into a quartz tube and the spectrum of the sample can be recorded. This technique thus retains the primary advantage of continuous flow, namely, the ability to record the entire spectrum of the sample at any given time, t, after mixing.

In addition, since the sample is temporally static, its behavior in the magnetic field can be more completely characterized. For example, the saturation properties of the sample can be determined over a range of temperatures down to that of liquid He. ENDOR spectra can be obtained, and, theoretically, even electron spin echo spectra could be generated from the freeze-quenched material.

As in continuous flow, the time, t, is varied by either changing the distance between mixing chamber and nozzle or by changing the flow rate. In practice, since changes in the flow rate alter the mixing characteristics of the system and the size of the frozen pellets obtained, it is best to vary t by using reaction tubes of varying length. The shortest time accessible is ∿ 5 msec although mixing itself is accomplished in ∿ 2 msec; added to this is the actual quench time of the cryogenic bath, which also is ∿ 5 msec. Much of the excellent kinetic data on xanthine oxidase transients discussed in Section V have been obtained using the rapid freeze technique and Ballou has detailed the design, construction, and use of the system now most commonly employed.[34,35]

V. APPLICATIONS

We have attempted to choose representative examples from the literature or our own work to discuss in this section. It is our hope that these examples will help clarify the various sections above by example. One area that is not given adequate attention here is the general one of metal replacement, even though we have noted its utility in earlier sections.

A. The B_{12} Coenzyme

Certainly one of the most versatile of the cofactors of enzymatic catalysis is vitamin B_{12}. This cobalt-containing coenzyme has been the subject of many reviews and its properties will not be elaborated upon here.[36-38] However, the mechanism of action of certain enzymes in which B_{12} serves as cofactor has been elucidated by the use of ESR. Thus, a discussion of the results of such experiments is relevant and serves to illustrate how ESR has been instrumental in these studies. These experiments have demonstrated the presence of organic free radicals during turnover generated via homolytic bond fission of a carbon-cobalt, or a C−H bond. The nature of these radicals has been determined by a combination of g value, linewidth and saturation analysis, and by assignment of hyperfine couplings. Isotopic substitution has also proved useful. Kinetic studies have indicated the competence of these radicals as intermediates. Thus, these systems provide a variety of examples of how ESR can be used in the study of biologically interesting systems.

ESR resonances have been observed in (frozen) solutions of dioldehydrase,[39-41] glyceroldehydrase,[42] ethanolamine ammonia lyase,[43,44] and the B_{12}-dependent ribonucleotide reductase.[45-47] The intensity of these resonances depends upon the concentration of available substrate(s). These spectra are not attributable to the coenzyme itself, since B_{12} or 5′-deoxyadenosylcobalamine (DBCC) is not paramagnetic. However, what is known as Co(II)-cobalamine, or $B_{12(r)}$, which lacks the carbon-cobalt bond of the native coenzyme does exhibit a characteristic ESR spectrum. Model studies have established the g value, hyperfine couplings, and saturation behavior expected for this species. A typical spectrum is shown in the low field portion of Figure 16. Minimally, therefore, the reactions catalyzed by these enzymes do involve rupture of the bond between the 5′-carbon of the deoxyadenosyl ligand and the cobalt; that a Co(II) species is formed rather than Co(I) requires that homolytic bond fission occurs, and consequently, a second unpaired electron must result. In fact, such a radical species has been detected as illustrated in Figure 16.

The properties of this spectral "doublet" that is observed in each of these four enzymes are summarized in Table 1.[48] Although an assignment of this radical species has been elusive, the strategies employed in an attempt to do so are instructive.

That the doublet is attributable to a single electron center is established readily by comparing the separation of the two resonances at different microwave frequencies. There is no difference in this parameter at 9 and 35 GHz; this frequency-independent nature shows that neither g-value anistropy nor two (unrelated) g values are responsible for the spectra observed. The linewidths, themselves, are suggestive of the unpaired electron being contained within an organic molecule rather than centered on a metal ion. The spin relaxation differences that are responsible for these linewidth effects can be gauged by saturation experiments. A typical result for ribonucleotide reductase is illustrated in Figure 7 of Reference 46. At 12°K, the power at half-saturation, $P_{1/2}$, for the two resonance lines is 23 μW (g = 2.036) and 2.5 mW (g = 1.965). The slopes of the lines in the saturation region are similar, indicating that the spin packets that make up the resonance envelopes have similar relaxation rates. The $P_{1/2}$ values for the doublet are normally a 1/10 of those needed to cause appreciable saturation of the Co(II) signal. Thus, the saturation study not only distinguishes between the two types of resonances, but also indicates that the doublet spectrum does not arise from a metal centered electron spin.

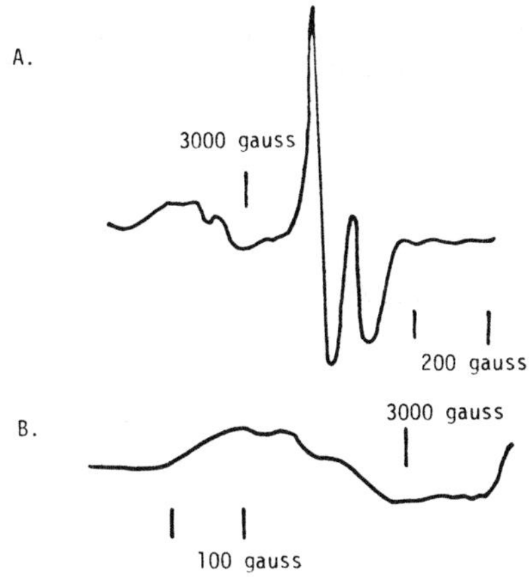

FIGURE 16.[44] EPR spectra of the enzyme coenzyme complex in the presence of L-2-aminopropanol. The reaction mixture contained 15.4 nmol of ethanolamine ammonia-lyase, 63 nmol of coenzyme, 1.0 μmol of L-2-aminopropanol·HCl, and 2.0 μmol of potassium phosphate buffer (pH 7.4) in a volume of 0.3 mℓ. The EPR spectra were taken at 98°K. Microwave frequency was 9.162 GHz, microwave power was 1.0 milliwatt, and the modulation amplitude was 10 G. Gain for the lower spectrum was 2 times the gain for the upper spectrum.

Isotopic substitution in the organic species that serve as cofactor or substrates in these reactions represents a rational means of locating the atoms with which this unpaired electron interacts. The doublet nature of the spectrum is suggestive of a hyperfine interaction with a single nuclear spin, I = ½, thus selective deuterium substitution is a reasonable way of elucidating the electron spin distribution. As noted in Table 1, the results have been inconsistent in that there is no single pattern of isotopic effect. This perhaps is not surprising, for although all four enzymes use B_{12} as cofactor, the reactions catalyzed are distinctly different, particularly the reductase one. As is obvious from the table, the assignment of the radical in this latter enzyme has proved difficult. This system will be discussed in more detail below. The data from the other three do indicate that the electron spin contributing to the doublet spectrum is at least in part localized in the substrate, but little ESR evidence exists that places spin on the deoxyadenosyl moiety. This inference illustrates the type of unique information provided by ESR. In these reactions, the stable intermediate formed following cleavage of the cobalt-deoxyadenosyl-5′-carbon bond results from transfer of a hydrogen atom from substrate to cofactor, i.e., that the reaction is a radical one, rather than a carbonium ion or carbanion one. This process can be pictured.[49]

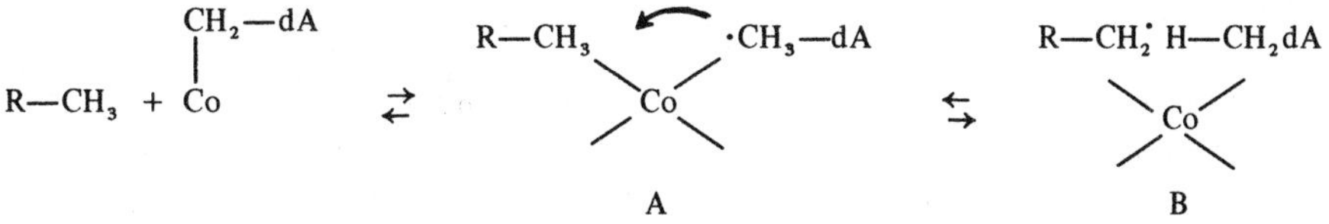

Table 1[48]
VITAMIN B-12-REQUIRING ENZYMES WHICH EXHIBIT THE DOUBLET EPR SPECTRUM

Enzyme:	Ribonucleotide reductase	Glycerol dehydrase	Ethanolamine ammonia lyase	Diol dehydrase
Source of enzyme:	*Lactobacillus leichmanii*	*Aerobactor aerogenes* (No. 572 PZH)	*Clostridium* (Uncl.)	*Aerobactor aerogenes* (AJCC 8724)
Substrate or inhibitor:	Nucleoside triphosphates	Propane diol	α-Amino propanol	Propane diol
Integrated intensity	8.5% of Co[a]	50% of Co[a]	60% of Co[a]	0.76[b]
Splitting (G)	110	154	70	151
g High field	1.965	1.944	1.99	1.96
Linewidth (G)	37	39	41	43.6
g Low field	2.032	2.035	2.04	2.05
Linewidth (G)	30	36	54.5	38.6
L/H intensity ratio	2:1	4.7:1	2.2:1 (est)	4.7:1
Microwave power	2.7 mW (n.s.)	Not given	1 mW (n.s.)	1 mW
Temperature dependence of signal	12°K + (splitting decrease)	Not studied	8-243°K (no effect)	100-214°K (no effect)
Optical evidence for B_{12r}		+	+	+
Kinetic competence	+	+	+	+
Spin density on substrate	−?	Not given	+	+

Note: n.m., not measurable; n.s., not saturated.

[a] Integration over doublet only.

[b] Integration over Co(II) plus doublet.

Presumably, B is the predominant species, based on the results summarized in Table 1.

As noted, ribonucleotide reductase is distinct from the other three B_{12} enzymes that exhibit ESR spectra in the presence of substrates. Aside from and related to experimental differences, is the fact that this enzyme's reaction consumes reducing equivalents in the form of the dithiol ↔ disulfide redox reaction. The biological reductant is thioredoxin, but various 1,3- and 1,4-dithiols can substitute (at higher concentrations); pre-reduced enzyme can also support turnover, suggesting that a dithio-disulfide redox pair exists in the enzyme (see below for further discussion of this feature). As shown in Table 1, isotopic substitution has not unambiguously determined the source of the doublet spectrum exhibited by this enzyme system. However, fast freeze kinetic experiments, as well as analog studies suggest that in this enzymic process, at any rate, the doublet species is not catalytically relevant.

Stopped-flow absorbance measurements that apparently record the production and consumption of Co(II)-cobalamin (B_{12r}) (Figure 17) can be correlated with the appear ance and disappearance of a different ESR resonance than those illustrated and discussed above. The X-band spectrum of this species is shown in Figure 1 of Reference 47. Aside from the loss of the hyperfine structure, there is no difference in the K-band spectrum indicating the resonance is attributable to a single paramagnetic species. Furthermore, saturation of this signal becomes apparent at microwave power in excess of

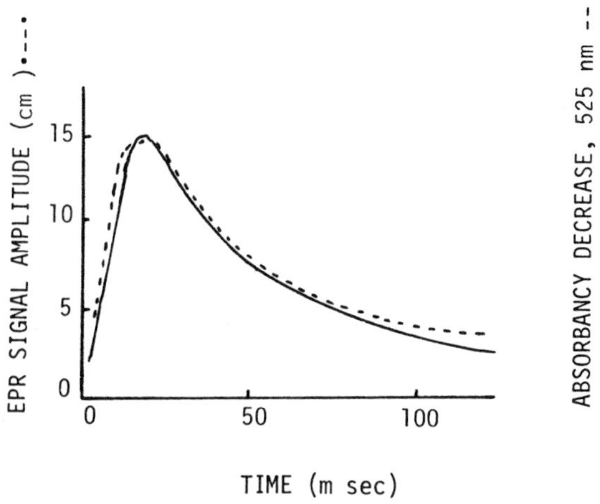

FIGURE 17.[47] Comparison of kinetics of the ESR signal amplitude changes with those of absorbance changes in stopped flow experiments.

30 mW, in contrast to the behavior of the doublet spectrum above. Estimates of the spin concentration in this system indicate that there are two unpaired electrons per enzyme molecule. As seen for the doublet spectrum, isotopic substitution had no effect on the spectrum of this freeze-quench species; this included use of 5′,5-dideuterio- and 5′-^{13}C-coenzyme, as well as D_2O as solvent.

Although the data do not permit a structural assignment of the radical to be made, the saturation behavior suggests that the paramagnetic species is "like" a transition metal. However, the fact that there are two unpaired electrons per molecule, as there must be if there is a homolytic bond fission, argues for this spin complex to be something other than B_{12r}, although the absorption spectrum is characteristic of this coenzyme form. Possibly, an organic radical (deoxyadenosyl?) is present, but its resonance is broadened by the Co(II). Kinetic studies using appropriate analogs of deoxyadenosine do suggest a temporal relationship between the "active" spectrum produced by freeze-quench and the probable "dead-end" doublet spectrum. Thus, the interpretation of the ESR spectrum of the former intermediates is of importance to the understanding of the mechanism of action of ribonucleotide reductase.

An interpretation of the doublet spectrum that may also be relevant to the freeze-quench resonance is that it is due to an exchange coupling of Co(II) and organic-centered unpaired electrons.[48] This can be compared to an "AB" system of coupled spins commonly encountered in NMR, two coupled nuclei that have similar chemical shifts ("g values"). This analysis can explain a number of puzzling and seemingly unrelated spectral phenomena, i.e., the $\sim$ 2:1 intensity ratio of the resonance lines of the doublet, the doublet splitting and intensity dependence on temperature, and perhaps, even the difference between the doublet and freeze-quench samples. Briefly, the typical doublet spectrum including the g $\sim$ 2.3 region may not represent simply B_{12r} and the "organic radical", but in fact, a system in which the two electrons experience an electrostatic exchange coupling, J, which is equivalent to the energy separation of the radical doublet. By taking into account the anisotropic nature of the cobalt g and hyperfine values, the doublet spectra of the four B_{12} enzymes discussed here can be simulated quite successfully.[48]

What is the significance of this analysis? First, exchange is temperature-dependent and thus the behavior seen for ribonucleotide reductase is not surprising.[46] Simple hyperfine coupling of an electron to a nucleus I = ½ would not exhibit such behavior. Secondly, since the doublet is not associated with such a nucleus, isotopic substitution (D for H, for example) is not a logical experiment. Lastly, the variability of the doublet intensities and splittings from enzyme to enzyme (and substrate to substrate) can reasonably be due to small differences in J and/or the angle between the two electron spin-containing orbitals. In an extreme situation, even the freeze-quench spectrum may be due to the same chemical species as that which exhibits the doublet spectrum, but which conformationally differs significantly enough to generate the spectrum observed. The extension of the spectral analyses to include this possibility should be the subject of future research.

B. The Iron-Containing Ribonucleotide Reductase

Escherichia coli contains a ribonucleotide reductase that does not depend upon B_{12} as a cofactor to support turnover. Thus, although both the B_{12} dependent and independent enzymes use thioredoxin as the hydrogen donor, the characteristics of the enzymes' active sites and corresponding catalytic intermediates must be different. In fact, a somewhat more precise picture of these intermediates in the *E. coli* reductase cycle has been obtained, primarily through the application of ESR techniques.

The enzyme is composed of two subunits, proteins B1 and B2. Both are involved in catalysis; B1 contains an active dithiol - disulfide pair(s) (up to three), while B2 contains two Fe(III) atoms.[50-52] In addition, B2 contains variable amounts of a paramagnetic doublet species.[53] A brief description of how the properties of the Fe(III) and radical centers were elucidated follows.

The X-band ESR spectrum of *E. coli* ribonucleotide reductase B2 protein in the absence of substrate is shown in Figure 18.[53] The saturation behavior at 88°K is given in the insert, and shows that at this and higher temperatures, saturation is apparent only above ∿ 2 mW. Below 77°K, the signal is more readily saturated. Although line broadening occurs as the temperature is increased, a spectrum is readily obtainable at 25°C in liquid solution; the doublet splitting persists under these conditions. Isotopic replacement with ^{57}Fe or ^{56}Fe, or D_2O from H_2O (for sample preparation only) has no effect. In the Q-band spectrum (ν = 35 GHz) the major doublet persists. The data all indicate that the resonance originates from an unpaired electron in an organic molecule, which is coupled to a nuclear spin, I = ½. The similarity to the B_{12} reductase signal, which appears only in the presence of substrate is remarkable (see Section V.A).

However, the ambiguity that remains in this latter system has been resolved for *E. coli* enzyme. The radical is not centered on the iron; Mössbauer spectroscopy and magnetic susceptibility measurements show that the two iron atoms are high spin Fe(III) centers which are antiferromagnetically coupled and thus diamagnetic (below 195°K).[54] The existence of the radical depends on the presence of the metal, however; also, a known free radical scavenger such as hydroxylamine while not removing the metal, eliminates the electronic properties associated with the radical including the ESR spectrum. Thus, there is little likelihood that the doublet spectrum arises from an exchange coupling as described for the B_{12} system(s).

To establish the identity of the radical, two elegant experiments have been performed.[55] A strain of *E. coli* that produces 5 to 10% of its soluble protein content as reductase was grown in D_2O and the ESR spectrum of the cells recorded; the comparison between the cells grown in H_2O and those grown in D_2O is shown in Figure 19. Significantly, the doublet collapses to a single line, as would be expected if $>\dot{C}$—-H were replaced by $>\dot{C}$—-D.

Note that because of the differences in g_N the hyperfine coupling of D should be ∿ 1/6 that of H, a coupling that would not be resolved. In a second experiment, cells were again grown in D_2O, but with the addition of specific non-deuterated amino

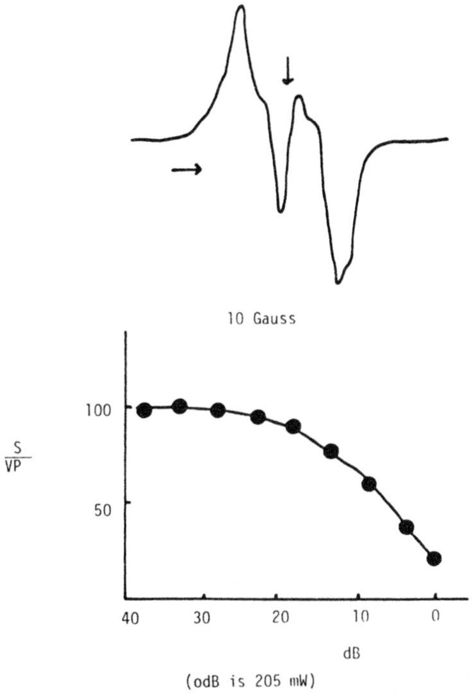

FIGURE 18.[53] Electron spin resonance spectrum of protein B2. Temperature, 88°K. The g value is 2.0047 at the zero passage indicated by the arrow. Cavity used: Varian V-4531, with gas flow Dewar insert. Field modulation frequency, 100 kHz, and amplitude, 2.4 gauss peak to peak; microwave frequency, 9.08 to 9.09 GHz, and power, 2 mwatts; field scanning rate, 25 gauss per min; time constant of detector, 1s. INSERT Microwave saturation of ESR signal from protein B2. Vertical axis, signal amplitude (S) divided by the square root of microwave power (P). Unsaturated level set equal to 100. Horizontal axis, microwave power attenuation in decibels. Temperature, 88°K.

acids; the "D_2O" spectrum persisted for all amino acids added with the exception of tyrosine. When tyrosine was added, the normal doublet spectrum was observed. In a complementary experiment, growth of cells in H_2O in the presence of specifically deuterated tyrosine localized the radical species still further. The collapse of the doublet is caused by incorporation into the reductase of β,β'-dideuteriotyrosine. Thus, the radical is formed from the homolytic cleavage of a benzylic C—H bond in the side chain of a tyrosine in the B2 subunit. These experiments are perhaps among the best reported that use isotopic substitution to characterize a radical center endogenous to a biological macromolecule. Although the B_{12}-dependent reductase is not similar to the *E. coli* enzyme except for the doublet ESR spectrum, the fact that the radical in the B_{12} system is not on the cofactor or substrate suggests that it might be found in the protein itself. This possibility bears further study.

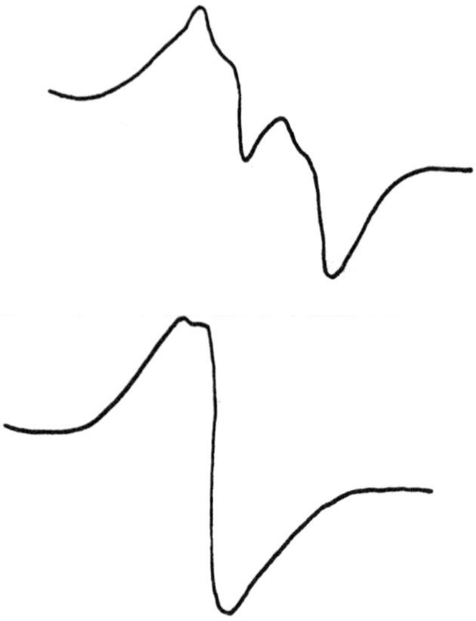

FIGURE 19.[55] (a) ESR spectrum at 77K of the cell culture KK546 of *Escherichia coli* grown in H_2O. The modulation amplitude was 1.5G. (b) ESR spectrum at 77K of the cell culture KK546 grown in D_2O. Instrument settings were as in a.

C. Xanthine Oxidase[56,57]

Perhaps no other biological macromolecule has been studied so extensively and effectively by ESR as has xanthine oxidase. This enzyme, commonly isolated from cows' milk, catalyzes the oxidation of xanthine to uric acid; the reaction is, in fact, the hydroxylation of the C8 position of the purine ring. What is so interesting to the ESR spectroscopist about xanthine oxidase is the fact that this protein contains a variety of redox active, and thus potentially paramagnetic, centers of different chemical types. A feature it shares in common with a few other enzymes (sulfite oxidase, aldehyde oxidase, nitrate and sulfate reductase) is the combination of nonheme iron, flavin, and molybdenum cofactors, which together are responsible for the enzymatic activity. Although not completely established as fact, the structure of xanthine oxidase appears best described as a dimer of identical subunits, each of which contains 1 Mo, 1 FAD, and 2 Fe_2S_2 clusters. These two subunits appear to function independently of each other. Thus, each active site is capable of accepting/donating a total of six electrons during catalysis. The elucidation of how these redox centers function in the enzymic reaction is a textbook example of the use of ESR in biology.

The Mo centers in xanthine oxidase are probably in the VI oxidation state in the completely oxidized enzyme; the enzyme is diamagnetic and exhibits no ESR spectrum in this form, as expected for the d° Mo(VI) ion.[56,57] The Fe_2S_2 clusters are typical in that they appear to contain antiferromagnetically coupled, high-spin Fe^{3+} (S = 5/2), and thus are also diamagnetic, as is the flavin, as FAD. The completely reduced enzyme exhibits an ESR spectrum characteristic of the reduced Fe_2S_2 cluster, namely a rhombic spectrum (inequivalent $g_\perp$ values) with g_x = 1.899, g_y = 1.935, and g_z = 2.002. Although observable in a broadened form at liquid nitrogen temperatures, the spectrum is sharpened considerably with the use of helium as coolant. In fact, below ∼ 25°K a second species is observed in the ESR spectrum of the reduced enzyme, also attributable to a rhombic Fe_2S_2 cluster with g_x = 1.91, g_y = 2.007 and g_z = 2.12. This is a unique Fe_2S_2 system since only one of the principle g values is below the free spin value (2.0023).[58,59]

Table 2[56,60]
ASSUMED RELATIVE REDOX POTENTIALS FOR THE GROUPINGS OF XANTHINE OXIDASE

Species	Relative redox potential (mV) Alone	Xanthine bound
Fe/S II_{ox}/Fe/S II_{red}	0	0
FE/I I_{ox}/Fe/S I_{red}	−24	−24
FAD/FADH	−60	−60
FADH/$FADH_2$	+60	+60
Mo(VI)/Mo(V)	−60	+120
Mo(V)/Mo(IV)	−31	+57

These two clusters, termed Fe/S I and Fe/S II, respectively, are each one electron redox centers. This can be determined by anaerobic titration of the oxidized enzyme with dithionite.[60] Since one of the principle g values for each cluster is found at a unique magnetic field (at a given frequency), that is, g_y = 1.95 for Fe/S I, and g_z = 2.12 for Fe/S II, the reduction of each cluster can be followed directly. Data can thus be obtained that show, qualitatively, that Fe/S II_{ox} has a more positive reduction potential than Fe/S I_{ox}.[60] Thus, as indicated by ESR the two Fe_2S_2 clusters in the active site of xanthine oxidase are both spectrally and chemically distinct, and therefore, must be structurally different, as well.

That the Mo does not contribute to the ESR spectrum of the fully reduced enzyme shows that it is probably in the IV oxidation state. A spin even system (S = 1), it would not exhibit a detectable resonance. The 2-electron reduction of the Mo (VI → IV) is expected to go through the paramagnetic (S = ½), V state. Indeed, reductive titration by dithionite does generate a signal attributable to this oxidation state.[60] It never accounts for more than 25% of the total Mo and thus must have a more positive potential than Mo (VI); both species show less electron affinity than the two Fe/S centers. The FADH· intermediate is also evident in ESR spectra of partially reduced enzyme (g = 2.0035), although in relatively small amounts. In the reductive titration, the semiquinone accounts for at most ∿ 5% of the total flavin. This shows clearly that this half-reduced cofactor has the most positive reduction potential of the six redox states (centers) in the active site. The relative electron affinity constants for these centers are given in Table 2, and the ESR spectra of the two Fe/S clusters and the FADH are shown in Figure 2 of Reference 56.

Of the four redox centers that exhibit ESR spectra, the most interesting and informative is the Mo(V). In addition to the redox behavior outlined above, the ESR spectra also help to characterize the enzyme-substrate interaction. First, the Mo(V) signal is not represented by a single spectrum; during turnover, there are two distinct Mo(V) species formed in sequence, termed the very rapid and rapid signals.[61] Early literature refers to these two as the γ, δ and α,β signals, respectively.[62] These two resonances differ in three important ways: rate of formation and decay; principle spin Hamiltonian parameters; the presence of superhyperfine splitting. That the two species are temporally related is indicated by their formation half-times, 5 and 15 msec; and decay half-times, 40 and 500 msec with maximal signal intensity at 15 and 65 msec, respectively.[56,60] These data are obtained by rapid freeze experiments and are illustrative of the use of this kinetic ESR technique.[60,63]

The differences in the spectra themselves can be dealt with on several levels. The enzyme as normally isolated contains a natural abundance mixture of I = 0 Mo isotopes (−94, −96, and −98) and I = 5/2 isotopes (−95 and −97). The former constitute 75% of the total, and thus the spectra are essentially devoid of metal hyperfine splittings, although the contribution from the latter isotopes can be detected at high signal levels. The g values for the very rapid and rapid signals are given in Table 3. Interesting is the similarity of these values to those for complexes between Mo(V) and various thiols. Although without other experimental evidence, this indicates that one or more cysteinyl residues in the protein could serve as ligands to the Mo.[56]

The analysis of the ESR spectra of the Mo(V) center and the characterization of this site was in large measure carried out by Bray and Meriweather who compared the "natural abundance" enzyme with an enzyme prepared from a cow that had received an injection of ^{95}Mo (as sodium molybdate).[64] For reference, 184 mg of ^{95}Mo were used for a 546 Kg cow. They included a tracer of ^{99}Mo, and by determining the specific activity of the xanthine oxidase could quantitate the isotopic composition of the enzyme independent of the ESR measurements. This stands as one of the most elegant experiments in isotopic substitution yet reported.

The analysis proceeds as follows. The spectrum of the very rapid or γ,δ signal from the oxidase containing only ∿ 25% isotopes with nuclear spin is given in Figure 1 of Reference 64. Three major resonances are seen, with some evidence of additional transitions. Although one might expect the contribution of these isotopes to be more in evidence, keep in mind that with I = 5/2, six hyperfine lines result. Thus, each line is effectively only $1/3 \times 1/6 = 1/18$ of the intensity of each of the three major transitions present. This same point must be kept in mind when inspecting the spectrum of the ^{95}Mo-enriched protein. With three g values, and with I = 5/2, a total of $3(2 \times 5/2 + 1) = 18$ lines would be expected. Furthermore, there should be no resonances at the normal g-field positions because of the even-splitting. Clearly, neither of these predictions are born out; in particular, the three transitions unsplit by electron-nuclear coupling remain clearly evident. As indicated, however, the isotopic purity of this sample was established by radiolabeling; it contained 74% Mo isotopes with nuclear spin. The three lines noted are due to the 1/4 Mo sites that exhibit no nuclear hyperfine coupling, and, as above, appear relatively intense because the resonance lines are not split. This illustrates the importance of the independent isotope analysis.

Actually, the residual Mo-even isotopes provide convenient field markers. The nuclear splitting of each of the three principle g tensors is seen readily by inspection. For example, A_z is easily resolved at the low field side, while A_x is measured from the separation of the two highest field lines. Using these two values, one can mark lines within the central complex of resonance, which correspond to the expected (six) transitions in both g_z and g_x. Lines remaining unmarked must belong to g_y and thus A_y can also be assigned. Spectral overlaps (and lack of resolution) can be expected to diminish the total number of transitions observed, as is the case here. This analysis was nicely summarized by a splitting or stick diagram,[64] and the g and A values are given in Table 4.

Significantly, the rapid, or α,β-, Mo(V) signal is different in detail from the one just described. As noted in Table 3, this Mo(V) is probably axially symmetric since $g_x = g_y$ within the limits of resolution at X-band frequencies and corresponding magnetic fields. However, as seen in the original spectrum, there are not the two lines expected for the "natural abundance" enzyme, but at least four dominant resonances are observed. The weaker lines can be ascribed to the Mo isotopes, I ≠ O; they generate a more intense resonance pattern in this spectrum since spin density is distributed into only two orientations. In particular, in $g_\perp$ the unpaired electron is not divided into

Table 3[56]
MOLYBDENUM EPR SIGNALS FROM REDUCED FORMS OF XANTHINE OXIDASE

Signal	Origin	Treatment required	g_{av}	g_z	$\|A_{av}\|(^1H)(G)$	Exchange of interacting ^1H with ^2H_2O	Notes
Very rapid	Active enzyme	Reduction for very short times, at high pH, with xanthine, only	1.977	2.025	None	—	—
Rapid	Active enzyme	Reduction with any substrate	1.973—1.974	1.989—1.994	12—14	Yes	Several related species
Inhibited	HCHO or CH_3OH treatment of active enzyme	None	1.973	1.953	3.9—5.6	No	Stable in air
Slow	Desulfo enzyme	Reduction for long periods (e.g., 20 min with dithionite)	1.965	1.956	16	Yes	—

Table 4[64]
SPIN HAMILTONIAN PARAMETERS

Molybdenum − γ, δ Signal (Molybdenum −95 Coupling)

$g_x = 1.951$	$A_x = 37G$
$g_y = 1.956$	$A_y = 24G$
$g_z = 2.025$	$A_z = 41G$

Molybdenum - α, β Signal (Molybdenum −95 Coupling)

$G_{\parallel} + 1.971$	$A_{\perp} = 28G$
$g_{\parallel} = 1.990$	$A_{\parallel} = 67G$
$A_{\parallel}(^1H) = 13G$	$A_{\perp}(^1H) = 16G$

two distinct g tensors. Thus, each line in this region is expected to be twice as intense as the corresponding line in the rhombic spectrum discussed above.

The origin of the doublet splitting of $g_\perp$ and $g_\parallel$ could be investigated by determining the frequency dependence of the magnitude of this splitting. However, shfs is generally not seen in 35 GHz spectra, for example, thus this approach often results in ambiguity. The second method is that used by Bray and Meriweather. The rapid signal generated using the 74% Mo (I ≠ O) sample exhibits splitting assignable to the metal. The lines in $g_\perp$ discussed above are clearly resolved. $2 \times (2 \times 5/2 + 1) = 12$ lines can be assigned to $g_\perp$ as indicated in the splitting diagram. This assumes the doublet splitting to be of shfs origin, presumably a single nucleus $I = \frac{1}{2}$. That this is undoubtedly correct is supported by the presence of a similar splitting of the $g_\parallel$ resonance lines, shfs clearly resolved on the lines at the extremes of magnetic field seen best at higher gain. Thus, a consistent set of spin Hamiltonian parameters can be derived from this qualitative analysis, which can then be used to construct a splitting diagram. The comparison is quite satisfying. In summary, oxidized enzyme reacts with xanthine within 15 msec to generate a Mo(V) species that is rhombic and lacks shfs (very rapid signal). This species converts more slowly to one that is axial and exhibits the shfs attributable to a nucleus, $I = \frac{1}{2}$, probably ^1H (rapid signal). A_H is approximately 15G (see Table 4).

The structural differences underlying the spectral differences between the very rapid and rapid species naturally are of great interest. That this conversion occurs only with xanthine, the specific substrate, and not with other purines, suggests that the substrate, while bound to or near the Mo is involved in this change. Isotope and model studies point to the solution to the puzzle. First of all, generation of these two species in D_2O results in the initial appearance of the typical rapid signal, which then loses the shfs structure contributed by the $I = \frac{1}{2}$ nucleus. This confirms that this nucleus is ^1H, and it is in exchange with solvent within the lifetime of the rapid species. By determining the extent of exchange at various pH values, the pKa of the conjugate acid involved was found to be ~ 8. Even more precise information is obtained when 8-deuteroxanthine is used as substrate; the $I = \frac{1}{2}$ splitting in the rapid signal is lost. Thus, the ^1H nucleus responsible is from the C-8 position of xanthine (the reaction center).

Where this proton is in the rapid species is still a matter of some debate. Two possibilities exist. First, the ^1H nucleus, as a hydride, could be coordinated directly to the Mo(V). However, A_H in $g_\parallel$ and $g_\perp$ are nearly the same; that is, the shfs is nearly isotropic. This would be unusual if the ^1H coordination, itself, were highly anisotropic. The second possibility is that the nucleus as a proton is bound to one of the protein ligands to the Mo(V) and in this way interacts with the unpaired spin density through a dipolar interaction or via spin delocalization into the ligand itself. Model studies tend to support this hypothesis and indicate that the ligand atom protonated is a nitrogen.

Stiefel and co-workers prepared the Mo(V) complex, $Mo(S_2CN(C_2H_5)_2)(SNHC_6H_4)_2$.[65,66] The ESR spectrum of this species has $A_H = 7.4$ G, similar to that seen in the rapid species. The proton is readily exchangeable. The N–D and $N{-}CH_3$ complexes lack the ^1H shfs. Interestingly, $A_N = 2.4$ for the model, significantly smaller than the ^1H nucleus. The lack of ^{14}N coupling in the enzyme spectrum has always been questioned; clearly, coupling of magnitude exhibited by the model would go unresolved in the larger linewidths characteristic of biological samples. Thus, it is quite probable that the shfs in xanthine oxidase is due to a substrate-derived proton, which is transferred to a nitrogenous ligand atom during catalysis. The catalytic implications of this are quite interesting and have been extensively reviewed.[66,67]

D. Galactose Oxidase — A Case History of the ESR Study

One of the most recent and most satisfying electron spin resonance studies on a

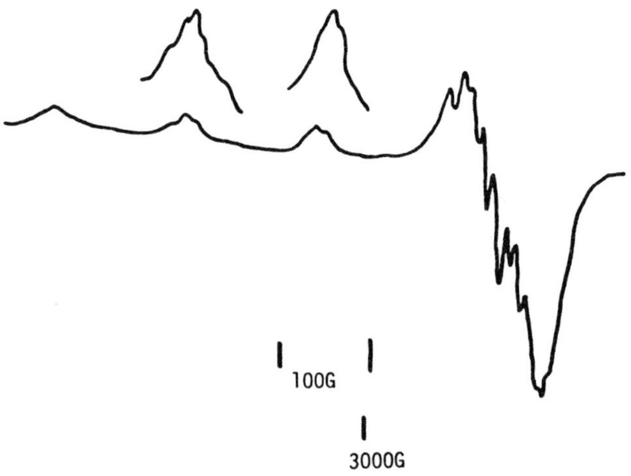

FIGURE 20.[73] ESR spectrum of $^{63/65}$Cu-galactose oxidase at 9.090 GHz. This figure represents an average of six scans. The insert shows the $M_I \pm \frac{1}{2}$ transitions at higher signal level.

biological system that has been carried out in our laboratories centered on the metalloprotein, galactose oxidase (GOase).[68] GOase, a single chain copper protein, is unique because it contains a single nonblue copper(II) atom and no other prosthetic groups. Thus, its spectral properties exemplify those characteristic of the nonblue copper(II) centers in metalloproteins without interference from other metal sites. As such it offered an opportunity to gain information about a large class of nonblue sites in multinuclear copper proteins.

Our initial studies were carried out on an enzyme that contained both copper-63 and copper-65 in natural abundance. The spectrum in Figure 20 is typical of those we obtained.[69] At that time, two alternative suggestions were offered for the interpretation of this spectrum. (1) The structure in the perpendicular region of the spectrum could be due to metal hyperfine, or (2) the structure could be due to ligand hyperfine. Of course, the possibility existed that both factors contributed to the complex structure. Our route to assigning the spectrum was to note that the $A_\parallel$ and $g_\parallel$ values corresponded to that expected for a CuN_2O_2 coordination environment and thus we might expect to find five hyperfine lines in the perpendicular region due to nitrogen superhyperfine splittings.[68,71] Since more than five were detected, a more complex or alternate interpretation was sought.

The quality of any spectrum of any enzyme will often vary with preparation procedures. As we obtained better enzyme, we were able to detect what clearly seemed to be a five line superhyperfine splitting pattern on two of the three well-resolved parallel lines (Figure 20). Such an observation required that some of the structure in the parallel region can be due to nitrogen superhyperfine splittings, since such structure in copper(II) systems is often nearly isotropic, that is, a similar magnitude of splitting should exist in the perpendicular region. As noted above, since more than five lines existed in this latter region, its assignment was still not clear. One alternative was that g value anisotropy existed, $g_{xx} \neq g_{yy}$, and that each g value envelope consisted of five superhyperfine lines and this overlapping pattern resulted in the complex perpendicular region.

As noted in Section II.A, this could be determined by examining a spectrum at a higher field strength since the g value differences would be accentuated under those conditions. We were surprised to observe the spectrum in Figure 21.[73] The perpendicular region exhibits little or no g value anisotropy. We then rediscovered that at X-

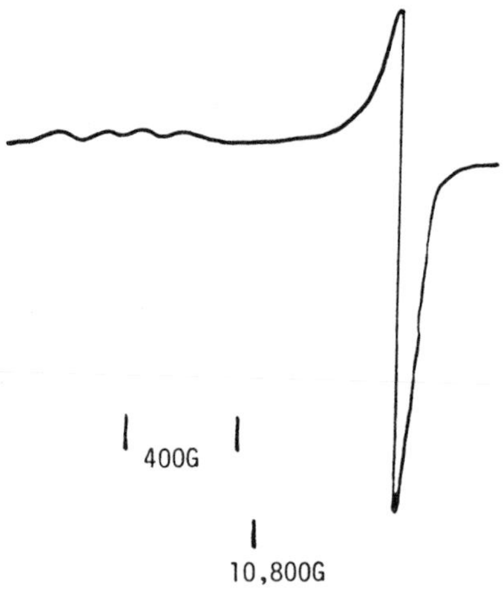

FIGURE 21.[73] ESR spectrum of $^{63/65}Cu$-galactose oxidase at 34.758 GHz.

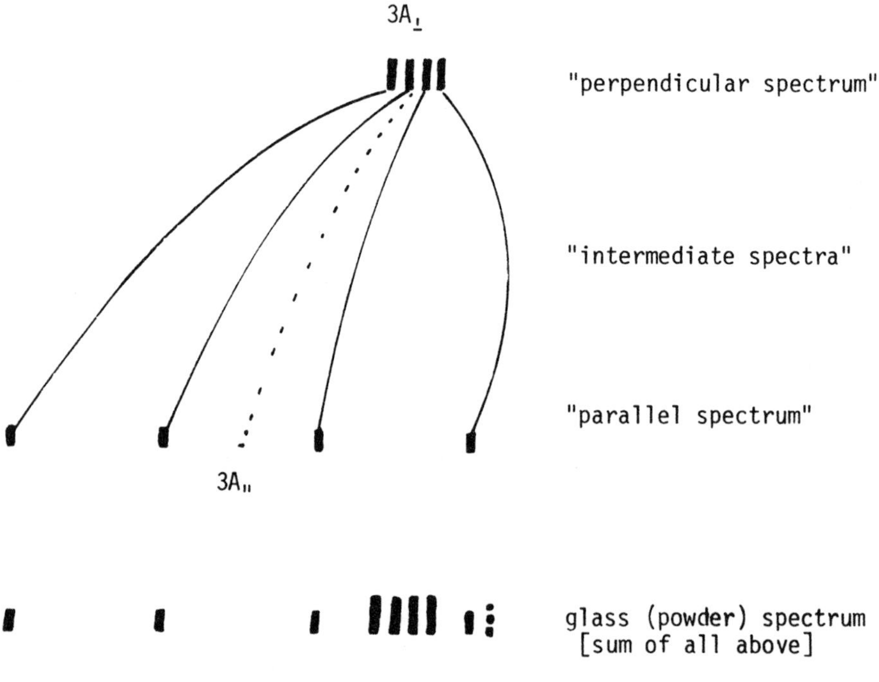

FIGURE 22. A stick diagram for a tetragonal Cu ESR spectrum showing how $g_{\perp} \rightarrow g_{\parallel}$ (....) as one proceeds through intermediate orientations. Note that the transition of $A_{\perp} \rightarrow A_{\parallel}$ (—) may change at a rate such that a false turnaround point * occurs outside the actual limits of the spectrum. This point will have features (*) intermediate between $M_I = -3/2(\parallel)$ and $M_I = 3/2\ (\perp)$. It always occurs (even though often overlooked or obscured) when we have $g_{\parallel} >> g_{\perp}$ and $A_{\parallel} >> A_{\perp}$. See also Figure 5.

band frequencies, an "extra" peak will occur at high fields in copper ESR spectra. This peak has been called an overshoot line and represents an intermediate orientation between the parallel and perpendicular orientation.[72,73] Figure 22 attempts to diagrammatically indicate how this feature occurs. Thus, the complex structure in the perpen-

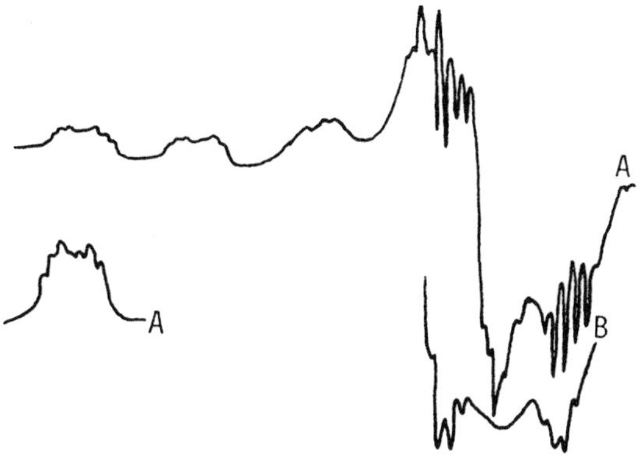

FIGURE 23.[73] ESR spectrum of ^{63}Cu-galactose oxidase-$^{19}F^-$ complex at (A) X- and (B) Q-band frequencies (perpendicular region only); [KF] = 0.22 M. The insert shows the M_I = + 3/2 transition at X-band at higher signal level.

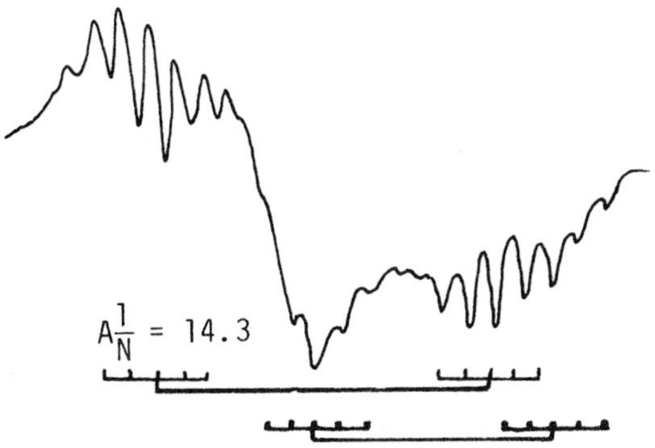

FIGURE 24.[73] Perpendicular regio of the ESR spectrum in 23A.

dicular region appeared to be both an overlap of hyperfine splittings due to nitrogen on the fourth parallel line, the $g_{\perp}$ region, and the overshoot region.

We were able to prove this quite nicely by actually making the spectrum more complicated![73] We knew from previous work that F^- was bound by the copper(II) atom in GOase in a 1:1 complex. Since the fluoride coupling constant should be large, it was expected that this effect would produce a spectrum that might complement the information sought on the native protein. Figure 23a presents the GOase-F^{19} spectrum where the ^{63}Cu protein has been employed to eliminate any extra features due to ^{65}Cu. Hyperfine coupling due to one $^{19}F^-$ (I = ½) is clearly evident, especially on the M_I = + 3/2 transition in the parallel region (see insert) indicating effective coordination to only one such ion ($A_F^{\parallel}$ = 41.0 G). Of particular value is the observation that the shfs due to nitrogen is better resolved in this case ($A_N^{\parallel}$ = 11.2 G) than in the native protein. This well-resolved pattern again could be confidently assigned to only two equivalent nitrogen atoms. In addition, a striking new feature consisting of two well-separated sets of five lines appear in the perpendicular region of the spectrum (Figure 24). This could have been due to a strong rhombic symmetry imposed by the F^- ion or to a strong dipolar coupling of the $^{19}F^-$ in perpendicular orientations ($A_F^{\perp}$ = 175.4 G). The 35 GHz spectrum is virtually superimposable on the X-band spectrum in this re-

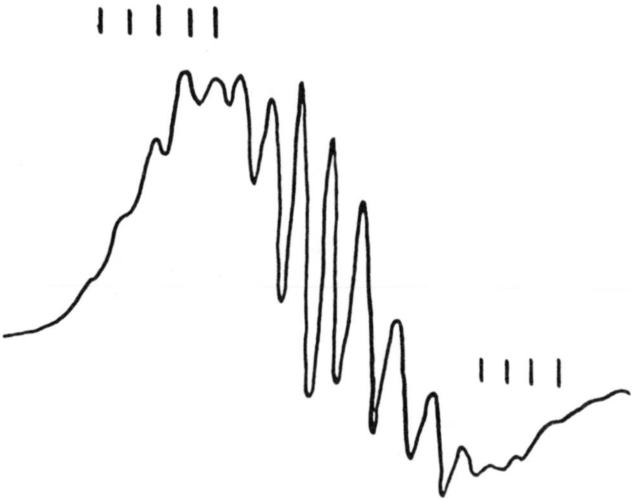

FIGURE 25.[73] Perpendicular region of the ESR spectrum of galactose oxidase-imidazole complex [Im] = 0.12M.

gion (Figure 23b), which rules out the first alternative. The overshoot line, which is also split by the $^{19}F^-$ again clearly shows a five-line pattern ($A_F{}^O$ (overshoot) = 128.1 G) with a hyperfine splitting value ($A_N{}^O$ = 13.3 G) intermediate between that in the parallel region ($A_N{}^{\parallel}$ = 11.2 G) and the perpendicular region ($A_N{}^{\perp}$ = 14.3 G). Note that the ^{19}F coupling in the overshoot line is of an intermediate magnitude as well.

Now the problem became one of attempting to identify the exact origin of the nitrogenous ligand groups to copper(II) in the protein. Again we employed ligand binding in an attempt to gain further information. We reasoned that if we were to add an exogenous nitrogen donor ligand in which the nitrogen donor atom was different than the endogenous ligands, then a very complicated superhyperfine pattern would result due to the inequivalent coupling expected between copper(II) and the two kinds of bonded nitrogens. At the same time, if we were to add a like donor molecule, we would only increase our splitting pattern from five to seven lines.

The exact identification of the endogenous nitrogen ligands was thus deduced in the following manner. Addition of imidazole as an exogenous ligand yielded a more complex hyperfine splitting pattern in the perpendicular region than that exhibited by the native enzyme (Figure 25). However, the 35 GHz spectrum (not shown) gave no new rhombic distortion (g_{xx} remained nearly equal to g_{yy}) nor the appearance of an apparent copper hyperfine in the perpendicular envelope. Thus, the changes observed at 9 GHz must have been due entirely to new superhyperfine and must be from only one additional ligand. Again, the overshoot line was a valuable guide in interpreting the spectrum (Figure 25). In this case, one can find four lines of what clearly is a seven-line pattern (the maximum intensity occurs at the fourth line centered at 3254 G) due to three nitrogen atoms ($A_N{}^O$ = 13.4 G). The interpretation of the rest of the complex perpendicular region first involves locating the fourth parallel line (centered at 3128 G) partially resolved on the low-field side of the perpendicular envelope. Three of the potential seven lines are evident ($A_N{}^{\parallel}$ = 12.1 G). Careful examination of the remaining perpendicular region leaves only seven well-resolved intense lines ($A_N{}^{\perp}$ = 15.7 G), which again could be assigned to the coupling of three equivalent nitrogen atoms. As expected, the magnitude of the nitrogen coupling on the overshoot line was intermediate between that associated with the parallel and perpendicular regions. However,

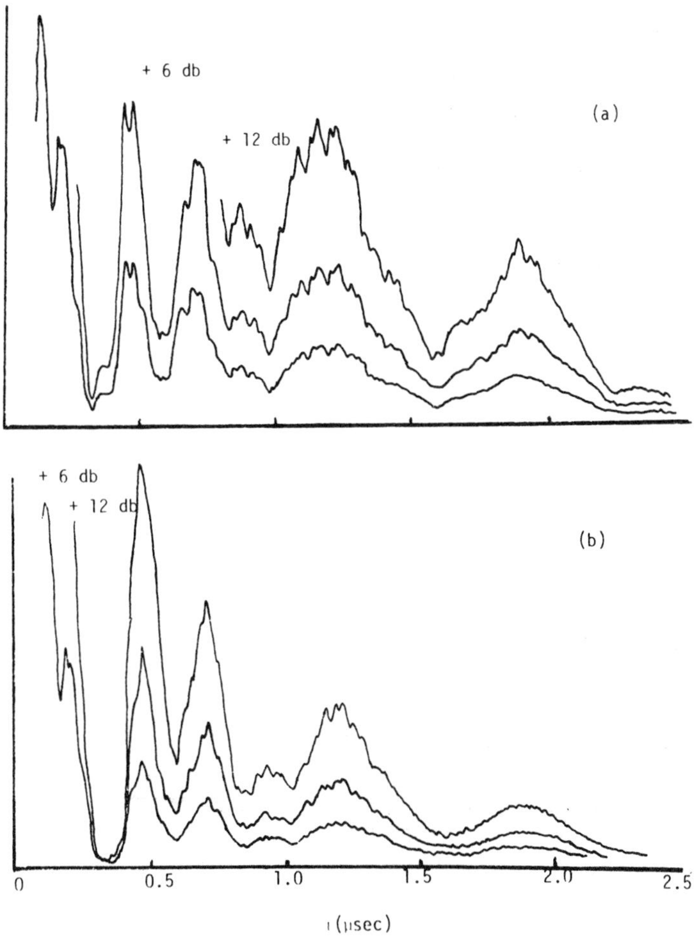

FIGURE 26.[74] Envelope modulation patterns for (a) galactose oxidase and (b) galactose oxidase-imidazole.

what was most significant was that this interpretation implied that the endogenous and exogenous nitrogen atoms were equivalent chemically and thus that the endogenous nitrogen structure is due to two histidine imidazole ligands. Any other amines considered to date (pyridine and methylimidazole) gave splittings of a more complex nature.

Coupled with this study was a simultaneous electron spin echo (pulsed electron spin resonance) study of GOase.[74] As noted in Section III.C, imidazole coordination leaves a distinctive "fingerprint" on the modulation envelope of a paramagnetic metal center.[20,21] The modulation envelopes for GOase and the GOase-imidazole complex (Figure 26) show the characteristic, low-frequency modulation attributable to the distal, noncoordinating pyrrole nitrogen in the imidazole ring. The "depth" of the modulation (the difference between the peaks and troughs) is related to the number of coordinating imidazoles; not surprisingly, this depth is more pronounced in the imidazole complex. Comparisons to modulation patterns of models (cf. Figure 14b) indicate that the copper(II) in GOase is coordinated to two (or three) protein imidazoles. The direct ESR data outlined above suggests strongly that there are only two.

VI. APPENDIXES

A. Sample Preparation and Obtaining the Spectrum

The ESR spectrometer in good condition can detect of 10^{-4} to 10^{-5} molar spins in a volume of 100 $\mu\ell$. Thus the total number of spins needed is approximately 10^{15} to 10^{16}. Yet, to obtain good quality, reproducible spectra, samples optimally should be about 10^{-3} molar.

Once a solution is available, both room temperature and frozen glass spectra should be determined. In general, the following data can be determined:

1. A solution spectrum
   Isotropic g and A_{iso} values (these are normally obtained with the aid of a "flat" cell to lower the dielectric loss of microwave signal). Practically, if the sample is in a solvent that has a significant dipole moment, a quartz flat cell will be required. In a nonpolar (assuming no H_2O contamination) solvent, a normal quartz ESR tube can sometimes be used.
2. A frozen solution (or glass spectrum)
   Anisotropic $g_{\|}$ (or g_{zz}), $g_{\perp}$ (or g_{xx} and g_{yy}), $A_{\|}$ (or A_{zz}) and $A_{\perp}$ (or A_{xx} and A_{yy}) values. Practically, any solvent that can be frozen quickly (so the solute is held trapped in the solvent lattice) can be used. The better the "glass" properties, the better the resolution. Glycerol:water mixtures yield excellent frozen glasses if one is assured that the glycerol does not interact with the sample. Samples are normally placed in 3 mm I.D. quartz sample tubes, which should be checked as blanks to be sure no impurity signal is present in the tube. Glass sample tubes should be avoided because of the impurity signals. One "trick" that can be employed is to change the insert used for producing the low temperature. This will cause minor changes in the microwave frequency and subtle changes in complex ESR spectra can be used to assign the origin of the features.
3. A diluted powder (parameters similar to the frozen solution)
   Often, when a paramagnetic complex is to be studied, especially for small molecule models, a similar diamagnetic complex can be prepared. If it is possible to prepare these together so that the paramagnetic one becomes an impurity in the diamagnetic one (3 to 5% impurity), similar results are obtained as the frozen glass. Practically these spectra can be obtained at room temperature without significantly lowering their resolution quality. At the same time, this procedure is more tedious than simply making up a solution of the complex to be studied. Normally this procedure is not employed unless (1) there is a question as to the identity of the species being studied in solution, or (2) solubility prevents frozen glass data from being obtained.

A variety of quartz tubes suitable for ESR are available from Wilmad Glass Co., Buena, N.J. Quartz flat cells for a variety of uses are available from James Scanlon, Solvang, Calif. Both types of quartz-ware are available from spectrometer manufacturers as well.

Temperature regulation can be provided in a variety of ways. Each one has its various advantages. In most cases, a gas, usually N_2 or He, is passed through a heat exchanger immersed in a cooling bath and then over a heating coil. A thermocouple placed between the heating coil and the sample regulates the heating of the gas stream and thus the temperature of the sample. Such regulated temperature ESR inserts are available for all spectrometers and generally can operate from 400°K to approximately

100°K with liquid nitrogen as coolant or to near 10°K with liquid helium as coolant (A special liquid helium transfer dewer from Air Products will allow temperature regulation and variation to near 4°K).

Another type of heat exchanger can be used for temperatures near that of liquid helium. In this one, a metal rod that is in contact with the sample is cooled by a stream of gaseous helium pumped off a liquid helium reservoir. The rate of boil off controls the temperature (7 to 30°K). This method requires a very good thermal contact between the sample and the rod.

Various closed cycle cryogenic refrigerators are also available with ESR inserts for temperature regulation. The low thermal capacity of these coupled with vibration at the sample resulting from the refrigerator motor has limited their early success in our hands. Very good reports of their success in other laboratories have recently appeared. These require a much higher initial investment, but do not require the constant expensive supply of liquid nitrogen or helium. They are capable of temperature down to near 10°K.

Spectra can also be obtained from samples that are immersed directly in a coolant. Liquid nitrogen and liquid helium dewers are available. Under the appropriate conditions, temperature to near 1°K (pumping on liquid helium) can be obtained. Liquid nitrogen insert dewers are inexpensive and easy to use. The "bumping" that results from nitrogen bubbling in the ESR cavity can be very annoying and can cause spiking on the recorded signal. A thin film of glycerol on the outside of the sample tube decreases this boiling effect. The liquid helium insert dewers are very expensive and require a good deal of cryogenic expertise. In addition, they tend to use a great deal of liquid helium and an afternoon's work can be expensive ($100 to $300).

A problem that often occurs when running samples below ambient temperatures is condensation of water in the ESR cavity between the dewer insert and the walls of the cavity. This is a major problem, but can be virtually eliminated by passing an anhydrous gas (N_2 or He) through the cavity during cooldown and data collection. Most commercial instruments have wave guides with a gas nipple for just this purpose.

B. Effects of Power Levels and Temperature

Saturation, the point at which the two energy levels are equally populated and thus no absorption occurs, is not a major problem in most biological systems. However, since maximizing signal intensity by optimizing instrument settings is always important, one should be aware of the dangers of saturation.

It is a straightforward procedure to test whether saturation is possible at any temperature of interest. If one plots the intensity of an ESR signal vs. microwave power, a straight line should be produced with a slope of 0.5, since signal height is proportional to the square root of microwave power. If a deviation occurs at higher microwave powers, then saturation is beginning to occur and a lower power setting should be selected.

The use of lower temperatures to increase resolution of an ESR signal can also induce saturation if the same power level is maintained throughout a temperature study. Again, saturation effects can be tested by plotting the area of the resonance of interest against 1/T, since the power absorbed by a sample is proportional to the difference in the two energy levels between which the transition takes place. This difference is governed by the Boltzmann equation, $e^{-h\nu/kT}$. Since the area under an absorption curve is proportional to total spins (and thus power absorbed) a graph of area vs. 1/T should result in a straight line.

The decrease in linewidth with decreasing temperature will normally reach a limit below which a temperature decrease does not increase resolution. This temperature is the one at which the spin-spin relaxation has taken over from the spin-lattice relaxation as the dominant relaxation process. A plot of linewidth vs. temperature can thus identify the "adequate" temperature for "low temperature" measurements.

C. Calculation of g and A Values

The actual determination of the magnitude of the A and g values of an ESR spectrum is straightforward, but as is often the case, an example calculation can be more informative than a lengthy discussion. Let us consider the equations that would be employed to calculate the position of <u>any</u> line in an isotropic spectrum.

$$H_o = H_m + AM_I + \frac{A^2}{2\,H_o}\,[I\cdot(I+1) - M_I^2] + \frac{A^3}{4\,H_o^2}\,[\,\ldots$$

In this expression, H_o represents the center of the particular set of lines of interest and it is this H_o value that will later be determined in order to calculate the correct g value. H_m represents the field position of the line associated with the spin orientation of the metal nucleus. A is the hyperfine coupling yet to be determined and I represents the spin quantum number of the nucleus to be considered. For example, the VO^{2+} ion has seen use as an ESR probe since it has a rich ESR spectrum. If the A value were approximately 100 G, then the last of the terms in the equation above would be only 0.1 G at X-band frequencies and is often neglected. We then can write a simple expression for the positions for all 8 lines in the isotropic ESR spectrum for VO^{2+}. We'll let the term $A^2/2H_o$ be equal to K.

$$H_{7/2} = H_o - 7/2A - 7/2K \quad (1)$$
$$H_{5/2} = H_o - 5/2A - 19/2K \quad (2)$$
$$H_{3/2} = H_o - 3/2A - 27/2K \quad (3)$$
$$H_{1/2} = H_o - 1/2A - 31/2K \quad (4)$$
$$H_{-1/2} = H_o + 1/2A - 31/2K \quad (4')$$
$$H_{-3/2} = H_o + 3/2A - 27/2K \quad (3')$$
$$H_{-5/2} = H_o + 5/2A - 19/2K \quad (2')$$
$$H_{-7/2} = H_o + 7/2A - 7/2K \quad (1')$$

Now, to obtain the A value, one can subtract pairs of these equations to yield these four independent determinations for the A value.

$$H_{1/2} - H_{-1/2} = A$$
$$H_{3/2} - H_{-3/2} = 3A$$
$$H_{5/2} - H_{-5/2} = 5A$$
$$H_{7/2} - H_{-7/2} = 7A$$

Note that the only pair of lines that are separated by A is the inner pair and often these two lines are the least accurately known.

By adding pairs of these equations, 1 and 1′, 2 and 2′, and so forth, and rearranging terms, equations can be obtained that are of the form:

$$4 + 4' = H_{1/2} + H_{-1/2}$$
$$= 2H_o - 62/2\,K$$

Thus,

$$\frac{H_{1/2} + H_{-1/2}}{2} = H_o - 62/4\,K$$

or

$$H_o = \frac{H_{1/2} + H_{-1/2}}{2} + \frac{31}{4}\frac{A^2}{H_o}$$

Now, we can represent $\frac{H_{1/2} + H_{-1/2}}{2}$ as the average of the two field positions in Gauss, $H_{4,5}$. Thus,

$$H_o = H_{3,6} + \frac{27}{4}\frac{A^2}{H_o}$$

$$H_o = H_{2,7} + \frac{19}{4}\frac{A^2}{H_o}$$

$$H_o = H_{1,8} + \frac{7}{4}\frac{A^2}{H_o}$$

The evaluation of H_o is thus an iterative process. One first assumes that H_o is $H_{a,b}$ and uses that H_o to calculate a new H_o and so forth. Normally, convergence will result after two or three cycles.

Often, we may not need to make these second order calculations for H_o. If A is large, i.e., 50 G, then we can estimate the error in H_o quickly at a field of 3000 G. Consider that the K term would be approximately 5 G under those considerations. That is

$$K = \frac{27}{4}\frac{A^2}{3000} \cong 5\text{ G}$$

Once a value for H_o has been determined, then the g value is obtained by substituting in the standard equation.

$$g = \frac{h\nu}{\beta H_o}$$

where $h\nu$ = microwave frequency in MHz ≈ 9,000; β = 1.39969; and $H_o \cong$ 3,000 G

$$\cong \frac{9000}{3000}(1.4) \cong 2$$

It is often convenient to report A values in cm^{-1} (or millikaisers, mK). This can be done by application of the following equation:

$$\frac{g}{2.1418}(A_G) \times 10^{+4} = A_{cm^{-1}}$$

Similar equations to all of those above can be written for $A_{\|}$ and

$$g_{\|}\left(K = \frac{A_{\|}^2}{2H_o}\right) \text{ and } g_{\perp} \text{ and } A_{\perp}\left(K = \frac{A_{\|}^2 + A_{\perp}^2}{4H_o}\right)$$

In order to more clearly demonstrate this calculation procedure, we have selected an example, obviously not of biological importance, but one that clearly demonstrates the value of second order calculations. The ESR frozen glass spectrum of NbO^{++} in an ethanol-solution that was saturated with HCl gas consists of 20 overlapping lines (10 parallel and 10 perpendicular; I = 9/2). The positions of the parallel lines are assigned as the center of the absorption peak, while the position of the perpendicular lines are assigned as the midpoint of the line tangent to the face of the first derivative line, regardless of the position of the apparent baseline. The positions of the 10 parallel and 10 perpendicular lines to 0.1 G are listed below (ν = 9112.77 MHz).

	$A_{\|}$	$A_{\perp}$
1	2172.4	2818.1
2	2419.3	2892.1
3	2672.1	2982.5
4	2929.2	3088.7
5	3188.0	3219.4
6	3466.7	3349.3
7	3740.3	3485.4
8	4019.2	3650.2
9	4302.5	3816.2
10	4588.3	4000.0

By subtracting pairs of the line positions and dividing by the appropriate number, $A_{\|}$ and $A_{\perp}$ can be determined five independent times.

	$A_{\|}$	$A_{\perp}$
$\frac{A_{1-10}}{9}$	268.4	131.3
$\frac{A_2-A_9}{7}$	269.0	132.0
$\frac{A_3-A_8}{5}$	269.4	133.5
$\frac{A_4-A_7}{3}$	270.4	132.2
$\frac{A_5-A_6}{1}$	278.7	129.9

Note that the $A_{5,6}$ is the least accurate of those determined. In this case, $A_{\|}$ and $A_{\perp}$ were taken as the average of the first four values. Now by adding pairs of lines and going through the iterative calculations to get H_o ($\|$) and H_o $(\perp)_x$, the following values of H_o result.

$\frac{H_m + H_{-m}}{2}$	1st Iteration	2nd Iteration	$\frac{H_m + H_{-m}}{2}$	1st Iteration	2nd Iteration
3380.3	3391.8	3391.8	3409.1	3438.7	3438.5
3360.9	3392.9	3392.8	3354.2	3427.9	3435.9
3345.6	3393.6	3392.9	3316.3	3441.7	3437.0
3334.7	3393.3	3392.2	3287.1	3440.9	3434.0
3327.3	3391.2	3390.0	3284.4	3452.0	3443.9
		3391.9			3436.3

In this case again, we averaged the first four $H_o(\|)$ to calculate $g_{\|}$ as 1.9194 and the first four $H_o(\perp)$ to yield $g_{\perp}$ as 1.8946. Note the tremendous error that would be introduced by taking the midpoint between the middle pair of lines as H_o and not applying second order calculations.

D. Calibration of an ESR Spectrum

Quite often, even in those laboratories where good ESR spectrometers are available, some supporting equipment such as gaussmeters and frequency counters with sufficient range to determine the microwave frequencies are not available.

The spectrometer can be calibrated using a dilute solution of vanadylacetylacetone or $VO(acac)_2$ in benzene (g using $H_{4,5}$ = 1.986) and solid diphenylpicrylhydrazyl or DPPH (g_{DPPH} = 2.0036). The field sweep can be calibrated in Gauss by assigning the peak separation between the fourth and fifth lines of the $VO(acac)_2$ as 108.0G. For example, for a 1,000G scan, the sweep we found was 25.7 G/cm. This was taken as an average for several runs and the instrument should be recalibrated periodically to insure stability. The g values of a sample can be assigned by the shift from the DPPH line. The absorption due to DPPH is located between the fourth and fifth line of the $VO(acac)_2$. The magnetic field of the g value can be calculated by the following relationship:

$$h\nu = g_{DPPH}\beta H_{DPPH} = g_{4,5}\,\beta H_{4,5}$$

where $g_{4,5}$ equals the g value calculated without second order considerations of fourth and fifth line of the $VO(acac)_2$ and has a value of 1.986. $H_{4,5}$ is calculated by measuring the distance it lies from the g_{DPPH}.

$$H_{4,5} = H_{DPPH} + \Delta H$$

Since g_{DPPH} = 2.0036, Equation 1 may be rewritten as

$$2.0036\ H_{DPPH} = 1.986(H_{DPPH} + \Delta H)$$

For this particular calibration ΔH equals 28.27G. This gives us a value of H_{DPPH} = 3190.0 G.

Measured values of H_{DPPH} for several calibrations will vary somewhat. This average value can be used when assigning a g value for a sample. Although fluctuations may seem large, in actual measurement all g values can be assigned using g_{DPPH} = 2.0036, and the distance which the g value of the sample lies from g_{DPPH}. Thus the actual assignment of H_{DPPH} becomes arbitrary. Errors induced by this method give very small errors in calculated g values (±0.0002).[a] Other standards for calibration have been tabulated and discussed.[75,76]

E. Determination of Spin Densities[75,76]

Except in the most favorable cases, it is generally not possible to use ESR spectra to determine the number of paramagnetic spins present. If a standard is chosen that has a similar spectrum, one can perform double integrations (by hand if need be) to yield results that are reliable to ± 20%. If a computer is available, which can accept the observed ESR signal, the various routines are available to carry out the double integration. We find that the error results from the choice of the baseline for the original spectrum.

Samples can be marked by placing a piece of tape with a small amount of DPPH around the outside of the ESR tube. The spectra are then recorde d and assignments of A and g values are made.

ACKNOWLEDGMENT

We are truly indebted to those many colleagues who have contributed to our appreciation of paramagnetic resonance. We particularly thank Robert Kurland, Robert Allendoerfer, Dennis Chasteen, Carl Brubaker, Jr., Max Rogers, Leon Stock, Lawrence Piette, Jack Peisach, and William Mims. We, of course, accept all responsibility for any shortcomings of this chapter. Our research has been supported by the National Science Foundation (BMS73-01248) and the Graduate School of the State University of New York at Buffalo. RDB has been a Fellow of the Camille and Henry Dreyfus Foundation.

REFERENCES

1. **Bersohn, M. and Baird, J. C.,** *An Introduction to Electron Paramagnetic Resonance,* W. A. Benjamin, New York, 1966.
2. **Poole, C., Jr.,** *Electron Spin Resonance — A Comprehensive Treatise on Experimental Techniques,* Wiley-Interscience, New York, 1967.
3. **Carrington, A. and McLachlan, A.,** *Introduction to Magnetic Resonance,* Harper & Row, New York, 1967.
4. **Pake, G.,** *Paramagnetic Resonance,* Harper & Row, New York, 1967.
5. **Alger, R.,** *Electron Paramagnetic Resonance Techniques and Applications,* Wiley-Interscience, New York, 1968.
6. **Wertz, J. and Bolton, J.,** *Electron Spin Resonance—Elementary Theory and Applications,* McGraw-Hill, New York, 1972.
7. **Ingram, D. J. E.,** *Biological and Biochemical Applications of Electron Spin Resonance,* Plenum Press, New York, 1967.
8. **Ehrenberg, A., Malmstrom, B., and Vanngard, T., Eds.,** *Magnetic Resonance in Biological Systems,* Pergamon Press, New York, 1967.
9. **Feher, G.,** *Electron Paramagnetic Resonance with Applications to Selected Problems in Biology,* Gordon and Breach, New York, 1970.
10. **Swartz, H. M., Bolton, J. R., and Borg, D. C., Eds.,** *Biological Applications of Electron Spin Resonance,* Wiley-Interscience, New York, 1972.
11. **Goodman, B. A. and Raynor, J. P.,** Electron spin resonance of transition metal complexes, *Chem. Br.,* 10, 254, 1974.
12. **Feher, G.,** Observation of nuclear magnetic resonances via the electron spin resonance line, *Phys. Rev.,* 103, 834, 1956.
13. **Feher, G.,** Electron nuclear double resonance (ENDOR) experiments, *Physica (The Hague),* 24, S80, 1958.
14. **Hyde, J. S.,** in *Magnetic Resonance in Biological Systems,* Pergamon Press, New York, 1967, chap. 17.
15. **Mims, W. B.,** Electron spin echoes, in *Electron Paramagnetic Resonance,* Geschwind, S., Ed., Plenum Press, New York, 1972, chap. 4.
16. **Mims, W. B.,** Envelope modulation in spin-echo experiments, *Phys. Rev. Sect. B,* 5, 2409, 1972.
17. **Mims, W. B.,** Amplitudes of superhyperfine frequencies displayed in the electron spin-echo envelope, *Phys. Rev. Sect. B,* 6, 3543, 1972.
18. **Swartz, H. M., Bolton, J. R., and Borg, D. C., Eds.,** *Biological Applications of Electron Spin Resonance,* Wiley-Interscience, New York, 1972, 61.
19. **Mims, W. B. and Peisach, J.,** Pulsed EPR studies of metalloproteins, submitted for publication.
20. **Mondovi, B., Graziani, M. T., Mims, W. B., Otzik, R., and Peisach, J.,** Pulsed electron paramagnetic resonance studies of types I and II copper of *Rhus vernicifera* laccase and porcine ceruloplasmin, *Biochemistry,* 16, 4198, 1977.
21. **Mims, W. B. and Peisach, J.,** The nuclear modulation on effect in electron spin echoes for complexes of Cu^{2+} and imidazole with ^{14}N and ^{15}N, *J. Chem. Phys.,* 69, 4921, 1978.
22. **Mims, W. B.,** Measurement of the linear electric field effect in EPR using the spin echo method, *Rev. Sci. Instrum.,* 45, 1583, 1974.

23. **Platt, J. R.,** Electrochromism, a possible change of color producible in dyes by an electric field, *J. Chem. Phys.*, 34, 862, 1961.
24. **Peisach, J. and Mims, W. B.,** Linear electric field-induced shifts in electron paramagnetic resonance: a new method for study of the ligands of cytochrome P-450, *Proc. Natl. Acad. Sci. U.S.A.*, 70, 2979, 1973.
25. **Mims, W. B. and Peisach, J.,** Linear electric field effect measurements of variant low-spin forms of ferric cytochrome C, *Biochemistry*, 13, 3346, 1974.
26. **Peisach, J., Orme-Johnson, W. R., Mims, W. B., and Orme-Johnson, W. H.,** Linear electric field effect and nuclear modulation studies of ferredoxins and high potential iron-sulfur proteins, *J. Biol. Chem.*, 252, 5643, 1977.
27. **Chance, B., Gibson, Q. H., Eisenhardt, R. H., and Lonberg-Holm, K. K., Eds.,** *Rapid Mixing and Sampling Techniques in Biochemistry*, Academic Press, New York, 1964.
28. **Piette, L.,** Continuous flow methods adopted for EPR apparatus, in *Rapid Mixing and Sampling Techniques in Biochemistry*, Academic Press, New York, 1964, 131.
29. **Borg, D. C.,** Continuous flow methods adopted for EPR apparatus, in *Rapid Mixing and Sampling Techniques in Biochemistry*, Academic Press, New York, 1964, 135.
30. **Bray, R. C.,** Sudden freezing as a technique for the study of rapid reactions, *Biochem. J.*, 81, 189, 1961.
31. **Bray, R. C. and Pettersson, R.,** ESR measurements, *Biochem. J.*, 81, 194, 1961.
32. **Bray, R. C.,** Quenching by squirting into cold immiscible liquids, in *Rapid Mixing and Sampling Techniques in Biochemistry*, Academic Press, New York, 1964, 195.
33. **Palmer, G. and Beinert, H.,** An experimental evolution of the Bray rapid freezing technique, in *Rapid Mixing and Sampling Techniques in Biochemistry*, Academic Press, New York, 1964, 205.
34. **Ballou, D. P. and Palmer, G. A.,** Practical rapid quenching instrument for the study of reaction mechanisms by EPR, *Anal. Chem.*, 46, 1248, 1974.
35. **Ballou, D. P.,** Instrumentation for the Study of Rapid Biological Oxidation-Reduction Reactions by EPR and Optical Spectroscopy, Ph.D. thesis, University of Michigan, Ann Arbor, 1971.
36. **Wood, J. M. and Brown, D. M.,** The chemistry of vitamin B_{12}-enzymes, *Struct. Bonding (Berlin)*, 11, 47, 1972.
37. **Babior, B. M., Ed.,** *Cobalamin*, John Wiley & Sons, New York, 1975.
38. **Abeles, R. H. and Dolphin, D.,** The vitamin B_{12} coenzyme, *Chem. Res.*, 9, 114, 1976.
39. **Finlay, T. H., Valinsky, J., Mildvan, A. S., and Abeles, R. H.,** ESR studies with dioldehydrase, *J. Biol. Chem.*, 248, 1285, 1973.
40. **Valinsky, J. E., Abeles, R. H., and Mildvan, A. S.,** ESR studies with dioldehydrase II, *J. Biol. Chem.*, 249, 2751, 1974.
41. **Valinsky, J. E., Abeles, R. H., and Fu, J. A.,** ESR studies on dioldehydrase III, *J. Am. Chem. Soc.*, 96, 4709, 1974.
42. **Cockle, S. A., Hill, H. A. O., Williams, R. J. P., Davies, S. P., and Foster, M. A.,** Detection of intermediates during the conversion of propane-1,2-diol to proprionaldehyde by glyceroldehydrase, *J. Am. Chem. Soc.*, 94, 275, 1972.
43. **Babior, B. M., Moss, T. H., and Gould, D. C.,** The mechanism of action of ethanolammonia lyase, *J. Biol. Chem.*, 247, 4339, 1972.
44. **Babior, B. M., Moss, T. H., Orme-Johnson, W. H., and Beinert, H.,** The mechanism of action of ethanolamine ammonialyase, *J. Biol. Chem.*, 249, 4537, 1974.
45. **Hamilton, J. A. and Blakley, R. L.,** ESR studies of ribonucleotide reduction catalyzed by the ribnucleotide reduction of *lactobacillus leichmaunia*, *Biochim. Biophys. Acta*, 184, 224, 1969.
46. **Hamilton, J. A., Tamao, Y., Blakley, R. L., and Coffman, R. E.,** ESR studies on cobalamin-dependent ribonucleotide reduction, *Biochemistry*, 11, 4696, 1972.
47. **Orme-Johnson, W. H., Beinart, H., and Blakley, R. L.,** Cobamides and ribonucleotide reduction, *J. Biol. Chem.*, 249, 2338, 1974.
48. **Schepler, K. L., Dunham, W. R., Sands, R. H., Fee, J. A., and Abeles, R. H.,** A physical explanation of the EPR spectrum observed during catalysis by enzymes utilizing coenzyme B_{12}, *Biochim. Biophys. Acta*, 397, 510, 1975.
49. **Sando, G. N., Blakley, R. L., Hogenkamp, H. P. C., and Hoffmann, P. J.,** Studies on the mechanism of adenosylcobalamin-dependent ribonucleotide reduction by the use of analogs of the coenzyme, *J. Biol. Chem.*, 250, 8774, 1975.
50. **Thelander, L.,** Physiochemical characterization of ribonucleotide diphosphate reductase from *E. coli*, *J. Biol. Chem.*, 248, 4591, 1973.
51. **Brown, N. C., Elliasson, R., Reichard, P., and Thelander, L.,** Spectrum and iron content of protein B2 from ribonucleotide diphosphate reductase, *Eur. J. Biochem.*, 9, 512, 1969.
52. **Thelander, L.,** Reaction mechanisms of ribonucleotide diphosphate reductase from *E. coli*, *J. Biol. Chem.*, 247, 4858, 1974.

53. **Ehrenberg, A. and Reichard, P.**, ESR of the iron-containing protein B2 from ribonucleotide reductase, *J. Biol. Chem.*, 247, 5485, 1972.
54. **Atkin, C. L., Thelander, L., Reichard, P., and Lang, G.**, Iron and free radicals in ribonucleotide reductase, *J. Biol. Chem.*, 248, 2464, 1973.
55. **Sjöberg, B. -M., Reichard, P., Gräslund, A., and Ehrenberg, A.**, Nature of the free radical in ribonucleotide reductase from *E. coli, J. Biol. Chem.*, 252, 536, 1977.
56. **Bray, R. C.**, Molybdenum iron-sulfur flavin hydroxylases and related enzymes, in *The Enzymes*, Vol. 12, 3rd ed., Boyer, P. O., Ed., Academic Press, New York, 1975, 299.
57. **Bray, R. C. and Swam, J. C.**, Molybdenum-containing enzymes, *Struct. Bonding (Berlin)*, 11, 107, 1972.
58. **Lowe, D. J., Lynden-Bell, R. M., and Bray, R. C.**, Spin-spin interaction between molybdenum and one of the iron-sulfur systems of xanthine oxidase and its relevance to the enzymic mechanism, *Biochem. J.*, 130, 239, 1972.
59. **Edmondson, D., Ballou, D., van Heuvelen, A., Palmer, G., and Massey, V.**, Kinetic studies on the substrate reduction of xanthine oxidase, *J. Biol. Chem.*, 248, 6135, 1973.
60. **Olsen, J. S., Ballou, D. P., Palmer, G., and Massey, V.**, The mechanism of action of xanthine oxidase, *J. Biol. Chem.*, 249, 4363, 1974.
61. **Bray, R. C. and Vänngård, T.**, "Rapidly appearing" Mo EPR signals from reduced xanthine oxidase, *Biochem. J.*, 114, 725, 1959.
62. **Palmer, G., Bray, R. C., and Beinert, H.**, Direct studies on the electron transfer sequence in xanthine oxidase by EPR spectroscopy, *J. Biol. Chem.*, 239, 2657, 1964.
63. **Olsen, J. S., Ballou, D., Palmer, G., and Massey, V.**, The reaction of xanthine oxidase with molecular oxygen, *J. Biol. Chem.*, 249, 4350, 1974.
64. **Bray, R. C. and Meriwether, L. S.**, Electron spin resonance of xanthine oxidase substituted with Mo-95, *Nature (London)*, 212, 467, 1966.
65. **Pariyadath, N., Newton, W. E., and Stiefel, E. I.**, Monomeric Mo(V) complexes showing ^{1}H, ^{2}H, and ^{14}N shfs in their EPR spectra. Implication for molybdenum enzymes, *J. Am. Chem. Soc.*, 98, 5388, 1976.
66. **Stiefel, E. I., Newton, W. E., Watt, G. D., Hadfield, K. L., and Bulen, W. A.**, Molybdeoenzymes, the role of electrons, protons, and dihydrogen, *Adv. Chem. Ser.*, 162, 353, 1977.
67. **Stiefel, E. I. and Garner, J. K.**, in *Proc. First Int. Conf. on the Chemistry and Uses of Molybdenum*, Mitchell, P. C. H., Ed., Climax Molybdenum Co., London, 1974.
68. **Bereman, R. D., Ettinger, M. J., Kosman, D. J., and Kurland, R. J.**, Characterization of the copper(II) site in galactose oxidase, *Adv. Chem. Ser.*, 162, 263, 1977.
69. **Giordano, R. S. and Bereman, R. D.**, Stereoelectronic properties of metalloenzymes I: a comparison of the coordination of copper(II) in galactose oxidase and a model system, N,N′-ethylenebis(trifluoroacetyl-acetoniminato)-copper(II), *J. Am. Chem. Soc.*, 96, 1010, 1974.
70. **Peisach, J. and Blumberg, W. E.**, Structural implications derived from the analysis of EPR spectra of natural and artificial copper proteins, *Arch. Biochem. Biophys.*, 165, 691, 1974.
71. **Giordano, R. S.**, The Stereoelectronic Properties of the Type 2 Cu(II) Enzyme Galactose Oxidase and Related Model Systems, Ph.D. thesis, State University of New York at Buffalo, 1976.
72. **Neiman, R. and Kivelson, D.**, ESR line shapes in glasses of copper complexes, *J. Chem. Phys.*, 35, 156, 1961.
73. **Bereman, R. D. and Kosman, D. J.**, Stereoelectronic properties of metalloenzymes 5. Identification and assignment of ligand hyperfine splittings in the electron spin resonance spectrum of galactose oxidase, *J. Am. Chem. Soc.*, 99, 7322, 1977.
74. **Kosman, D. J., Peisach, J., and Mims, W. B.**, Pulsed EPR studies of the Cu(II) site in galactose oxidase, *Biochemistry*, 19, 1304, 1980.
75. **Bolton, J. R., Borg, D. C., and Swartz, H. M.**, in *Biological Applications of Electron Spin Resonance*, Wiley-Interscience, New York, 1972, chap. 2.
76. **Fee, J. A.**, Transition metal EPR related to proteins, *Methods Enzymol.*, 49, 512, 1978.

Chapter 3

NUCLEAR RELAXATION STUDIES OF LIGAND-ENZYME INTERACTIONS

Thomas Nowak

TABLE OF CONTENTS

I. INTRODUCTION

The phenomenon of nuclear magnetic resonance (NMR) is a well-known process in physics and a commonly used technique in organic chemistry. The applications of this phenomenon to biochemical problems are rapidly increasing, as reflected in the growing number of papers in the literature dealing with the applications of NMR. A portion of the recent popularity of NMR can be attributed to the interests of scientists trained in other disciplines (physics, physical chemistry, physical organic chemistry, inorganic chemistry, etc.) to biochemical problems. However, a more thorough development of NMR theory and recent technical developments have made the study of biochemical problems even more favorable over the past few years. This chapter will attempt to deal primarily with the study of ligand-protein interactions by NMR techniques. More specifically, the use of paramagnetic probes such as cations will be stressed. Emphasis will be placed on the understanding of the phenomena, the techniques involved, the information to be gained, and the limitations of the techniques. Several recent texts have dealt with more general applications of NMR to the study of biochemical systems including proteins.[1,2] These texts serve as excellent references for additional detailed analyses and descriptions of NMR phenomena.

The magnetic resonance phenomena contain three principle parameters that reflect useful information; the chemical shift, coupling constants, and relaxation rates. It is the latter phenomenon that will be emphasized in these applications. The judicious applications of nuclear relaxation rates can result in kinetic, thermodynamic, and structural information concerning the formation of enzyme-ligand complexes. A description of the NMR phenomena can be provided via a quantum mechanical approach or a classical mechanical description. Since most of the phenomena can adequately be described via a classical theory that may be easier to visualize, this will be the preferred approach used in this chapter.

II. NMR PHENOMENON

Nuclei can be characterized by several properties that include mass and charge. A number of nuclei (and the electron) also exhibit the property of a magnetic moment, M_o. Among the nuclei exhibiting such a property are ^1H, ^2H, ^{13}C, ^{15}N, ^{31}P, and ^{23}Na. When a nucleus that exhibits a magnetic moment is placed in a static magnetic field, H_o, the magnitude of M_o is given by

$$M_o = \chi_o H_o \tag{1}$$

where χ_o is the magnetic susceptibility. Upon interaction with the magnetic component of the electromagnetic field, the nucleus also exhibits the property of spin along the component of H_o. The precession frequency, ω_o of the nucleus is proportional to H_o by the magnetogyric ratio γ

$$\omega_o = \gamma H_o \tag{2}$$

where ω_o is in radians sec^{-1}. The number of energy levels of a nucleus of spin I in a magnetic field is 2 I + 1 and the energy levels are separated (ΔE) by

$$\Delta E = \frac{\mu H_o}{I} \tag{3}$$

where μ is the nuclear magnetic moment and is given by Equation 4

$$\mu = \frac{\gamma h I}{2\pi} \tag{4}$$

where h is Planck's constant. The frequency of radiation (μ_0) that induces a transition between energy levels is thus related to Equation 2.

$$\nu_o = \frac{\omega_o}{2\pi} = \frac{\gamma H_o}{2\pi} \text{ (cycles per sec (cps) or Hz)} \tag{5}$$

Thus, in Figure 1 we can visualize a nucleus in the presence of a magnetic field, aligned with the field along the z axis (standard nomenclature). The nucleus precesses about the z axis. A source of radiofrequency (rf) energy (H_1) is directed along the x axis and when the frequency obeys the resonant condition

$$\nu_{rf} = \nu_o \tag{6}$$

the magnetic moment absorbs energy from H_1 and tips to a new angle θ' in the x-y plane. A receiver in the y axis detects the component in the y plane and translates the y vector into an NMR signal (Figure 2).

Under equilibrium conditions nuclei are distributed among the energy levels in a normal Boltzmann distribution.

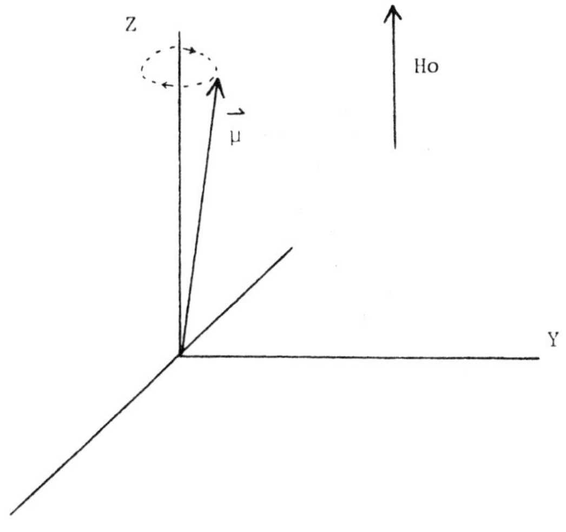

FIGURE 1. Precession of a magnetic moment in a static magnetic field, H_o. The nucleus precesses about the z axis, parallel to the magnetic field with the Larmor angular frequency ω_o.

$$\frac{N_{upper}}{N_{lower}} = e^{-\Delta E/kT} \tag{7}$$

Here, N_{upper} and N_{lower} refer to the population of nuclei in the upper and lower energy states, respectively. A disruption of this distribution by some physical process such as the addition of a magnetic field or addition of an rf field, perturbs this equilibrium. The return back to equilibrium in its environment (lattice) is an exponential relaxation process characterized by a time constant T_1. This relaxation time is called a *spin-lattice* or a *longitudinal relaxation time.* Spin lattice relaxation occurs from the interaction of the nuclear dipoles with random fluctuating magnetic fields provided by surrounding dipoles within the lattice which have a frequency that of resonance frequency (Equation 5).

A second relaxation process, T_2 is called a *spin-spin* or *transverse relaxation time.* This process characterizes the loss of phase coherence of a group of spins in the xy plane. If a strong rf source is imposed along the x axis, the total magnetization is flipped into the xy plane. A removal of this rf source allows the nuclei to precess about the z axis. Since each dipole has its own microenvironment, they will all begin to precess with their own frequency, thus phase coherency will be lost. The loss of net magnetization is an exponential process with a time constant T_2. Spin-spin relaxation occurs via the same mechanisms that cause T_1 relaxation, however additional processes also contribute to T_2 relaxation. Random nonfluctuating magnetic fields also contribute to the T_2 process. Such local random fields cause shifts in resonance frequencies of the nuclear spins in the sample. The greater the variation, the wider the resonance peak in the nmr spectrum. The T_2 relaxation rate is thus related to the linewidth at half-height ($\nu_{1/2}$) by Equation 8

$$1/T_2 = \pi \nu_{1/2} \tag{8}$$

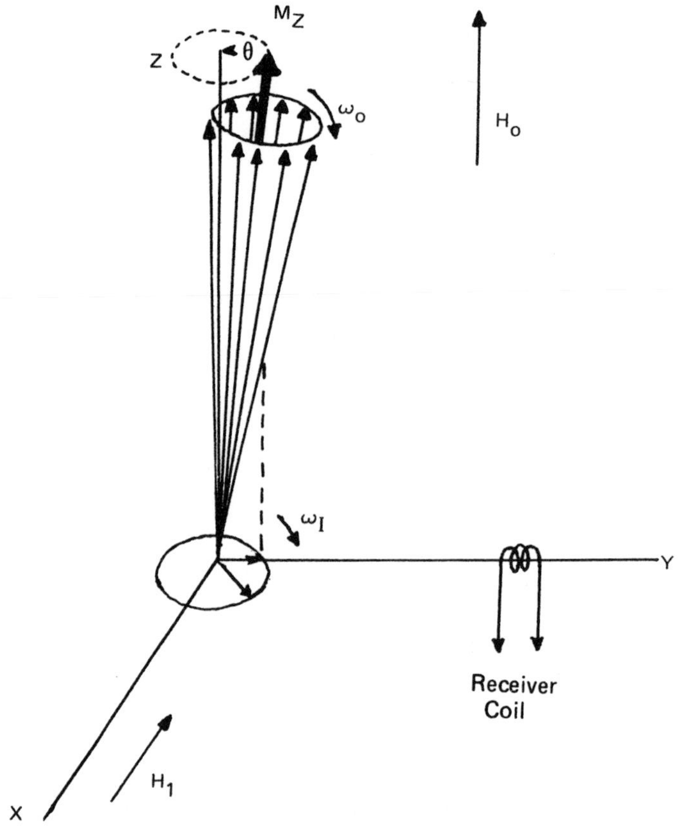

FIGURE 2. The generation of a nuclear magnetic resonance signal. The net ensemble of excess nuclear moments provides a resultant vector, M_z in the z direction. The addition of an oscillator frequency (radiofrequency) field, H_1 is applied perpendicular to H_o. When the precession frequency; ω_1 of the H_1 satisfies the resonance condition ($\nu_{rf} = \nu_o$) energy is transferred to the nuclei and the resultant vector is tipped at an angle θ into the y axis. The resultant vector wobbles about the z axis at ω_o and provides an alternating field that induces a current in the receiver generating an NMR signal.

III. RELAXATION RATE MEASUREMENTS

These relaxation processes are perhaps best visualized by describing the methods used to measure T_1 and T_2. To describe such processes it is helpful to change the reference in the coordinate system. Instead of using the coordinate system of the laboratory, the coordinate system, rotating in the same direction and frequency as the nuclear moments precess, will be used. The rotating system is referred to as the rotating frame of reference. This reference frame is analogous to our treatment of the rotation of the earth as a rotating frame. Before a person begins to drive his car, we treat the motion as zero, or fixed, and not as the motion of the rotating earth. The rotating frame coordinate system will be designated as the x′, y′, and z′ axes, respectively, each rotating with the frame at a frequency ω.

Figure 3 demonstrates the application of an rf field, H_1 at resonance frequency along the x′ axis in the presence of H_o along z′. The application of H_1 causes the net magnetization vector M to tip via an angle θ to the y′ axis to induce an NMR signal. The angle θ is determined by the strength of H_1 or the length of its application, t_{app} by Equation 9

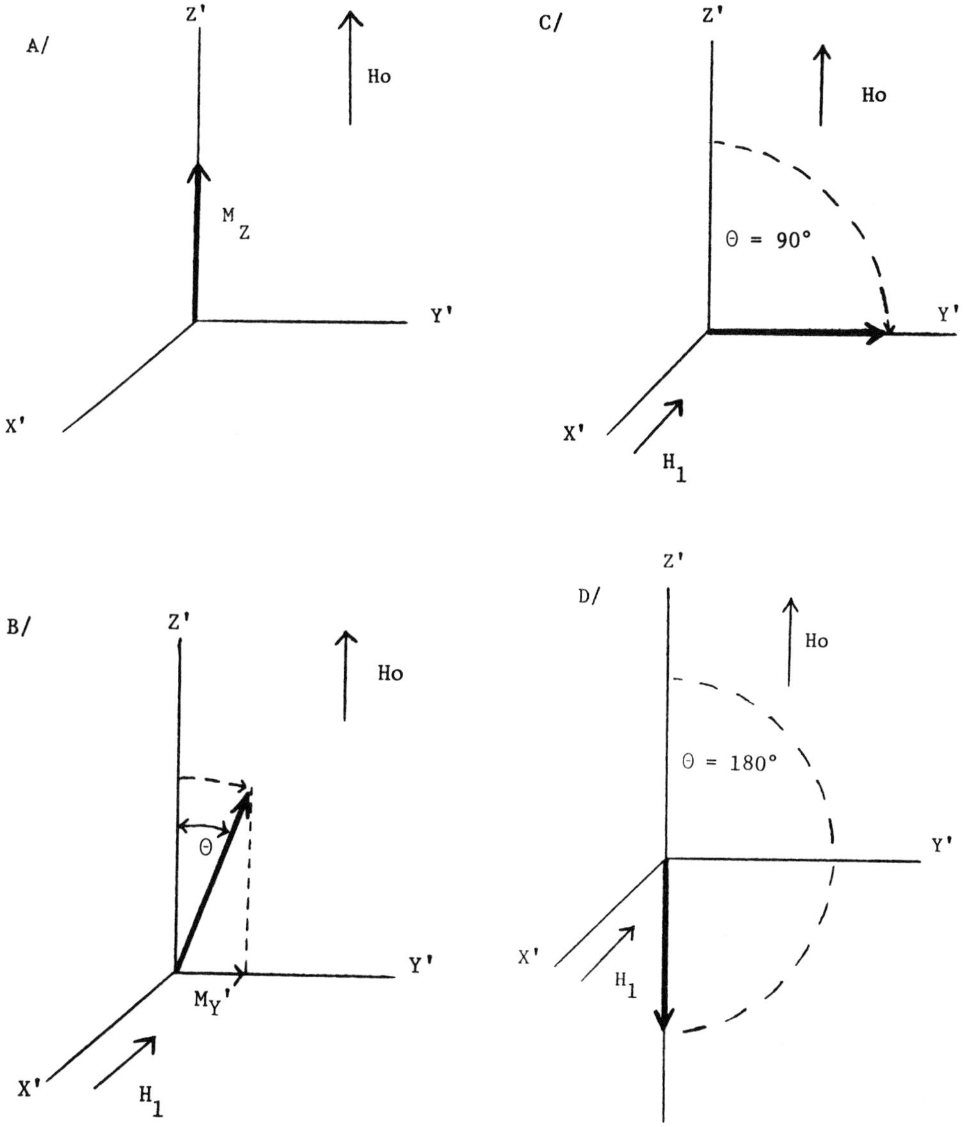

FIGURE 3. The application of an rf field, H_1 and its effect on the resultant magnetic vector M in the rotating frame of coordinates x', y', and z'. (A) The static magnetic field H_o induces a net magnetic vector in the z coordinate M_z. (B) The application of an rf field under resonance conditions, $\nu_{rf} = \nu_o$ induces a "tipping" of the vector by the angle θ into the y' coordinate. (C) The application of H_1 for a sufficient length of time or at a sufficient value of γH_1 can induce a 90° tilt to obtain a maximum signal intensity. (D) At a sufficient value of γH_1, the time of application of H_1 used in (C) can be doubled to obtain a 180° tilt, which results in no observed NMR signal.

$$\theta = \gamma H_1 t_{app} \tag{9}$$

Pulsed NMR techniques are usually accomplished by using a fixed H_1 and the time of application for an rf pulse pulse is varied. Thus a 90° pulse is obtained by varying the length of t_{app} until M is tipped $\pi/2$ or 90° into the y' axis (Figure 3C). This induces a maximum NMR signal. A 180° degree or π pulse (Figure 3D) is determined by varying the time t_{app} until the first zero signal (past 0°) is obtained.

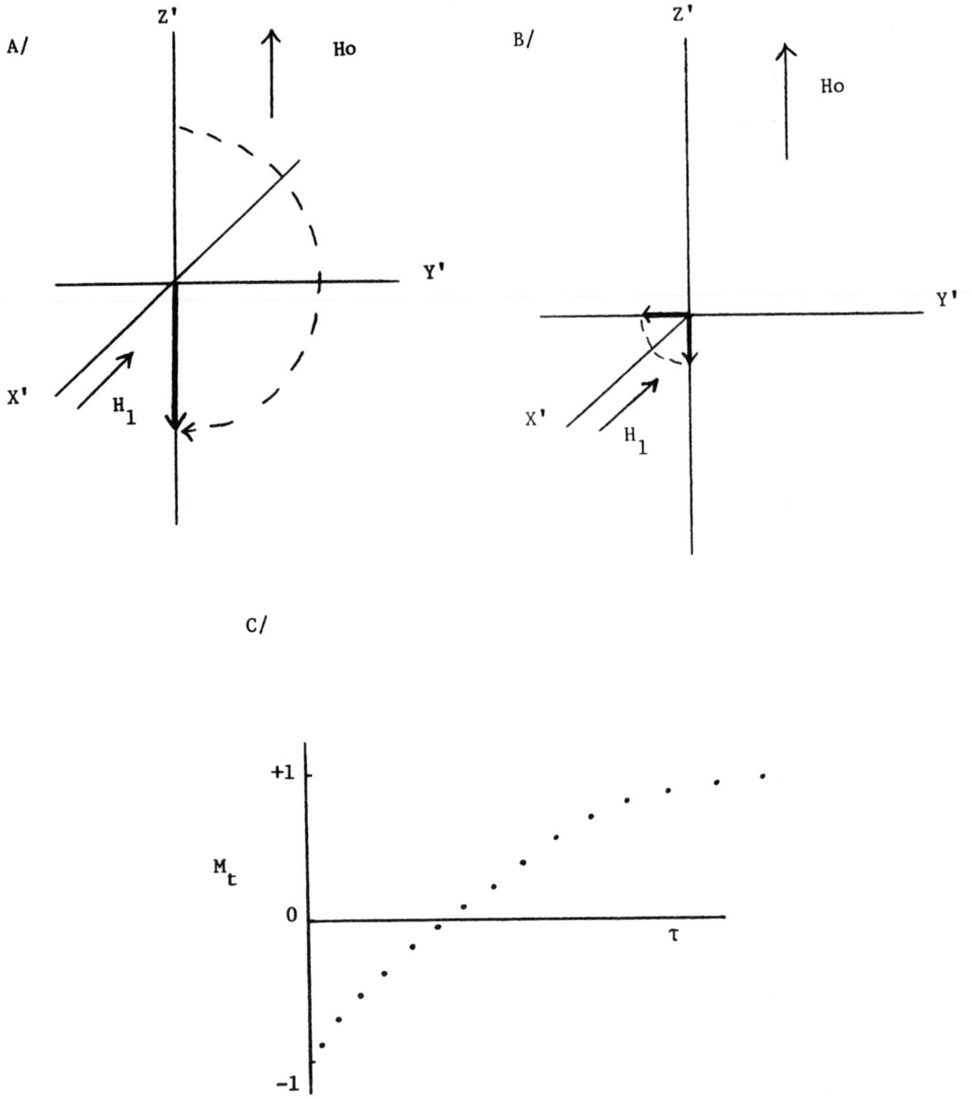

FIGURE 4. The determination of T_1 by a 180°, τ, 90° pulse sequence. (A) M is tipped 180° by a pulse at time 0. After time τ, when M has partially relaxed a 90° pulse is applied and M is tipped into the y′ axis to generate a signal of amplitude M_t (B). The initial signal amplitude, M_t is plotted as a function of τ (C).

Only the most common and instructive relaxation rate measurements will be described here. T_1 measurements are most commonly made by a series of pulses in a 180°-τ-90° sequences (Figure 4).[3,4] The net magnetization is tipped 180°. This is followed by a waiting time, τ where the net magnetization begins to relax back to equilibrium with a time constant T_1. This is followed by a 90° pulse, which then places the magnetization into the y′ axis and induces an NMR signal. The intensity of M_t reflects the decay of M back to equilibrium. The decay of M_z is given by the Block equations where

$$\frac{dM_Z}{dt} = \frac{(M_Z - M_O)}{T_1} \tag{10}$$

and upon integration with $M_z = -M_o$ at t = 0 we obtain

$$M_z = M_o\,(1 - 2\,e^{-t/T_1}) \qquad (11)$$

Since one normally measures M_z as signal amplitude (with most pulsed experiments) or the integral of the absorption peak, the useful value of M_z is thus signal amplitude or area. A practical form of Equation 11 is in the form

$$\ln(M_\infty - M_t) = \ln 2\,M_\infty - \frac{t}{T_1} \qquad (12)$$

A plot of ln $(M_\infty - M_t)$ vs. t gives a slope of $-1/T_1$. Care must be taken to fulfill two criteria; M_∞ must be measured accurately by choosing sufficiently long values for t, and the repetition of these pulse sequences must be performed with an adequate waiting time. This waiting time must allow the net magnetization to return to equilibrium before it is again tilted 180°. If this does not occur, abnormally short T_1 values will be obtained. If a waiting time of four T_1 values is used 96.3% of equilibrium is reached: six T_1 values allow 99.5% of equilibrium to be reached. The requirement of M_∞ can be minimized if a fit of the data of M_t vs. t can be obtained with a unique value of T_1. Another method of determining T_1 is to determine the value of t when M_t is zero. It can be seen from Equation 12 that when $M_t = \frac{1}{2} M_\infty$, when magnetization is halfway back to equilibrium, then $t = \ln 2\,(T_1)$ or 0.693 T_1 and $1/T_1 = 0.693/t_{1/2}$. This is identical to the measurement of the half-time of a first order reaction. The accuracy of this method is dependent upon the accuracy in the measurement of M_t. Most values of T_1 for protons of small molecules in liquid solution are between 0.1 to 5 sec.

As noted (Equation 8), T_2 measurements can be estimated from the linewidths of a high resolution spectrum if the resonance of the nucleus in question can be resolved. Since most common spectrometers have a resolution limit of 0.2 to 0.5 Hz, the upper limit of an estimation of T_2 to 1.6 to 0.64 sec can be obtained. A T_2 of longer than 0.6 to 1.6 sec results in a narrower natural linewidth and is thus limited by homogeneity of the instrument. Thus a T_2 measured from a narrow resonance line is often T_2^* where T_2^* is an apparent relaxation time and

$$T_2^* = T_{2\ \text{actual}} + \frac{1}{\pi\,(\nu_{1/2})_{\text{instr.}}} \qquad (13)$$

where $(\nu_{1/2})_{instr.}$ is the instrumental inhomogeneity resulting in the amount of line broadening $\nu_{1/2}$.

Values for T_2 can be determined by pulsed methods using spin echoes first introduced by Hahn[5] then modified by Carr and Purcell[6] and by Meiboom and Gill.[7] These methods are called spin-echo techniques and can be used to overcome the problems of inhomogeneity. In the Hahn method a 90°-τ-180° sequence is applied and an echo is observed at time 2 t. This method is depicted in the rotating frame in Figure 5. The 90° pulse tips M into the y′ axis. As each nucleus responds to its own microenvironment and precesses about z′, those nuclei that precess faster than average appear to have a positive motion, and those that precess slower appear to have a negative motion. At time t a 180° pulse causes the nuclei to rotate about the x′ axis, but they continue to rotate in the same direction. At time 2 t, the nuclei rephase along the −y′ axis resulting in an echo of the spins. As the value of t increases, the amplitude of the echo, M_{2t} decays in an exponential fashion

$$M_{2t} = M_o\,e^{-2t/T_2 - 2/3\,\gamma^2 G^2 \mathcal{D} \tau^3)} \qquad (14)$$

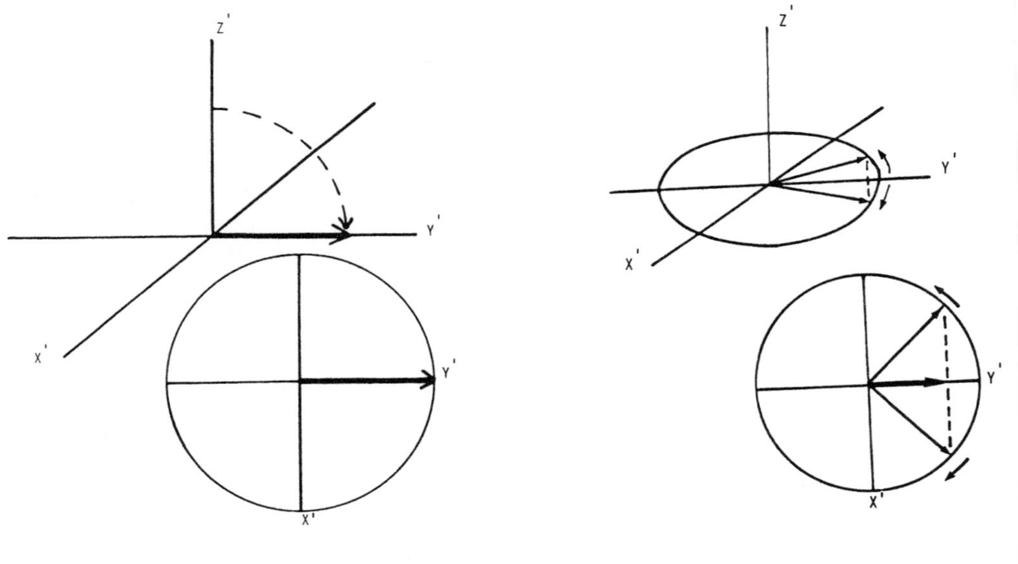

A B

FIGURE 5. The Hahn pulsed spin echo experiment to determine T_2. (A) A 90° pulse is applied with H_1 at time 0 and M is tipped into the y′ axis. (B) As the nuclei precess about H_o, they begin to precess out of phase. This dephasing process can be illustrated by the nuclei that precess faster than ω_o having a positive (clockwise) motion, and the nuclei that precess slower having a negative (counterclockwise) motion. (C) After time τ, a 180° pulse rotates all of the nuclei about the x′ axis. The nuclei with clockwise motion still move in a clockwise fashion. (D) The nuclei rephase along the negative y′ axis to yield an echo at 2τ. (E) A series of experiments performed by varying τ leads to echoes that diminish with increasing τ. The decrease in amplitude reflects the loss of phase coherence and thus a T_2 effect.

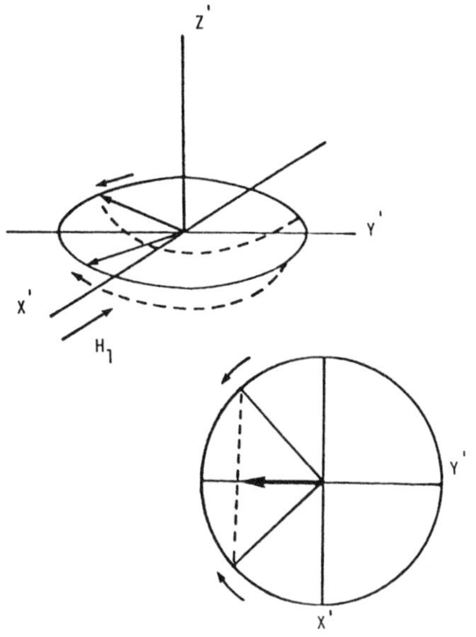

FIGURE 5C.

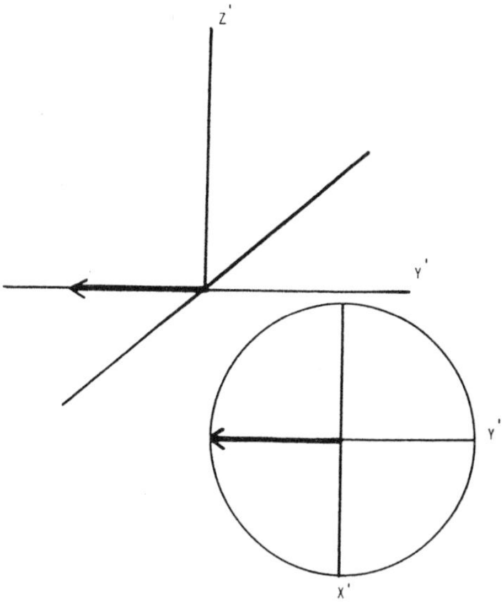

FIGURE 5D.

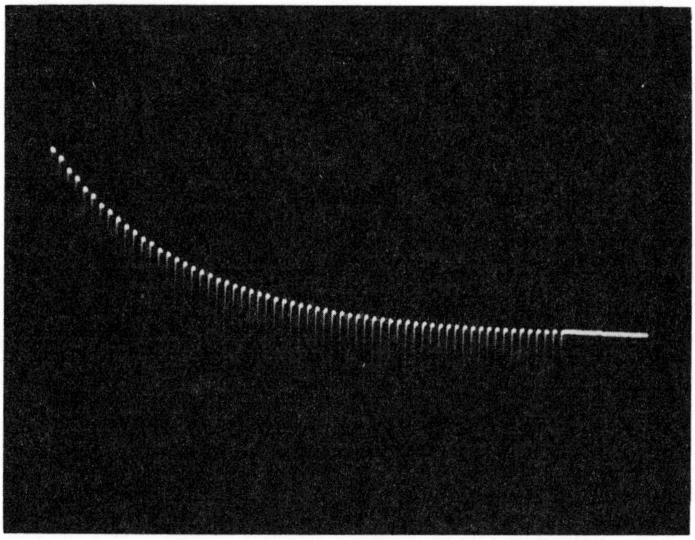

FIGURE 5E.

where G is the magnetic field gradient caused by inhomogeneity in H_o and $\mathcal{D}$ is the self-diffusion coefficient of the molecules in the sample. The diffusion term becomes pronounced at long values of t thereby effecting long values of T_2. The Carr-Purcell sequence is a simple modification that can shorten the time required to measure T_2 and diminish the effect of diffusion. This method is a 90°, t, 180°, 2 τ, 180°... repetitive sequence. The result is analogous to Figure 5 except that as the magnetization is rephased at 2 t and allowed to dephase again until 3 τ, another 180° retips the magnetization to the $+y'$ axis. As this sequence is repeated a series of echoes will be measured alternating in sign. Consequently a train of n echoes can be obtained from one sequence of measurements rather than n measurements with a waiting time of at least 6 $\times$ T_1 between measurements. Also, a choice of short values of t can be used to minimize the effect of self-diffusion since it is the time during 2 t when this process becomes important. A practical difficulty in these determinations is the ability to obtain a correct pulse. Incorrect 180° and 90° pulses result in incomplete rephasing and thus, error. The Meiboom-Gill modification of the Carr-Purcell technique can minimize most of the problems associated with T_2 measurements. The same pulse sequence as the Carr-Purcell technique is used except that the 180° pulses are applied at a 90° phase difference to the initial 90° pulse. This is analogous to applying a 180° pulse along the y' axis. As seen in Figure 6, magnetization is refocused along the y' axis resulting in all positive echoes. If the 180° pulse is not entirely correct, every other echo will be somewhat diminished in amplitude, but not in a cumulative fashion.

The measurements of relaxation rates of an individual nucleus (i.e., ^1H of H_2O) by pulsed techniques are easier for T_1 values rather than for T_2 values. For high resolution spectra in which individual nuclei of a ligand or ligands can be measured, the linewidths, therefore T_2^* values, can be measured easier than T_1 values. Normally, T_1 values are easier to interpret than T_2 values.

IV. NUCLEAR RELAXATION

The applications of nuclear relaxation rates to ligand-protein interactions are most easily interpretable with the use of paramagnetic ions as a relaxation probe. These cations can frequently be used as activating cations in metal-requiring enzymic systems. Therefore they are not perturbants of the system as required for many other spectroscopic techniques, i.e., fluorescence, EPR. It is this application that will be stressed.

The dominant mechanism for T_1 relaxation is a dipole-dipole interaction. These dipolar effects can occur from the interaction of the nucleus I_1 with the magnetic nucleus of I_2 where I_2 may be on the same molecule (intramolecular) or on a nearby molecule (intermolecular). Both effects can be important. The dipole-dipole interaction is modulated by a correlation time which, in this case, is the rotational correlation time of the molecule, τ_r. For a two spin system of unlike spins where I_2 has γ_2, the dipolar contribution to I_1 is given as

$$\frac{1}{T_1} = \frac{3\,\gamma_1{}^2\gamma_2{}^2 h^2 I_2 (I_2 + 1)}{r^6}\,\tau_r \qquad (15)$$

where r is the distance between the two nuclei. For intramolecular effects, τ_r is the rotational correlation time of the molecule whereas if intermolecular effects play a significant role a translational correlation time will become important. The r^{-6} dependence indicates these effects are modulated by magnetic dipoles principally from nearest neighbor atoms.

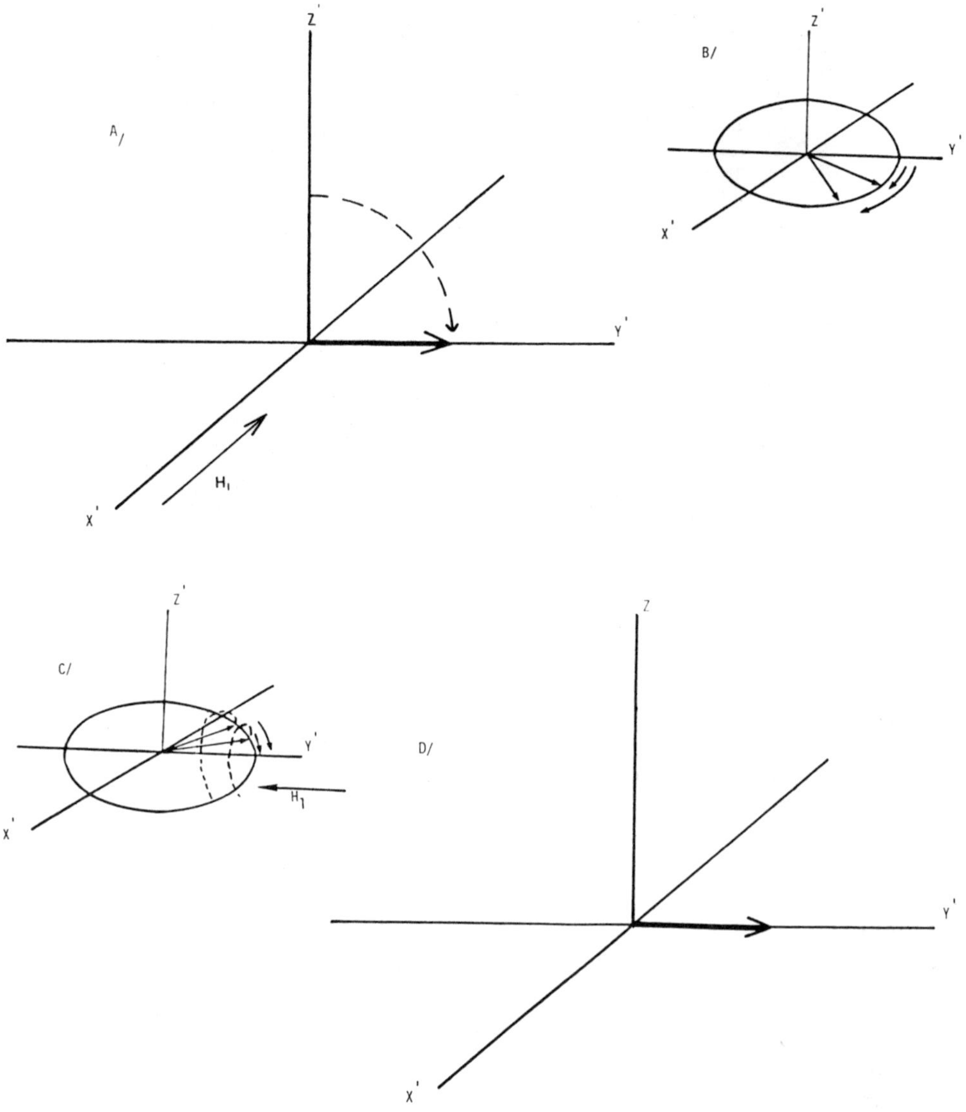

FIGURE 6. The Meiboom-Gill modification of the Carr-Purcell, pulsed spin echo technique for T_2 measurements. The general principles utilized are those demonstrated in Figure 5. (A) A 90° pulse tips the net magnetization into the y′ axis. (B) The nuclei begin to precess about the z′ axis with a loss of phase coherence. For illustrative purposes the slower moving nuclei are shown to lag behind the faster moving nuclei in contrast to the method of presentation in Figure 5. (C) After time τ a 180° pulse is now given 90° out of phase with the x′ axis. Effectively the 180° pulse is along the y′ axis. The result of this pulse is to place the faster moving nuclei behind the slower moving nuclei and allows them to "catch" the slower ones. (D) At time 2τ the nuclei all realign at the y′ axis giving an echo. The sequence of 180° pulses energy 2τ can be repeated yielding a train of echoes in the same positive y′ axis. (E) A series of experiments performed by varying τ leads to echoes which diminish with increasing time. The decrease in the amplitude reflects the loss of phase coherence and thus a T_2. The experiments are a Meiboom-Gill modification of the Carr-Purcell sequence for a T_2 measurement of the protons of distilled water (I) and of the protons of water in the presence of $10^{-3} M\, MnCl_2$ (II). The scale of the oscilloscope traces are shown in the figures.

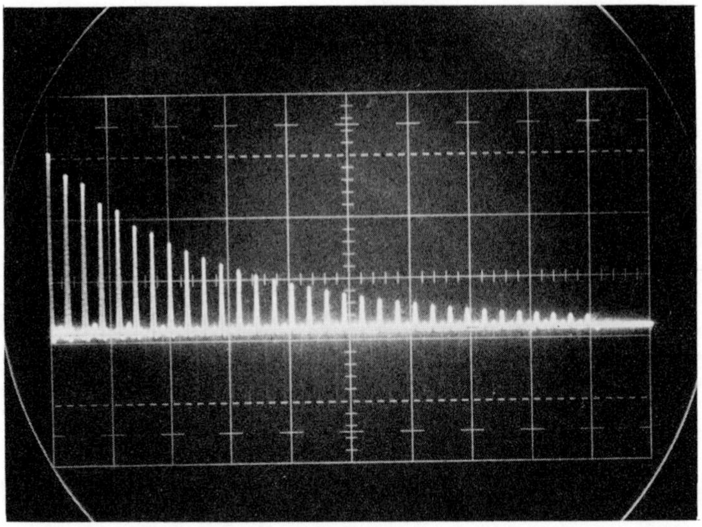

FIGURE 6E.

Paramagnetic ions can be extremely useful probes in studying enzyme-metal and enzyme-metal-ligand interactions. The magnetic moment of an electron is approximately 657 times greater than the magnetic moment of a proton. Thus an unpaired electron is much more efficient (657^2) than a proton in causing relaxation and can provide the dominant relaxation mechanism. The electron-nuclear interaction can contribute two effects to relaxation — a dipole-dipole effect and a scalar coupling effect. The dipole-dipole effect is a "through space" effect and depends upon the distance between the dipoles analogous to Equation 15. The scalar coupling is dependent upon the spin density of the electron at the nucleus and is a "through bond" effect. The effect of the paramagnetic ion on the relaxation rates of nuclei in the environment of the ion is given by the Solomon-Bloembergen equations,[8,9] which quantitate these effects.

$$\frac{1}{T_{1M}} = \frac{2\,S(S+1)\gamma^2 g^2 \beta^2}{15\,r^6}\left(\frac{3\,\tau_c}{1+\omega_I^2\tau_c^2} + \frac{7\,\tau_c}{1+\omega_s^2\tau_c^2}\right) + \frac{2\,S(S+1)}{3}\left(\frac{A}{h}\right)^2\left(\frac{\tau_e}{1+\omega_s^2\tau_e^2}\right) \qquad (16)$$

$$\frac{1}{T_{2M}} = \frac{S(S+1)\gamma^2 g^2 \beta^2}{15\,r^6}\left(4\,\tau_c + \frac{3\,\tau_c}{1+\omega_I^2\tau_c^2} + \frac{13\,\tau_c}{1+\omega_s^2\tau_c^2}\right) + \frac{S(S+1)}{3}\left(\frac{A}{h}\right)^2\left(\tau_e + \frac{\tau_e}{1+\omega_s^2\tau_e^2}\right) \qquad (17)$$

The terms $1/T_{1M}$ and $1/T_{2M}$ are the relaxation rates of the nuclei of spin ½ that reside at the paramagnetic site. The term S is the electron spin quantum number, g is the electronic "g" factor, β is the Bohr magneton, ω_I and ω_s are the Larmor precession frequencies of the nuclear and electron spins ($\omega_s = 657\ \omega_I$), respectively, r is the electron-nuclear distance, and A/h is the electron-nuclear hyperfine coupling constant in cycles per sec. The terms τ_c and τ_e are the dipolar and scalar correlation times, respectively. The first term of Equations 16 and 17 are the dipolar terms. The correlation times are given by

$$\frac{1}{\tau_c} = \frac{1}{\tau_r} + \frac{1}{\tau_s} + \frac{1}{\tau_m} \tag{18}$$

where τ_r is the rotational correlation time, τ_s is the electron spin relaxation time (generally the T_1 of the electron), and τ_m is the lifetime of the nucleus at the environment of the paramagnetic ion. The second term of Equations 16 and 17 is the scalar term and the correlation time is given by

$$\frac{1}{\tau_e} = \frac{1}{\tau_s} + \frac{1}{\tau_m} \tag{19}$$

The shortest of the processes involved in the modulation of the electron-nuclear interaction is thus the dominant process and the pertinent correlation time. This phenomenon is important in making a choice of paramagnetic ions. For example, in a simple metal-ligand complex in solution, τ_c is usually τ_r, which is approximately 10^{-11} sec (dependent upon the size of the ligand). In the formation of an enzyme-metal-ligand complex, τ_r increases by several orders of magnitude ($10^{-7} - 10^{-9}$ sec). The τ_c increases and another process may then become the important modulation process. The greater the increase in τ_c the larger the relaxation rate of the nuclei of the bound ligand. In these cases a *rate enhancement* is observed.

In nearly every application of paramagnetic ions for the investigation of enzyme-ligand interactions, the situation encountered is one in which the concentration of species of interest (ligand bound to the enzyme-metal complex) will be much less than the concentration of ligand in bulk solution. The observations thus made are the relaxation rates of the average of two ligand species designated as L_b and L_f for bound ligand and free ligand, respectively. This average is modulated by the lifetime of L_b to the enzyme-metal complex, τ_m.

The effect of chemical exchange between L_b and L_f can be manifested on the chemical shift $\Delta\omega$ of the nucleus and on the observed relaxation rates ($1/T_1$ and $1/T_2$). The question of "fast" exchange vs. "slow" exchange, as modulated by the exchange rate, $1/\tau_m$, depends upon the relative changes in the two phenomena. If the case exists in which the environment of L_b is different from L_f and a chemical shift difference exists then, in principle, two separate resonance peaks exist. If the rate of chemical exchange between the two environments is slower than the difference in chemical shifts

$$\left(\frac{1}{\tau_m} << \Delta\omega\right)$$

then the two resonance peaks should be observable. As the exchange rates increase, the resonance peaks broaden and coalesce to a single peak. The single peak is an average of both resonances in the region of *fast exchange.*

$$\left(\frac{1}{\tau_m} >> \Delta\omega\right)$$

The single resonance frequency ω_{obs} is related to the fraction of ligands bound (P_b) and free ($P_f = 1 - P_b$).

$$\omega_{obs} = P_b\omega_b + P_f\omega_f \tag{20}$$

The values ω_b and ω_f are the chemical shifts of the nuclei of the bound and free ligands, respectively. In the case where the fraction of $L_b << 1$, the observed change in chemical shift for L_f ($\omega_{obs} - \omega_f$) thus simplifies to

$$\omega_{obs} - \omega_f = P_b \omega_b \tag{21}$$

In the consideration of chemical exchange on relaxation rates, two relationships have been described:[10,11]

$$\frac{1}{T_1\ obs} = \frac{1}{T_{1f}} + \frac{P_b q_b}{(T_{1M} + \tau_m)} \tag{22}$$

$$\frac{1}{T_2\ obs} = \frac{1}{T_{2f}} + \frac{P_b q_b}{\tau_m} \left[\frac{\frac{1}{T_{2M}} \left(\frac{1}{T_{2M}} + \frac{1}{\tau_m} \right) + \Delta\omega^2}{\left(\frac{1}{T_{2M}} + \frac{1}{\tau_m} \right)^2 + \Delta\omega^2} \right] \tag{23}$$

where P_b is the mole fraction of bound ligands and q_b is the number of ligands bound. The value $\Delta\omega$ is the change in chemical shift. For enzyme-metal complexes containing manganous ion, but not all paramagnetic ions, the chemical shifts are negligible and the relationship for spin-spin relaxation then becomes analogous to Equation 21:

$$\frac{1}{T_2\ obs} = \frac{1}{T_{2f}} + \frac{P_b q_b}{T_{2M} + \tau_m} \tag{24}$$

Equations 22 and 24 can be written in a more general form to include the effects of the paramagnetic ion on ligand nuclei not in the primary sphere of the cation. These effects are outer sphere effects and are additive in nature. The outer sphere effects are normally small in comparison to primary effects, but may become substantial if the primary relaxation effects observed become limited by phenomena such as chemical exchange.

$$\frac{1}{T_1\ obs} - \frac{1}{T_{1f}} = \frac{1}{T_{1p}} = \frac{P_b\ b}{T_{1M} + \tau_m} + \frac{1}{T_1\ o.s.} \tag{25}$$

$$\frac{1}{T_2\ obs} - \frac{1}{T_{2f}} = \frac{1}{T_{2p}} = \frac{P_b\ b}{T_{2M} + \tau_m} + \frac{1}{T_2\ o.s.} \tag{26}$$

The terms $1/T_{1p}$ and $1/T_{2p}$ represent the paramagnetic effect on the longitudinal and transverse relaxation rates, respectively, and $1/T_{1o.s.}$ and $1/T_{2o.s.}$ represent outer sphere effects on relaxation rates.

V. ENZYME-CATION COMPLEXES

The characterization of an enzyme-metal complex can be performed, under ideal circumstances, to obtain a wealth of information concerning this binary complex. A prudent choice of cations must first be made. The first criterion is the ability of the cation to activate the enzyme or serve the normal function of the "physiological" metal. For example, it is often found that Mn^{+2} substitutes for Mg^{+2} and Co^{+2} substitutes for Zn^{+2} in metal-enzyme systems. In some cases a cation with the physical properties required for NMR studies cannot activate the enzyme. In such cases an inhibitory cation can be used, although the onus resides with the investigator that the cation occupies the same binding site as the "physiological" cation. A second criterion is that the cation remains a stable, defined species at neutral pH (or at the pH where enzyme studies are being performed). The third criterion is that the relaxation rates that are affected by the cation be distinguished when the cation binds the enzyme under conditions of the experiment. As we will see, the cations that fit the third criterion best of all are Mn^{+2}, Gd^{+3}, Eu^{+2}, and Cr^{+3}. Cations such as Fe^{+3}, Co^{+2}, and Cu^{+2} have also been shown to be useful in some systems.

The bound paramagnetic cation can perturb the relaxation rates of the solvent water protons. The first observation of such effects with proteins were made by Cohn,[12] who has pioneered the applications of paramagnetic probes in enzymology. A measurement of the relaxation rates, primarily the longitudinal proton relaxation rates (PRR), can yield information concerning the number of metal sites, n; the dissociation constant, K_d; the rate of water exchange from the metal, k_{off}; the hydration number of the bound metal, q; and the correlation time of the cation-water proton interaction, τ_c.

The interaction of a cation M to an enzyme E reversibly forms the binary E-M complex:

$$E + M \rightleftharpoons EM \tag{27}$$

The cation-water proton interaction in solution is normally dominated by τ_r, which is approximately 3×10^{-11} sec. For cations with a short τ_s ($\tau_s \leqslant 10^{-11}$ sec), $\tau_c \approx \tau_s$. The formation of a binary E-M complex in rapid exchange is then described by the Swift-Connick equation (Equation 22). Two changes are then elicited upon the bound metal. First the value of q for water decreases from q_f to q_b upon substitution by ligands of the enzyme. This change would result in a decrease in $1/T_{1p}$. The second change that occurs is an increase in τ_r to the value approaching τ_r of the enzyme if the cation is immobilized upon binding. Values of τ_r depend upon the shape and size of the enzyme. Globular proteins normally give isotropic rotation and have values ranging from 10^{-6} to 10^{-9} sec depending on their molecular weight. Therefore if τ_c for the free cation is τ_r, the value for τ_c may increase by several orders of magnitude to yield an increase in $1/T_{1p}$. The increase in τ_c depends upon the process or processes that determine τ_c. In cases (i.e., Ni^{+2}) where τ_s is very short ($\tau_s \approx 10^{-12}$ sec) and if τ_s is invariant upon binding to a macromolecule, little or no change in $1/T_1$ of water is observed. Cations such as Mn^{+2} and Gd^{+3}, which have τ_s values $\sim 10^{-9}$ sec cause a substantial increase (an enhancement) in $1/T_{1p}$ upon binding to an enzyme. This enhancement is due to the increase in τ_c.

To study the interaction of an enzyme with a paramagnetic cation the simplest measurements to be made are the $1/T_1$ values of the solvent protons. The $1/T_1$ values measured in the presence of the enzyme will be designated with an asterisk (*). Thus when metal is added to a solution containing enzyme:

$$\left(\frac{1}{T_{1\,obs}}\right)^* = \left(\frac{1}{T_{1p}}\right)^* + \left(\frac{1}{T_{1o}}\right)^* \tag{28}$$

where $(1/T_{1o})^*$ is the diamagnetic solution containing enzyme (buffer, salt, etc.) without the paramagnetic ion. The paramagnetic effect $[(1/T_{1p})^* = (1/T_{1\,obs})^* - (1/T_{1o})^*]$ is dependent upon both bound metal and the aquocation complex:

$$\left(\frac{1}{T_{1p}}\right)^* = \left(\frac{1}{T_{1p\,free}}\right) + \left(\frac{1}{T_{1p\,bound}}\right) \tag{29}$$

and using Equation 22.

$$\left(\frac{1}{T_{1p}}\right)^* = \left(\frac{pq}{T_{1M} + \tau_{m\,f}}\right) + \left(\frac{p^*q^*}{T_{1M} + \tau_{m\,b}}\right) \tag{30}$$

The value of $1/T_{1p}$ can be compared in the presence and absence of the enzyme to determine the relative effect, or enhancement, ε, upon binding to the enzyme

$$\epsilon^*_{obs} = \frac{\left(\frac{1}{T_{1p}}\right)^*}{\frac{1}{T_{1p}}} \tag{31}$$

where $(1/T_{1p})^*$ includes the effects of bound and free cation. Since the Mn^{+2} cation is the most applicable ion for such studies, more specific application will be discussed. For Mn^{+2} binding to an enzyme, an increase in ε^* is usually observed. Two general approaches can be used to study the formation of the binary E-Mn^{+2} complex:

1. Titrate Mn^{+2} with enzyme. The $(1/T_{1p})^*$ or ε^* is measured as a function of enzyme (Figure 7). Values for ε^* are extrapolated to obtain the value for enhancement due only to the binary E-Mn complex (ε_b). The data can be fit, or recalculated to yield the information in the form of a Scatchard plot.[13] Thus information for K_d and n, the number of Mn^{+2} sites per enzyme, can be obtained. This approach is useful only if all of the metal sites are equivalent.
2. Direct measurement of free Mn^{+2}. It has been shown that the binding of Mn^{+2} to a ligand in solution results in a large increase in the linewidth of the EPR spectrum of Mn^{+2}.[14] In most cases, the spectrum of bound Mn^{+2} is invisible compared to hexaaquo Mn^{+2}. This phenomenon is useful in the measurement of Mn^{+2} bound to a ligand. A titration of Mn^{+2} by enzyme results in a decrease in the EPR spectrum amplitude of the first derivative spectrum. The amplitude is directly proportional to free Mn^{+2}. A measure of Mn^{+2}_f and therefore a calculation of Mn^{+2}_{bound} is performed by EPR with a measurement of $(1/T_{1p})^*$ (or ε^*) for each titration point. The ε^* is related to free and bound Mn^{+2}:

$$\epsilon^* = \frac{[Mn^{+2}]_f}{[Mn^{+2}]_{Total}}\,\epsilon_f + \frac{[Mn^{+2}]_b}{[Mn^{+2}]_{Total}}\,\epsilon_b \tag{32}$$

where the enhancement of free Mn^{+2} (ε_f) is, by definition, unity. Equation 32 can also be written as

$$\frac{1}{T_{1p}}^* = \frac{[Mn^{+2}]_f}{[Mn^{+2}]_T}\left(\frac{1}{T_{1p}}\right)_f + \frac{[Mn^{+2}]_b}{[Mn^{+2}]_T}\left(\frac{1}{T_{1p}}\right)_b \tag{33}$$

A Mn^{+2} binding study using NMR and EPR measurements can detect nonequivalent sites directly. Thermodynamically nonequivalent sites, observed as a biphasic Scatchard plot, must be fit by the more general equation for x sites each having $(K_d)_x$. Values for n_x and $(K_d)_x$ are thus obtained.

The value measured for ε_b for a particular E-M binary complex contains information concerning the structure of that complex in terms of (1) the hydration number of bound Mn^{+2} (q^*), (2) the rate of water exchange ($1/\tau_m^*$), and (3) the correlation time, τ_c for the bound Mn^{+2} — H_2O interaction. From Equations 31 and 25

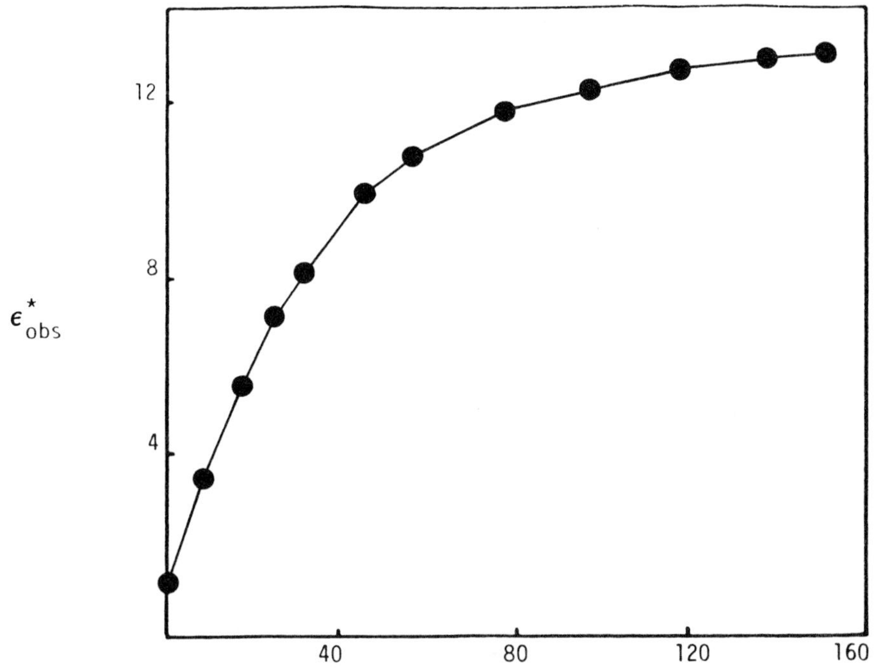

FIGURE 7. Titration of Mn^{+2} with the enzyme enolase. The $1/T_{1p}$ of a sample of 40 μM $MnCl_2$ in 50 mM tris·Cl pH 7.5 in a volume of 0.050 mℓ was measured. This solution was titrated with a solution containing an identical concentration of Mn^{+2} and buffer and included 283 μM enolase sites. Upon each addition of enzyme $1/T_{1p}$ was measured. The ε_{obs} was calculated for each point in the titration curve. Extrapolation to infinite concentration of enolase yields a value of 13.7 ± 0.3 for ε_T identical to the values obtained from combined PRR and EPR studies. The curve through the data is a "best fit" calculated using a value of K_d = 10 ± 1 μM. Rearranging the data in terms of a Scatchard plot yields a value of n = 2.3 ± 0.1 sites per dimer and a K_d = 13 ± 2 μM assuming identical sites. EPR studies are consistent with these observations.

$$\epsilon_b = \frac{\dfrac{q^*}{T_{1M}^* + \tau_m^*} + \left(\dfrac{1}{T_{1p\ o.s.}}\right)^*}{\dfrac{q}{T_{1M} + \tau_m}} \qquad (34)$$

If $\varepsilon >> 1$, $(1/T_{1p})_{o.s.}{}^*$ is negligible. In all cases $q^* < q$; for $Mn^{+2}{}_f$ q = 6 and q* is expected to be an integer, 1,2,3. (Since the protons of water re being measured q* can actually be a value of 0.5 representing a hydroxyl group in which the proton of bound water is hydrogen bonded. Likewise, only a single proton of water on bound Mn^{+2} may be in rapid exchange with the solvent). Hence, q*/q may be 1/6, 1/3, or 1/2. For $Mn(H_2O)_6{}^{+2}$, $\tau_m << T_{1M}$ therefore if $\varepsilon > 1$ is observed, $(T_{1M}{}^* + \tau_m{}^*) < T_{1M}$. The enhancement must then be caused by a change in relaxation $T_{1M}{}^*$.

A further analysis of ε_b for values of q*, $T_{1M}{}^*$, (τ_c), and $\tau_m{}^*$ requires additional experimentation. A brief analysis requires a measure of $(1/T_{2p})^*$. Since $T_{2M} < T_{1M}$ always occurs, if $(1/T_{1p})^* = (1/T_{2p})^*$, then it is likely that $(1/T_{1p})^* = pq/\tau_m{}^*$. Two additional experiments that would substantiate this conclusion are a study $(1/T_{1p})^*$ as a function of temperature and as a function of frequency. If $(1/T_{1p})^*$ represents the exchange rate of H_2O from bound Mn^{+2}, Arrhenius behavior is observed and $(1/\tau_m)^*$ (k_{off} for bound H_2O) and $\Delta E^{\neq}{}_{act.}$ for this process are obtained. The exchange rate is

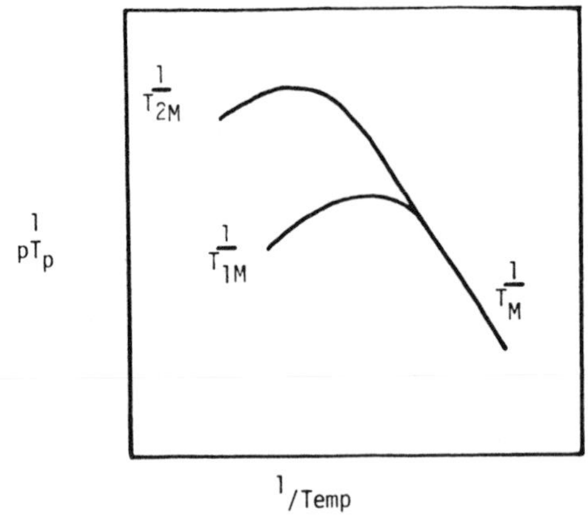

FIGURE 8. Temperature dependence of the paramagnetic effect upon relaxation rates. The case is illustrated where a minor population of the nuclei is bound in the vicinity of a paramagnetic center. A negative slope is indicative of "slow" exchange where the exchange rate is dominant. A region of intermediate exchange is observed as the slope changes and leads to the region of "rapid" exchange where nuclear relaxation is dominant.

also independent of frequency. Values for T_{1M}* hence q* and τ_c cannot be obtained under these conditions; an upper limit to T_{1M}* can be obtained.

If $(1/T_{2p})^* > (1/T_{1p})^*$ then $(1/pT_{2p})^* = q^*/T_{2M}^* + \tau_m^*$ and $(1/pT_{1p})^* = q^*/T_{1M}^*$. To determine if $(1/pT_{2p})^*$ is dominated by T_{2M}^* or τ_m^* a temperature dependence is performed. If τ_m^* is dominant a negative slope is obtained and k_{off} and $E^{\neq}_{act.}$ is obtained. If T_{2M}^* is dominant, a flat or positive dependence is normally observed (Figure 8) and $k_{off} > 1/pq^*T_{2p}^* = 1/q^*T_{2M}^*$. When T_{2M} is dominant, a frequency dependence of T_2 is expected. A determination of $(1/pT_{1p})^*$ allows the possible evaluation of q* and τ_c. The value $(1/pT_{1p})^*$ is related by the Solomon-Bloembergen equation (16) to q* and τ_c by

$$\left(\frac{1}{pT_{1p}}\right)^* = q^* K f(\tau_c) \tag{35}$$

where K is the collection of constants given in Equation 16. An assumption required is that the value of r for the Mn-water proton distance is known and is the same as for a simple Mn-H_2O complex in solution. The term $f(\tau_c)$ is the correlation function and must be the term that increases. From Equation 18,

$$\frac{1}{\tau_c} = \frac{1}{\tau_m} + \frac{1}{\tau_r} + \frac{1}{\tau_s}$$

estimations of these parameters may be performed. Since q* and K are independent of ω_I, and $f(\tau_c) = [(3\tau_c/1 + \omega_I^2\tau_c^2) + (7\tau_c/1 + \omega_s^2\tau_c^2)]$, $(1/T_{1p})^*$ should be frequency dependent if $\omega_I^2\tau_c^2 \approx 1$ or if $\omega_s^2\tau_c^2 \approx 1$. Since $\omega_s = 657\ \omega_I$, for a cation such as Mn^{+2} $\omega_s^2\tau_c^2 >> 1$ and the second term is negligible. Therefore, $T_{1p}^* \propto \omega_I^2$. A value (or upper limit) for $1/\tau_m^*$ is evaluated from $(1/T_{2p})^*$ vs. 1/Temperature. A value for τ_r can be estimated as τ_r of the protein, calculated from Stokes law

$$\tau_r = 4\pi R^3\eta/3kT \qquad (36)$$

in which R is the radius of the enzyme, assumed to be a sphere undergoing isotropic motion in a medium of viscosity η. The term k is the Boltzmann constant. The term τ_r has a positive temperature dependence

$$\tau_r = \tau_r^0 e^{E_R/RT} \qquad (37)$$

The term τ_s can have either a positive or negative temperature dependence

$$\tau_s = \tau_s^0 e^{\pm E_S/RT} \qquad (38)$$

The electron relaxation rate $1/\tau_s$ has an analogous description to the nuclear relaxation rate[15]

$$\frac{1}{\tau_s} = B\left[\frac{\tau_\nu}{1 + \omega_s^2\tau_\nu^2} + \frac{4\tau_\nu}{1 + 4\omega_s^2\tau^2_\nu}\right] \qquad (39)$$

The value B is a constant that relates the resultant electronic spin S and the zero field splitting of Mn^{+2}, and is a function of the anisotropy of bound Mn^{+2}. The term τ_ν is a correlation time for symmetry distortions at the metal attributed to solvent impact that modulates $1/T_{1e}$.

$$\tau_\nu = \tau_\nu^0 e^{E_\nu/RT} \qquad (40)$$

The Bloembergen-Morgan equation (Equation 39) indicates that τ_s is frequency dependent.

The evaluation of τ_c for the enzyme-bound Mn^{+2}-H_2O interaction, and thus an estimation of q* is a difficult process and can be approached by several methods.

1. A value for τ_c can be calculated from estimates or measurements of τ_m, τ_r, and τ_s. A value for (or lower limit for) τ_m can be measured by a temperature dependence of $(1/T_{2p})^*$. A value for τ_r can be estimated from Equation 36 assuming a globular protein and isotropic tumbling. A value for $1/T_{2e}$ (spin-spin relaxation rate of the electrons) for Mn^{+2} can be measured from the spectral linewidth of the EPR spectrum of bound Mn^{+2}. The assumption that $1/T_{2e} = 1/T_{1e}$ can be made.[16] From these values τ_c can be calculated using Equation 18. The relationship described in Equation 35 can thus be used with the calculated value of $f(\tau_c)$, K, and the measured values for $(1/pT_{1p})^*$ to estimate q*.

2. A more rigorous approach is to measure $(1/T_{1p})^*$ as a function of ω_I. If $(1/T_{1p})^*$ yields $(1/T_{1M})^*$ for the Mn^{+2} bound water, then τ_c and thus q* can be estimated. In order to demonstrate that $(1/T_{1p})^*$ yields $(1/T_{1M})^*$ (Equation 25) measurements of $(1/T_{1p})^*$ can be made as a function of temperature; $(1/T_{2p})^*$ can be measured; or a frequency dependence of $(1/T_{1p})^*$ can be performed. A frequency dependence of $(1/T_{1p})^*$ for Mn^{+2} — macromolecule complexes have been performed.[17] By a combination of Equation 22 and the dipolar portion of Equation 16, T_{1M}^* (pT_{1p}^*) is directly proportional to ω_I^2:

$$T_{1M}^* = \frac{qK}{3\tau_c} + \frac{qK\tau_c\omega_I^2}{3} \qquad (41)$$

where K is a combination of constants (Equation 16). A plot of T_{1M}* against ω_I^2 is expected to give a linear plot with a slope of $qK\tau_c/3$ and intercept of $qK/3\tau_c$. A value for τ_c can be obtained from their ratio. In some cases, nonlinearity between T_{1M} and ω_I^2 are observed if a sufficient range of ω_I is chosen.[18] This nonlinearity demonstrates that τ_c is also frequency dependent. A frequency dependence of τ_c can only occur if τ_s is the dominant correlation time. Thus to evaluate the required parameters from the data, Equation 41 must be substituted with Equation 18 for τ_c and τ_s must be substituted with Equation 39. A "best fit" to the data must then be performed choosing values for B and τ_v for Equation 39, estimating values for τ_r and τ_m (Equation 18) and calculating T_{1M} as a function of ω_I. A best fit to the data requires a proper choice of q* and τ_c. Thus these parameters can be obtained. A major difficulty with this approach is that the evaluation of q* and τ_c depends upon a fit to experimental data assuming values for parameters that are not clearly defined, nor independently easily measurable.

3. An approach that dispenses with several of the difficulties of the preceding method is to measure the effect of the paramagnetic species on $1/T_1$ of protons and of deuterons in a mixed H_2O/D_2O solution.[19] The paramagnetic effect of the cation on the $1/T_1$ of both nuclei is measured at a fixed magnetic field (H_o). Quadrupole relaxation of ^2H is not significant under these conditions ($\omega_I^2\tau_c^2 << 1$). From Equations 16 and 22

$$pT_{1M}^* = \frac{1}{q^*}(K')\left(\frac{1+\omega_I^2\tau_c^2}{\omega_I^2\tau_c}\right)H_o^2 + \frac{\tau_M}{q^*} \qquad (42)$$

the values for $(pT_{1M})^*$ are measured for H_2O and D_2O at the same value of H_o (different values of ω_I) and the results yield two equations and three unknowns (q*, τ_c, and τ_m) assuming values for r (part of constants arranged in K′) are those obtained crystallographically. Since τ_c should be frequency dependent, a measure of $(1/T_{1M})^*$ for ^1H and ^2H at two values of H_o yield four equations with four unknowns q*, τ_m, $\tau_{c(H_o)_1}$, and $\tau_{c(H_o)_2}$

VI. ENZYME-CATION-LIGAND COMPLEXES

The preceding section described the approach to the characterization of an enzyme-Mn^{+2} complex in terms of K_d, n, and ε_b. A more detailed study of ε_b leads to an evaluation of q*, τ_m, and τ_c. The formation of a ternary enzyme-Mn^{+2}-ligand complex can be studied by PRR experiments to identify the types of complexes formed and their thermodynamic constants. Relaxation rate studies of the nuclei (^1H, ^{19}F, ^{31}P, ^{13}C...) of the ligands that are bound to the enzyme complex can yield structural and kinetic information concerning these complexes. These data are important to the biochemist in describing the formation, structure, and the nature of the enzyme-ligand complexes.

The formation of a ternary complex by the addition of a ligand to an enzyme-Mn^{+2} complex can, in general, cause a perturbation in two parameters of E-Mn^{+2}, K_d and ε. The formation of an enzyme-ligand complex can change the affinity of the protein for the cation either in a positive or negative fashion. This possible change in affinity can be investigated analogous to the studies outlined using PRR and EPR to describe the interaction of Mn^{+2} to the enzyme. The success of such a study partially depends upon the relative affinity of the enzyme and the ligand for Mn^{+2}. If the enzyme has a much greater affinity for the cation than does the ligand ($K_{d\ Enz} < 100 \times K_{d\ Lig}$), and a relatively high affinity for the ligand itself (easy to saturate E with L), then results of such a study are usually unambiguous. The greater the competition of the enzyme and free ligand for the Mn^{+2} the greater the apparent number of metal sites.[20] In ideal cases a value for the enhancement of the ternary enzyme-Mn-ligand complex, ε_T is obtained from this study.

A perturbation of ε_{obs} upon titration of the binary complex with ligand results from a change in q*, τ_m, or τ_c upon formation of ternary complex. Such titrations are usually performed by the sequential addition of a solution containing enzyme, Mn^{+2} and ligand to a solution containing identical concentrations of enzyme and Mn.$^{+2}$ The value $1/T_{1p}$ or ε_{obs} is then treated as a simple phenomenon that changes upon addition of a ligand. Three observations are possible: (1) No change in ε_{obs} is observed. This could result from one of three possibilities; (a) The ligand may not bind to form a ternary complex, (b) the ligand binds at a site remote from the cation and causes no perturbation about the bound cation, or (c) compensating changes in q*, τ_c, and/or τ_m are such that fortuitously there is no change in ε. The first possibility can be tested by independent binding studies such as equilibrium dialysis, Hummel-Dreyer, or UV difference spectral changes. If independent studies show that binding does indeed occur, possibilities (b) or (c) can be tested experimentally. These possibilities are tested by performing the titration studies at a different magnetic field (vary ω_I) or by a study of the observed enhancement as a function of temperature. A comparison of the data for the case of E + Mn^{+2} vs. E + Mn^{+2} + ligand complexes would indicate whether changes in q*, τ_c, or τ_m for the Mn^{+2}-H_2O interaction have occurred upon addition of ligand. (2) An increase in ε_{obs} is observed. (3) A decrease in ε_{obs} is observed. Observation number 3 is more frequently observed and the results are more easily rationalized. In observation 2 or 3 the change in ε can be treated as a simple phenomenological change, and the titration data can be used to obtain a dissociation constant, K_3 for the ligand from the E-Mn^{+2} complex. A value for the enhancement of the ternary E-Mn-ligand complex, ε_T, is a reflection of the structure of the ternary complex and contains structural and kinetic information (in terms of q*, τ_c, and τ_m) analogous to ε_b. The observed enhancement is now a more complex function than that of the binary case defined by Equation 32. The observed enhancement value with ligand present is described by Equation 43:

$$\epsilon_{obs} = \frac{[Mn^{+2}]_f}{[Mn^{+2}]_T} + \frac{[Mn-L]}{[Mn^{+2}]_T}\epsilon_a + \frac{[E-Mn]}{[Mn^{+2}]_T}\epsilon_b + \frac{[E-Mn-L]}{[Mn^{+2}]_T}\epsilon_T \tag{43}$$

where ε_a is defined as the enhancement of the binary Mn-ligand complex. This value can be independently determined along with the dissociation constant K_1 analogous to the method described for determining ε_b and K_d for the Mn-enzyme binary complex. The following equilibria with their respective dissociation constants and enhancement values must be considered:

$$K_1 = \frac{[Mn][L]}{[MnL]} \qquad \epsilon_a \tag{44}$$

$$K_2 = \frac{[E][MnL]}{[EMnL]} \qquad \epsilon_T \tag{45}$$

$$K_d = \frac{[Mn][E]}{[EMn]} \qquad \epsilon_b \tag{46}$$

$$K_3 = \frac{[EMn][L]}{[EMnL]} \qquad \epsilon_T \tag{47}$$

$$K_S = \frac{[E][L]}{[EL]} \tag{48}$$

$$K_{A'} = \frac{[EL][Mn]}{[EMnL]} \qquad \epsilon_T \tag{49}$$

From thermodynamic definitions, note that

$$K_1K_2 = K_dK_3 = K_{A'}K_S \tag{50}$$

Equation 43 can be rewritten as:

$$\epsilon_{obs} = \frac{\frac{K_2}{[E]}\epsilon_a + \frac{K_3}{[L]}\epsilon_b + \frac{K_dK_3}{[E][L]} + \epsilon_T}{\frac{K_2}{[E]} + \frac{K_3}{[L]} + \frac{K_dK_3}{[E][L]} + 1} \tag{51}$$

A determination of values for K_2, K_3, and ε_T are desirable from these experiments. Two methods of approach can be used to approximate the parameters. The data can be fit by generating a theoretical curve[21] or by graphical analysis. In cases where $K_1 >> K_3$ and $K_S >> K_3$ graphical analysis yields values in good agreement with the values obtained from computer fit. The ε_T values can be obtained from the end point of titration curves. When $K_1 \approx K_d$, K_3 the graphical analysis fails to give satisfactory values for K_3 and ε_T. The computational analysis performed initially by Reed[21] involves solving simultaneous equations accounting for all possible equilibria (Equations 44 to 49) using a series of approximations in an iterative process. In more difficult cases (i.e., $K_1 \approx K_3$) where several "best fits" can be obtained leaving no unique fit to the data, limits can then be set for dissociation constants and ε_T. Cases where $\varepsilon_T > \varepsilon_b > \varepsilon_a$ usually give unique fits to the data (case 2 situations).

Graphical solutions may be performed by attempting to utilize Equation 51. Values of ε_{obs} are plotted vs. [L] or $1/\varepsilon_{obs}$ vs. 1/[L] to obtain values of ε_{obs} extrapolated to infinite concentrations of ligand. Equation 51 now simplifies to:

$$\epsilon_{obs} = \frac{\frac{K_2}{[E]}\epsilon_a + \epsilon_T}{\frac{K_2}{[E]} + 1} \tag{52}$$

Repeating the titration at several concentrations of enzyme should allow an extrapolation to infinite enzyme, thus a value for ε_T can be obtained. From Equation 52 where K_2 can be obtained, and from the relationships in Equation 50 where K_d is known, K_3 can be calculated. This parameter, K_3, can also be estimated by determining the concentration of free ligand that produces a half-maximal effect on ε_{obs}. This method is similar to the first order treatments used in obtaining dissociation constants from other techniques used to study binding (i.e., UV difference). Such studies may also require titrations at increasing enzyme concentrations and extrapolation to infinite enzyme (where K_S is assumed to be $> K_3$). A corollary to the first graphical approach is to extrapolate ε_{obs} to infinite enzyme concentration first where Equation 51 simplifies in terms of ε_{obs} as a function of [L] and K_3. Parameters estimated by graphical analysis should be used as "initial guesses" to values used in computer-assisted fits to the data.

The empirical results obtained from the study of PRR for binary E-Mn and ternary

Table 1
SUGGESTED COORDINATION SCHEME OF TERNARY COMPLEX DERIVED FROM EMPIRICAL DATA

Coordination scheme		Empirical behavior	Type
A. Substrate bridge	E-S-M	$\varepsilon_b < \varepsilon_t$[a]	I
B. Metal bridge	E-M-S (E–M<S)	$\varepsilon_b > \varepsilon_T$	II
C. Enzyme bridge	M-E-S	$\varepsilon_b = \varepsilon_T$	III

[a] Negligible formation of binary complex and $\varepsilon_{b(obs)} \approx 1$.

E-Mn-L complexes have been used to suggest the general coordination scheme for the ternary complex.[22] The general scheme is outlined in Table 1. Such possible schemes are based on indirect experiments. Variations of enhancement behavior with the simplified coordination schemes outlined can be envisioned and exceptions to the above scheme have been observed. Confirmatory experiments such as the measurement of relaxation rates of ligand nuclei are required to substantiate the scheme suggested from PRR data.

Following an evaluation of K_3 and ε_T for a ternary E-Mn-L complex, structural and kinetic data (q^*, τ_c, and τ_m) can be evaluated by temperature and frequency studies as outlined for the binary complexes. This information rigorously defines the parameters that have been perturbed resulting in a change in ε. A simple interpretation of type II behavior can be the assumption of simply a decrease in q^* upon ligand binding. A ratio of $\varepsilon_b/\varepsilon_T$ is expected to reflect q^*_b/q^*_T, the ratio of H_2O ligands displaced by the titrating ligand. A low ($\varepsilon_T < 2$) compared to ε_b is suggestive of the paramagnetic effect being an outer sphere effect or slowly exchanging water on the bound Mn^{+2} as a result of ternary complex formation.

Higher complexes, of the nature E-Mn-L_1-L_2 where substrate and modifier, two substrates, or a dead-end complex may be formed can also be investigated by PRR studies. The ε_{obs} must be expanded from Equation 43 to take into account two additional equilibria. The quantitative evaluation of the data becomes much more complex than evaluating ternary complex formation. However empirical observations and graphical analysis can yield qualitative information concerning quaternary or higher ordered complexes.

VII. STRUCTURAL AND KINETIC INFORMATION IN BINARY, TERNARY, AND HIGHER COMPLEXES

The theoretical section has described the relationship between $1/T_{1M}$ and $1/T_{2M}$ values of a nucleus and the distance r between the paramagnetic center and the nucleus of the ligand in question. The calculation of r first depends upon relating the measured values $1/T_{1p}$ and $1/T_{2p}$ to $1/T_{1M}$ and $1/T_{2M}$, respectively. Since $1/T_{1p}$ and $1/T_{2p}$ are a function of the distribution of paramagnetic species in solution, the species in question must either be shown to be dominant experimentally or the species distribution calculated from independent data, i.e., from PRR experiments, or from experiments performed at varying ligand concentrations or varying magnetic fields. In the study of binary Mn-L complexes where the relaxation rates of the ligand (i.e., ^1H, ^{19}F, ^{31}P, ^{13}C,

...) are measured as a function of increasing Mn^{+2}, the experiments must be performed where [L] > K_1. Thus all Mn^{+2} added is as the Mn-L complex. If this is not the case, the distribution of Mn^{+2} must then be calculated using K_1. The experiment where $1/T_1$ (and $1/T_2$) is measured as a function of Mn^{+2} can be performed at various concentrations of L to determine saturation of Mn^{+2}. Under these conditions the value p (Equations 22 and 24) is simply $[Mn^{+2}]_T/[L]_T$ assuming a 1:1 complex (q = 1). Thus

$$\frac{1}{pT_{ip}} = \frac{1}{T_{iM} + \tau_m} \quad i = 1,2 \tag{53}$$

Analogous care must be taken in defining $1/pT_{ip}$ for nuclei when ternary (or higher) complexes are formed. In cases where K_d and $K_3 << K_1$ this is not a substantial problem providing [L] >> K_3. Controls are performed in which $(1/T_{io})^*$ is the diamagnetic control for the experiment using a diamagnetic cation, (i.e., Mg^{+2}, for Mn^{+2}) to account for possible diamagnetic effects on relaxation. Usually such effects are negligible and $(1/T_{1o})^*$ measured in the absence of cation may serve as a proper control.

In ternary E-Mn-L complexes, care must be taken that $1/T_{ip}$ reflects the effect of enzyme-bound Mn^{+2} on relaxation. Experimentally, this is accomplished by measuring $1/T_{ip}$ as a function of Mn^{+2} at several levels of enzyme to ensure that Mn^{+2} is enzyme bound. The level of [L] used must be such that the ternary complex is saturated. A calculation of the distribution of paramagnetic species should demonstrate such conditions. If these conditions cannot be met, i.e., if $K_3 \approx K_1$, the distribution of species containing paramagnetic ion must be made and the contribution by the ternary complex can then be determined. Independent determination of $1/pT_{1p}$ for the Mn-L species must be obtained.

The distribution of paramagnetic species between the ternary enzyme-Mn-ligand complex and binary Mn-ligand complex can be determined experimentally. Such an evaluation is most important when the value $1/pT_{1p}$ for the binary complex is comparable to or greater than $1/pT_{1p}$ for the ternary complex. In these cases the observed, normalized relaxation rate is a function of the mole fraction, n of Mn^{+2} in the ternary complex

$$\left(\frac{1}{pT_{1p}}\right)_{obs} = n\left(\frac{1}{T_{1M}}\right)_{EML} + (1\text{-}n)\left(\frac{1}{T_{1M}}\right)_{ML} \tag{54}$$

where $(1/pT_{1p})_{obs}$ is the normalized, observed relaxation rate and $(1/T_{1M})_{EML}$ and $(1/T_{1M})_{ML}$ are the relaxation rates of the nuclei of the ligand L in the ternary and the binary complexes, respectively. To rigorously determine n, $(1/pT_{1p})_{obs}$ must be determined under the same experimental conditions at three frequencies (ω_{I_1}, ω_{I_2} and ω_{I_3}). When Mn^{+2} is the paramagnetic cation and L is a small molecule such that for the binary ML complex $\omega_I^2\tau_c^2 <<1$ $(1/T_{1M})_{ML}$ is invariant at ω_{I_1}, ω_{I_2} and ω_{I_3}. Using the simplest assumption that τ_c for the Mn-ligand interaction in the ternary complex is also frequency invariant over this range, the mole fraction n can be shown to be a function of ω_{I_1}, ω_{I_2}, ω_{I_3}; $(1/T_{1M})_{ML}$ and $(1/pT_{1p})_{obs}$ $(1/pT_{1p})_{obs_1}$, $(1/pT_{1p})_{obs_2}$, and $(1/pT_{1p})_{obs_3}$ (Equation 55). The values

$$n = 1 + \cfrac{\omega_{I_1}^2 \left[\left(\frac{1}{pT_{1p}}\right)_{obs_1} \left(\frac{1}{pT_{1p}}\right)_{obs_3} - \left(\frac{1}{pT_{1p}}\right)_{obs_1} \left(\frac{1}{pT_{1p}}\right)_{obs_2} \right] +}{\left(\frac{1}{T_{1M}}\right)_{ML} \left\{ (\omega_{I_1}^2 - \omega_{I_2}^2) \left[\left(\frac{1}{pT_{1p}}\right)_{obs_2} - \left(\frac{1}{pT_{1p}}\right)_{obs_3} \right] + \cfrac{\omega_{I_2}^2 \left[\left(\frac{1}{pT_{1p}}\right)_{obs_1} \left(\frac{1}{pT_{1p}}\right)_{obs_2} - \left(\frac{1}{pT_{1p}}\right)_{obs_2} \left(\frac{1}{pT_{1p}}\right)_{obs_3} \right] +}{(\omega_{I_2}^2 - \omega_{I_2}^2) \left[\left(\frac{1}{pT_{1p}}\right)_{obs_2} - \left(\frac{1}{pT_{1p}}\right)_{obs_1} \right]} \right\}}$$

$$\underline{\omega_{I_3}^2 \left[\left(\frac{1}{pT_{1p}}\right)_{obs_2} \left(\frac{1}{pT_{1p}}\right)_{obs_3} - \left(\frac{1}{pT_{1p}}\right)_{obs_1} \left(\frac{1}{pT_{1p}}\right)_{obs_3} \right]} \qquad (55)$$

$(1/pT_{1p})_{obs}$ are those measured at ω_I, ω_{I_2}, and ω_{I_3}, respectively. From such an evaluation, n can thus be determined and consequently $1/T_{1M}$ for the EML complex at three frequencies. The three values of $1/T_{1M}$ can thus be used to determine a value for τ_c for the ternary complex as discussed.

The experimental values for $1/pT_{1p}$ and $1/pT_{2p}$ can be related to $1/T_{1M}$ and $1/T_{2M}$, respectively when relaxation is demonstrated to be in fast exchange $1/\tau_m >> 1/T_{1M}$, $1/T_{2M}$. Criteria to demonstrate these conditions have been discussed. The relationship between $1/T_{1M}$, $1/T_{2M}$ and r, the metal-nucleus distance has been described (Equations 16 and 17). This relationship is simplified if the relaxation rate is dominated by the dipolar term and the hyperfine term is negligible. This is frequently true for $1/T_{1M}$, but not for $1/T_{2M}$. When $1/pT_{1p}$ has been shown to be in the slow exchange region, $1/pT_{1p}$ is a lower limit to the value of $1/T_{1M}$ and an upper limit to r can be calculated. Under these conditions $1/pT_{1p} \approx 1/pt_{2p} = 1/\tau_m$. Thus the value for the first order rate constant for ligand departure, k_{off} is determined. Under conditions where $1/pT_{1p} = 1/T_{1M}$ and $1/pT_{2p} = 1/\tau_m$, the value for r can be calculated from $1/T_{1M}$ and the value for k_{off} can be determined from $1/pT_{2p}$. Under conditions where $1/pT_{1p} = 1/T_{1M}$ and $1/pT_{2p} = 1/T_{2M}$, r can be calculated from $1/T_{1M}$ and the hyperfine coupling constant, (A/h) can be determined from $1/pT_{2p}$. A lower limit to $1/\tau_m$ can be estimated.

The calculations of r from the cation to the nuclei of the ligand that were measured can thus be calculated using two assumptions. From Equation 22, the value q, the number of ligands bound in the environment of enzyme-bound Mn^{+2} is normally assumed to be one. This is a reasonable assumption for most ligands (substrates, inhibitors, modifiers, etc.) and can usually be independently tested by performing direct ligand binding studies. If more than one ligand binds, it is the ligand closest to the paramagnetic center that provides the dominant contribution to $1/T_{1p}$ because of the inverse sixth power of r. The second criterion to be met to calculate r from $1/T_{1M}$ is the value of τ_c. A simplification of Equation 16 to calculate r is given as:

$$r\ (\text{in Å}) = X \left[T_{1M} \left(\frac{3\tau_c}{1 + \omega_I^2 \tau_c^2} + \frac{7\tau_c}{1 + (657\omega_I)^2 \tau_c^2} \right) \right]^{1/6} \qquad (56)$$

where X is a collection of constants. The value of X depends upon the paramagnetic cation and the nucleus under investigation. Thus for Mn^{+2}, X = 812 for ^1H, 796 for

^{19}F, 601 for ^{31}P, and 512 for ^{13}C. The value for τ_c can be determined by a number of approaches. Several have previously been discussed. For simple binary L-Mn^{+2} complexes where τ_c is τ_r, an estimate of τ_r can be obtained by the assumption that τ_r increases in proportion to the molecular weight of the complex. Since τ_r for $Mn(H_2O)_6^{+2}$ is 2.9×10^{-11} sec substitution of L for H_2O causes a small change in τ_r. The most rigorous approach is to determine $1/T_{1M}$ for the nuclei at various values of ω_I. If τ_c is dominated by τ_s, which may be frequency-dependent, this assumption can be used to estimate the limits of τ_c. The value for τ_c for the ion-ligand interaction can be obtained by a study of the ion-H_2O interaction in the same complex. The assumptions made are that one can measure τ_c for the ion-H_2O interaction in the ternary (or higher) complex. The same process that modulates the ion-H_2O interaction also modulates the ion-ligand interaction in the same complex. When τ_c is dominated by τ_r or τ_s (or both) this assumption is reasonable. An alternative method is to calculate τ_c from the ratio of $(1/T_{2M})/(1/T_{1M})$. The assumption of this approach is that there are no scalar contributions to both relaxation rates. This is a reasonable assumption for $1/T_{1M}$, but not for $1/T_{2M}$. Such scalar effects are sometimes observed even for carbon-bound protons or for second sphere complexes. Therefore $(1/T_{2M})/(1/T_{1M})$ can provide an upper limit to τ_c.

The measurement of the relaxation rates of the nuclei of a ligand that interacts with an enzyme-Mn^{+2} complex enables the calculation of the distances of the various atoms studied to the bound ion. The coordination scheme and geometry of the resultant ternary (and higher) complexes can then be determined. The rate constant for ligand departure of these ligands can also be obtained. An estimation of the rate of complex formation from the relationship

$$K_d = \frac{k_{off}}{k_{on}} \tag{57}$$

can be estimated assuming a simple second order process for binding occurs.

The utilization of nuclear relaxation techniques using a paramagnetic probe can be used to study specific ligand-protein interactions. Such studies can be performed by indirect measurements-observing the paramagnetic effect on the solvent, or by direct measurements of the ligand itself. The data obtained contains information concerning the thermodynamics, kinetic, and the structure of such complexes. These techniques are the only methods of observing ligand binding where such information is possible. Care must be taken — as with any technique — to recognize the information that can be obtained and the limitations of these techniques. The more sophisticated the question asked, the more difficult the experiments to be performed and the more one must consider the applications and limits of the theory involved.

REFERENCES

1. **Dwek, R. A.,** *Nuclear Magnetic Resonance in Biochemistry,* Clarendon Press, Oxford, 1973.
2. **James, T. L.,** *Nuclear Magnetic Resonance in Biochemistry,* Academic Press, New York, 1975.
3. **Vold, R. L., Waugh, J. S., Klein, M. P., and Phelps, D. E.,** Measurement of spin relaxation in complex systems, *J. Chem. Phys.,* 48, 3831, 1968.
4. **Allerhand, A., Doddrell, D., Glushko, V., Cochran, D. W., Wenkert, E., Lawson, P. J., and Gurd, F. R. N.,** Conformation and segmental motion of native and denatured ribonuclease A in solution. Application of natural abundance carbon-13 partially relaxed Fourier transform nuclear magnetic resonance, *J. Am. Chem. Soc.,* 93, 544, 1971.

5. **Hahn, E. L.**, Spin echoes, *Phys. Rev.*, 80, 580, 6.
6. **Carr, H. Y. and Purcell, E. M.**, Effects of diffusion on free procession in nuclear magnetic resonance experiments, *Phys. Rev.*, 94, 630, 1954.
7. **Meiboom, S. and Gill, R.**, Modified spin-echo method for measuring nuclear relaxation times, *Rev. Sci. Instrum.*, 29, 688, 1958.
8. **Solomon, I.**, Relaxation processes in a system of two spins, *Phys. Rev.*, 99, 559, 1955.
9. **Bloembergen, N.**, Proton relaxation times in paramagnetic solutions, *J. Chem. Phys.*, 27, 572, 1957.
10. **Swift, T. J. and Connick, R. E.**, NMR-relaxation mechanisms of O^{17} in aqueous solutions of paramagnetic cations and the lifetime of water molecules in the first coordination sphere. *J. Chem. Phys.*, 37, 307, 1962.
11. **Luz, Z. and Meiboom, S.**, Proton relaxation in dilute solutions of cobalt(II) and nickel(II) ions in methanol and the rate of methanol exchange in the solvation sphere, *J. Chem. Phys.*, 40, 2686, 1964.
12. **Cohn, M. and Leigh, J. S., Jr.**, Magnetic resonance investigations of ternary complexes of enzyme-metal-substrate, *Nature (London)*, 193, 1037, 1962.
13. **Scatchard, G.**, The attractions of proteins for small molecules and ions, *Ann. N. Y. Acad. Sci.*, 51, 660, 1949; Deranleau, D. A., Theory of the measurement of weak molecular complexes. I. General considerations. II. Consequences of multiple equilibria, *J. Am. Chem. Soc.*, 91, 4044, 4050, 1969.
14. **Cohn, M. and Townsend, J.**, A study of manganous complexes by paramagnetic resonance adsorption, *Nature (London)*, 173, 1090, 1954.
15. **Bloembergen, N. and Morgan, L. O.**, Proton relaxation times in paramagnetic solutions. Effects of electron spin relaxation, *J. Chem. Phys.*, 34, 842, 1961.
16. **Koenig, S. N.**, Note on the distinction between transverse and longitudinal relaxation times obtained from nuclear relaxation studies, *J. Chem. Phys.*, 56, 3188, 1972.
17. **Peacocke, A. R., Richards, R. E., and Sheard, B.**, Proton magnetic relaxation in solutions of *E. coli* ribosomal RNA containing Mn^{+2} ions, *Mol. Phys.*, 16, 177, 1969.
18. **Reuben, J. and Cohn, M.**, Magnetic resonance studies of manganese(II) binding sites of pyruvate kinase, *J. Biol. Chem.* 245, 6539, 1970.
19. **Burton, D. R., Dwek, R. A., Forsén, S., and Karlström, G.**, A novel approach to water proton relaxation in paramagnetic ion-macromolecule complexes, *Biochemistry*, 16, 250, 1977.
20. **Nowak, T. and Lee, M. J.**, Reciprocal cooperative effects of multiple ligand binding to pyruvate kinase, *Biochemistry*, 16, 1343, 1977.
21. **Reed, G. H., Cohn, M., and O'Sullivan, W. J.**, Analysis of equilibrium data from proton magnetic relaxation rates of water for manganese-nucleotide kinase ternary complexes, *J. Biol. Chem.*, 245, 6547, 1970.
22. **Cohn, M.**, Magnetic resonance studies of metal activation of enzymic reactions of nucleotides and other phosphate substrates, *Biochemistry*, 2, 623, 1963.

GENERAL REFERENCES

General Principles of NMR

Abragam, A., *Principles of Nuclear Magnetism*, Oxford University Press, New York, 1961.

Bovey, F. A., *Nuclear Magnetic Resonance Spectroscopy*, Academic Press, New York, 1969.

Farrar, T. C. and Becker, E. D., *Pulse and Fourier Transform NMR*, Academic Press, New York, 1971.

Applications of NMR to Enzymology

Mildvan, A. S. and Cohn, M., Aspects of enzyme mechanisms studied by nuclear spin relaxation induced by paramagnetic probes, *Adv. Enzymol. Relat. Areas Molec. Biol.*, 33, 1, 1970.

Dwek, R. A., Williams, R. J. P., and Xavier, A. V., Application of paramagnetic probes in biochemical systems, in *Metal Ions in Biological Systems*, Vol. 4, Siegel, H., Ed., Marcel Dekker, New York, 1974.

Sykes, B. D. and Scott, M., Nuclear magnetic resonance studies of the dynamic aspects of molecular structure and interaction in biological systems, *Annu. Rev. Biophys. Bioeng.*, 1, 27, 1972.

Mildvan, A. S. and Engle, J. L., Nuclear relaxation measurements of water protons and other ligands, *Methods Enzymol.*, 26C, 654, 1972.

Chapter 4

MAGNETIC RESONANCE STUDIES OF MEMBRANES

J. Ellis Bell

TABLE OF CONTENTS

I. INTRODUCTION TO PROBLEMS ASSOCIATED WITH MEMBRANE SYSTEMS

There are many problems associated with membranes that confront biochemistry today. It has become increasingly apparent that to understand many of the important functionings of the cell, we must understand the roles membranes and membrane-associated molecules play in both the structure of the cell and in regulation within the structure. We must understand how membranes within the cell are involved in compartmentation, in the maintenance of intracellular concentration gradients, and in the multitude of intracellular functions that apparently involve membrane-bound systems. Over the last ten years, many enzymes once thought to be soluble have been found to be associated with membranes in some way. What effect (if any) does membrane localization have on the catalytic or regulatory functions of such enzymes? In addition to intracellular phenomena, a variety of extracellular phenomena also involve membranes in some way. Hormones have receptors on the plasma membranes of cells and many exert effects at the level of the membrane by changing the permeability of the membrane. Many intercellular phenomena involve membrane-membrane interactions. How do cells communicate with one another at the level of membrane contacts?

Thus we have a variety of questions that can be asked at the level of membrane structure and function. How for instance do proteins affect the properties of the mem-

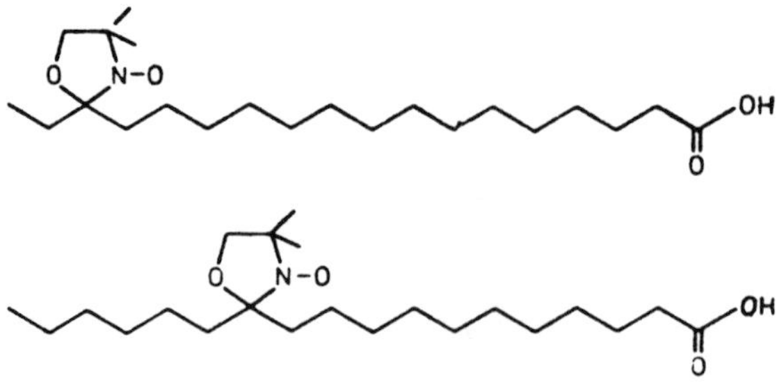

FIGURE 1. Structures of spin-labeled fatty acid derivatives I and II.

brane lipids and vice versa? How do molecules pass through membranes? Can membrane conformation affect the properties of membrane-associated proteins?

II. TYPES OF INFORMATION OBTAINABLE

Many of the experimental approaches that have been used to give information on membrane structure give static information. While crystallography, electron microscopy, and freeze fracture techniques have given much to the current ideas of membrane structure, they do not give dynamic information of the type required to answer many of the current problems in membrane biochemistry. Spectroscopic approaches however, have the potential to give this dynamic information.

Electron spin resonance studies, as with the case of fluorescence studies discussed in Volume I, Chapter 5, depend on the introduction of a specific spin probe. As we will see in the next section, ESR studies have been used to give information on the mobility of membrane lipids and on the interaction of membrane proteins with membrane lipids. The information obtainable from ESR studies depends in large part on the specific location of the introduced spin probe. Nuclear magnetic resonance studies have however, far greater potential. Not only do they frequently make use of either natural nuclei, or isotope replacements (which removes the perturbation of the membrane associated with the introduction of a fluorescence probe or a spin probe), but they also have the potential to give precise structural information as well as the more qualitative information associated with fluorescence or ESR measurements.

III. ELECTRON SPIN RESONANCE

A. ESR Studies Using Spin-Labeled Lipids

Spin-labeled analogs of fatty acids and steroids have been used to give information concerning their own environment in lipid bilayers and in "native" membranes. The spin label can be incorporated into various positions of, for example, stearic acid (Figure 1), and thus report on the environment of several regions of the fatty acid in the membrane. Marsh and Barrantes[1] used the above spin label derivatives of stearic acid to examine the lipid environment of acetylcholine receptor-rich membranes from *Torpedo marmorata*. In addition, they made use of the steroid spin-labeled derivative:

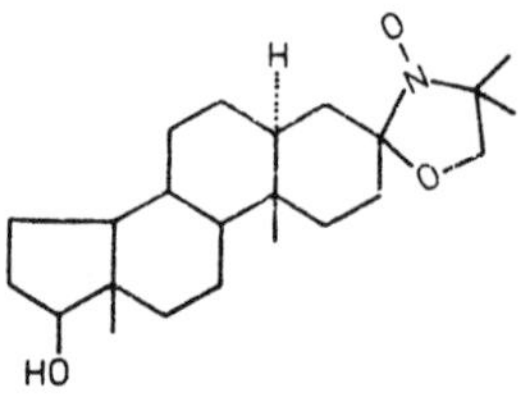

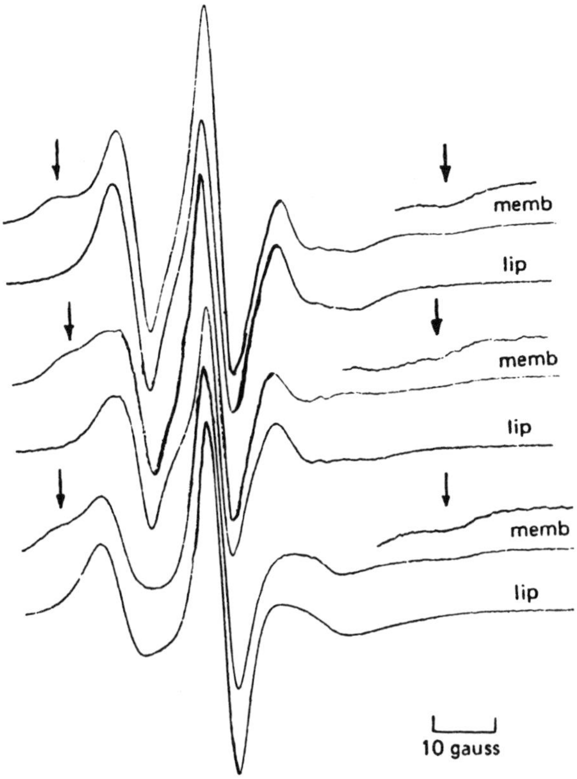

FIGURE 2. ESR spectra of lipid spin labels in AcChoR-rich membranes from *T. marmorata* and aqueous bilayer dispersions of the extracted lipids. The upper spectrum of each pair is from the membrane and the lower from the lipid bilayers. Upper two spectra: spin label II at −4°; middle two spectra: spin label I at 34°; lower two spectra: spin label III at 14°.

ESR spectra of these three probes in membranes, and in aqueous bilayer dispersions prepared from chloroform: methanol (2:1, v/v) solutions of extracted lipids containing 1% spin label indicate (Figure 2) that in the extracted lipid spectra for each label, a single component spectrum is obtained, indicating that the spin label is experiencing rapid anisotropic motion.[2] However, in the membranes each spectrum also contained a well-resolved component whose large anisotropy is characteristic of immobilization. This immobilized component is seen at high- and low-field extremes of the spectra. This component of the spectrum is best observed after the highly mobile component is subtracted from the membrane spectrum. Such difference spectra are shown in Figure 3. The motional correlation times for these spin labels can be calculated from the empirical equations:

$$T_r = a(1 - Azz/A^{R}zz)^{6}$$

$$T_r = a_m'(\Delta Hm/\Delta Hm^{R} - 1)^{6'm}$$

where Azz and ΔH for the high field or the low field region, have the values indicated in Figure 3. ΔH^R and Az^Rz are the values expected for a rigidly immobilized spin label.

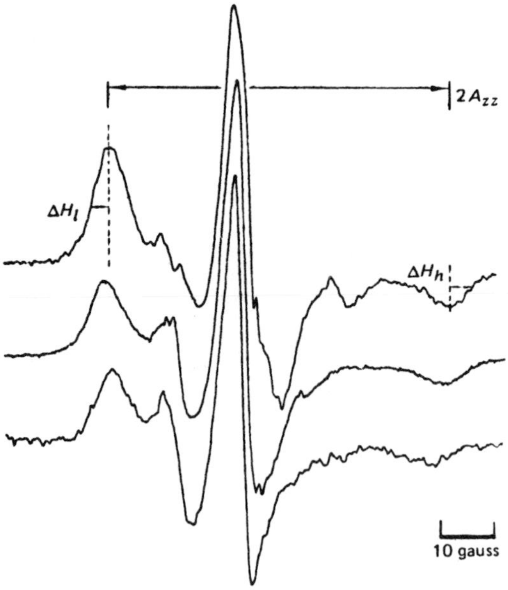

FIGURE 3. ESR difference spectra of lipid spin labels in AcChoR-rich membranes, obtained by subtracting the bilayer spectra of Figure 2 to yield the immobilized lipid component. Upper spectrum: spin label III in membranes at 14°; middle spectrum: spin label II in membranes at 4°; lower spectrum: spin label I in membranes at 34°.

The calibration constants, a and b, are obtained by spectral simulation. Thus rotational correlation times for the spin labels in the immobilized lipid regions of the membranes can be calculated from either the low-field linewidth, the high-field linewidth, or the outer splitting. It might be noted that the physical significance of these rotational correlation times varies with the nature of the probe. With the stearic acid spin labels, the direction of the maximum hyperfine splitting lies along the axis of the molecule, and the correlation time is for angular rotation of the long axes. However, with the steroid analog the direction of maximum hyperfine splitting is perpendicular to the long axis of the steroid, and thus the correlation time is for rotation around the long axis of the molecule.

This type of experiment shows that in the membrane there exists a class of immobilized lipids that appear to be associated with membrane proteins, since the immobilized signal is not seen in aqueous bilayers of the lipids alone. It is unlikely that this immobilized lipid is specifically associated with membrane proteins, since it is sensed by two different types of mobility probes — fatty acid analogs, and the steroid analog.

Fatty acid analogs have also been used to monitor mobility changes in membrane lipids induced by protein interaction with the membrane. Antibody-sensitized sheep erythrocytes were labeled with nitroxy-stearic acid analogs, and isosmolar ghosts prepared. Resealed isosmolar ghosts showed an almost identical spectrum to that obtained with whole erythrocytes (Figure 4). When these antibody-sensitized isosmolar ghosts were treated with complement, significant changes in the spectra are seen (Figure 5), indicating that complement protein-lipid interaction decreases the fluidity of the membranes.[3]

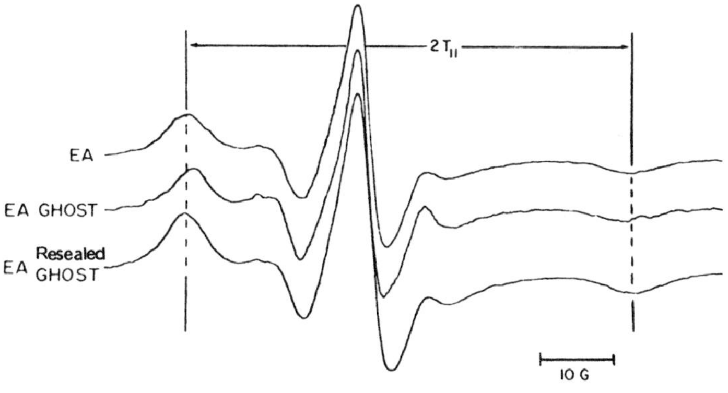

FIGURE 4. ESR spectra of sheep erythrocytes spin-labeled with 5-mitroxystearic acid. All spectra were measured at 22°. (From Dahl, C. E. and Levine, R. P., *Proc. Natl. Acad. Sci. U.S.A.*, 75, 4930, 1978. With permission.)

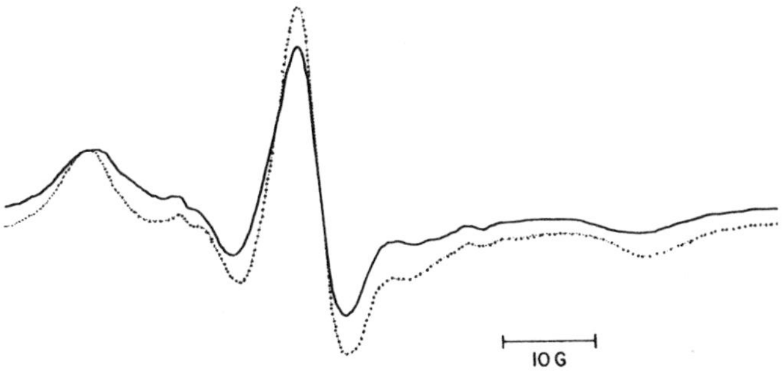

FIGURE 5. ESR spectra of 5-nitroxystearic acid in GA (solid line) and GA treated with complement (dotted line). Spectra were recorded at 22°. (From Dahl, C. E. and Levine, R. P., *Proc. Natl. Acad. Sci. U.S.A.*, 75, 4930, 1978. With permission.)

In the above, and many other cases, ESR has been used to indicate mobility changes in membrane lipids as a result of protein binding, changes in lipid content, or in various abnormalities compared to normal behavior of membrane lipids. However, a severe limitation of ESR in studying membrane fluidity is illustrated by examining results of membrane fluidity studies in erythrocyte membranes using fatty acid spin label analogs to study "normal" membranes compared to those isolated from Duchenne muscular dystrophy patients. Although conventional ESR using maleimide spin label had suggested that there were conformational differences in membrane proteins of Duchenne muscular dystrophy patients,[4] studies with fatty acid spin labels[5] indicated that there were no changes in membrane fluidity. The problem lies in the fact that classical ESR is only sensitive to motion corresponding to correlation times of 10^{-6} sec or less. Thus it is quite possible that motion, or changes in motion on much slower time scales would not be observed in classical ESR experiments. However, the introduction of saturation transfer ESR[6] has extended the time scale of sensitivity of ESR to molecular motions occurring with correlation times as long as 10^{-3} sec.

B. Saturation Transfer ESR

This technique is referred to as saturation transfer ESR, because it depends on the diffusion of saturation through the spectrum of a nitroxide spin label. The diffusion originates from rotational diffusion that modulates anisotropic magnetic interactions. The difference in time resolution of saturation transfer ESR (ST-ESR) and classical ESR (ESR) can be explained by considering the general electron magnetic resonance phenomenon.

The rate at which unpaired electron spins precess around the external Zeeman field is determined both by the magnitude of the Zeeman field, and by the molecular magnetic configuration of the unpaired electron. When the frequency of the incident microwave field matches that of the precessing electron, absorption occurs. In order to improve the signal to noise ratio by reducing the low frequency component of the i/f noise of the crystal detector, the external Zeeman field is time modulated. An ESR experiment involves detecting the out of phase component of the time dependent spin response wih respect to the first harmonic of the incident microwave, which is in phase with the first harmonic of the Zeeman modulation frequency. The power of the incident microwave is selected to ensure that the ESR signal intensity is a linear function of the microwave amplitude.

In contrast, ST-ESR uses a microwave amplitude higher than in the linear regime, with detection out of phase with the applied Zeeman field. This increases the dependence of the spin response on molecular processes that modulate energy transfer between the electron spin and the molecular lattice (that is I/T_1 processes), and between different portions of the resonant spectrum such as rotational motion and spin exchange. The detection of the response out of phase with the applied Zeeman modulation helps assure that the observed signal does indeed depend significantly on these saturation transfer processes. As a result, ST-ESR follows events that are competitive with T_1 rather than events occurring at a rate competitive with T_2, which is the case in ESR. Since the spin-spin relaxation rate, $1/T_2$ is usually much more than an order of magnitude faster than $1/T_1$ for nitroxide radicals, ESR is limited to faster processes than is ST-ESR.

That ST-ESR is indeed sensitive to processes that ESR cannot detect is shown in Figure 6. In studies with long chain fatty acid, maleimide spin label derivatives,[7] it was found that ESR could not distinguish between various states of the boundary layer in retinal rod outer segment membranes, whereas ST-ESR could. These experiments in fact show that there is an intermediate viscosity between the original viscosity of the discs and the viscosity of the final, highly delipidated membrane. In the ST-ESR spectra, spectrum d, especially in the high-field region, is not a linear combination of b and h, where b is obtained with free spin label in the membrane and h is obtained with maleimide-anchored spin label in highly delipidated membrane. This indicates that in the intact membrane (spectrum d) there is a component not due to either free spin label, or the highly immobile maleimide-anchored spin label.

Retinal rod outer segment membranes have been extensively studied using ST-ESR.[7,8] When rod outer segment membranes are labeled with 3-malemido-2,2,5,5-tetramethyl-1-pyrorolidinyloxyl (MSL) an ST-ESR spectrum as in Figure 7a is obtained. In Figure 7b is the spectrum obtained from MSL in a 50:50 mixture of glycerol and water. When this spectrum is subtracted from part a, the spectrum of the highly immobilized MSL-labeled protein is obtained (spectrum c). That the mobile component in spectrum a can be simulated by a 50:50 glycerol-water mix indicates that this mobile component corresponds to "membrane-associated" MSL and not to totally free MSL. It can be shown that the ST-ESR spectra reflect mobility of the probe within the membrane, and not rotation of the vesicles themselves by sonication of the isolated vesicles, which reduced their size from 5000 to 500 A, as judged by electron microscopy. No modifications in the spectra resulted from such treatment, eliminating vesicle motion

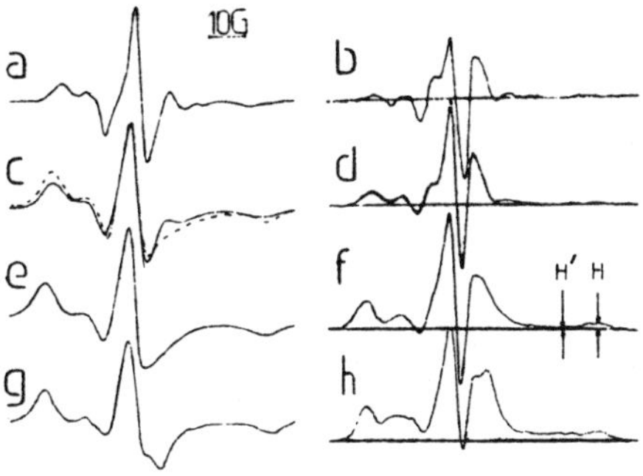

FIGURE 6. Comparison of the conventional EPR and saturation transfer EPR with (10,3) MSL in membranes progressively delipidated at 20°C. Spectra on the left correspond to the first harmonic, in phase, 10 mW, 2-G modulation. Spectra on the right correspond to the second harmonic, 90° out of phase, 5-G modulation, 32 mW. Spectra a and b: (10,3) fatty acid in intact membranes. Spectra c and d: (10,3)MSL in intact membranes. The dotted line spectra superimposed correspond to the signal due to the protein - bound label only (see text). Spectra e and f are obtained with partially delipidated membranes (30% delipidation). Spectra g and h correspond to highly delipidated membranes (over 70% delipidation). H and H′ correspond to the signal amplitude at the indicated positions. (From Favre, E., Baroin, A., Bienvenue, A., and De-Vaux, P. F., *Biochemistry,* 18, 1156, 1979. With permission.)

as the cause of the spectral shapes. The ST-ESR spectrum of MSL in disc membranes is sensitive to light. Illumination at 37°C for 45 min resulted in the changes shown in Figure 8; no changes are observed in ESR spectra obtained under similar circumstances. These spectral changes suggest that the correlation time for the label changes from 10 to 20 μsec to 70 to 80 μsec upon illumination. While ST-ESR spectral changes can be interpreted in terms of changes in correlation times, it is impossible at present to distinguish changes in r from changes in orientation of the label, which would give similar effects. However, the most likely explanation, in terms of changes in r is that on illumination, the MSL labeled protein aggregates. Saturation transfer spectra of MSL in disc membranes are also affected by lipid depletion of the membranes. (Again ESR spectra are *not* affected in similar experiments). When lipids are removed by phospholipase A_2 treatment the ST spectra (Figures 9a and 9b) show increased immobilization of the label. The final spectrum obtained (Figure 9c) when 60 to 70% of the lipids have been removed can be simulated by cross-linking proteins in native membrane with 5% glutaraldehyde, again suggesting that the proteins aggregate under these conditions.

It was mentioned earlier that ESR studies with fatty acid spin label derivatives could detect no differences between erythrocyte membranes from controls or from Duchenne muscular dystrophy patients. However, ST-ESR[9] has shown considerable differences using the same spin label derivative, 2-(3-carboxypropyl)-4,4-dimethyl-2-tridecyl-3-oxazolidinyl-oxyl (5-NS) as used in the ESR study.[5] When erythrocytes from a Duchenne-MD patient and a control were labeled with 5-NS for either a short time period or a

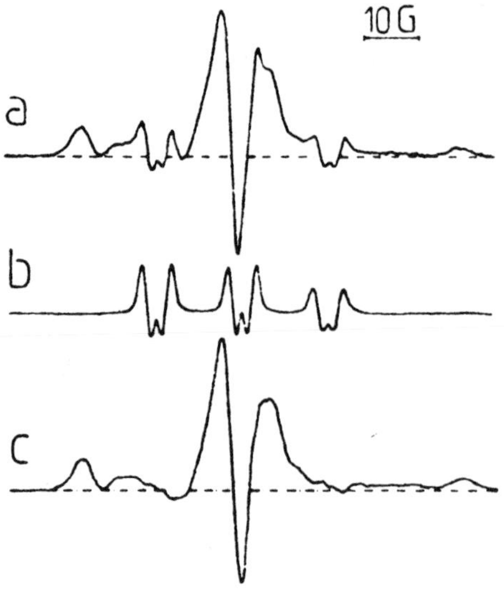

FIGURE 7. Saturation transfer spectra of spin-labeled rhodopsin (second harmonic, 90° out of phase, microwave power 32mW, modulation 5G, modulation frequency 50 kHz). (a) MSL bound to rhodopsin in disc membrane fragments at 20°C; (b) MSL in a mixture of glycerol-water (50:50) at 20°C; (c) combination of a and b, so as to minimize the contribution of the weakly immobilized component on Spectrum a. The electronic manipulation enables one to obtain the pure saturation transfer spectrum of MSL tightly bound to rhodopsin. This procedure is probably safer for the membranes than any combination of chemical preincubation followed by repeated extensive centrifugations of the membranes. (From Baroin, A., Bienvenue, A., and DeVaux, P. F., *Biochemistry*, 18, 1151, 1979. With permission.)

longer time period, distinct differences were seen (Figure 10). The differences were far greater in the early time point than the later time point. When the signal intensity (Figure 11) is monitored as a function of time after labeling, it can be seen that the control increases only marginally while the intensity in the Duchenne-MD patient doubled. There are two possible explanations for this type of observation. Changes in the polarity of the spin label insertion site could affect the magnetic tensor values, resulting in a shift of the peak positions, as well as modify the ability of the unpaired electron to exchange energy with its environment. Increased polarity would lead to increased exchange, which in the case of ST-ESR (which is dependent on lattice relaxation) would cause a decrease in signal intensity. Alternatively, spin exchange can give rise to this sort of effect. This occurs when the spin concentration is high enough to allow interactions between unpaired electrons, which can alter spectral shapes or decrease intensities. When such an exchange occurs with a frequency different from the ESR sensitivity, but such that it competes with $1/T_1$ processes, ST-ESR will respond to the process. Studies with varied concentrations of 5-NS indicated that spin exchange mechanisms most probably account for the observations in this case, suggesting that in erythrocytes

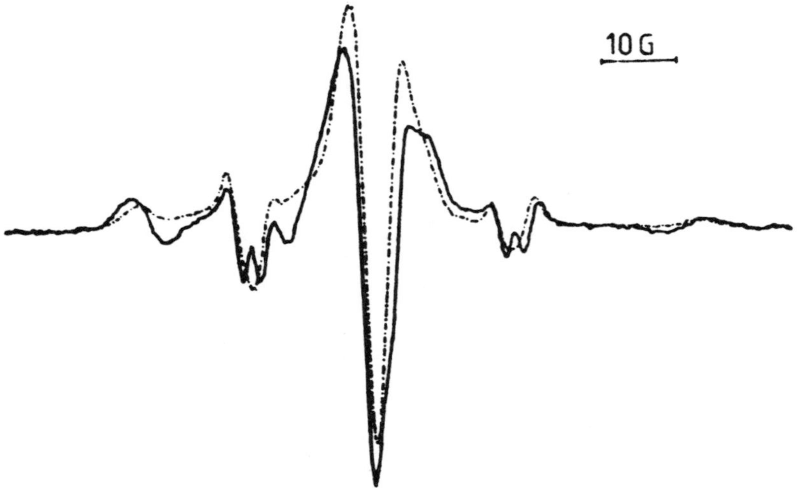

FIGURE 8. Effect of light on the saturation transfer spectra of MSL in disc. membranes, at 37°C. (Full line). Sample in the dark; (dotted line) sample illuminated continuously, temperature maintained at 37°C. The passage from the first spectrum to the second spectrum requires about 30 min after the beginning of the illuminations. The concentration of rhodopsin is approximately 300 μM. One spectrum is recorded in 8 min. (From Baroin, A., Bienvenue, A., and DeVaux, P. F., *Biochemistry*, 18, 1151, 1979. With permission.)

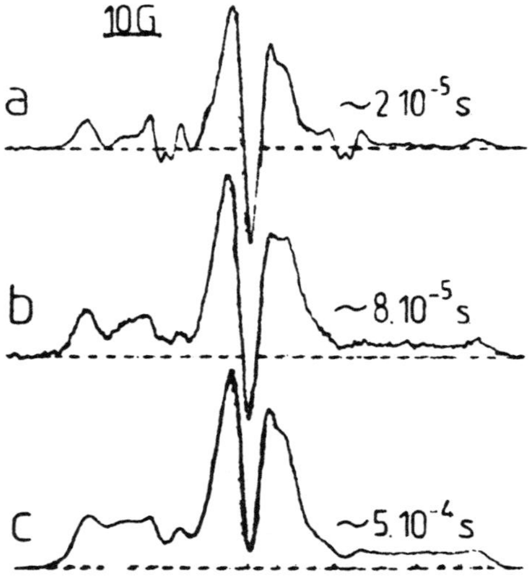

FIGURE 9. Effect of lipid depletion on the saturation transfer spectra of MSL in disc membranes. Temperature is 20°C. Concentration of rhodopsin, 200 μM. (a) Native membranes; (b) membranes from which 30 to 40% phospholipids have been removed; (c) 60 to 70% lipids removed. Approximate values of the rotational correlation times indicated on the spectra correspond to the average values obtained from a line — shape analysis of the type proposed by Thomas et al. (1976) for isotropic samples. (From Baroin, A., Bienvenue, A., and DeVaux, P. F., *Biochemistry*, 18, 1151, 1979. With permission.)

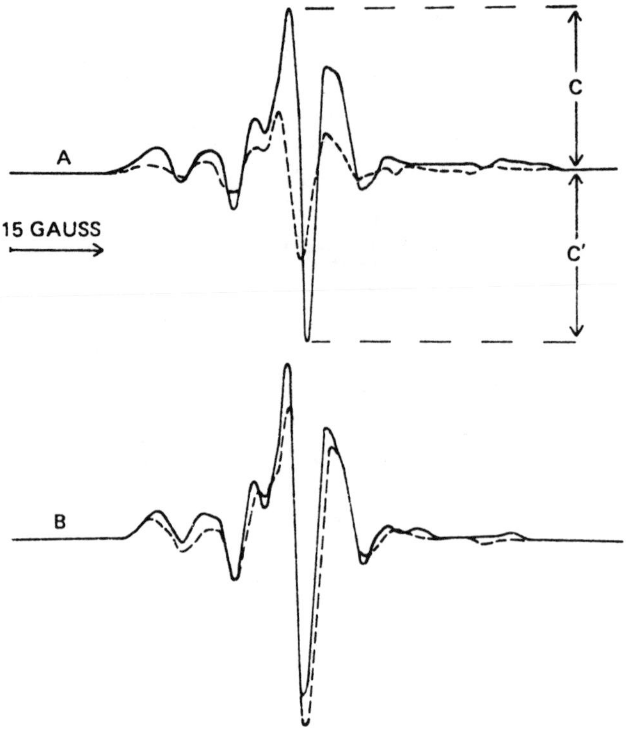

FIGURE 10. Tracings of representative ST-EPR spectra for erythrocytes from a DMD patient (---) and a control (-). (A) Early time point. (B) Late time point. (From Wilkerson, L. S., Perkins, R. C., Roelofs, R., Swift, L., Dalton, L. R., and Park, J. H., *Proc. Natl. Acad. Sci. U.S.A.*, 75, 838, 1978. With permission.)

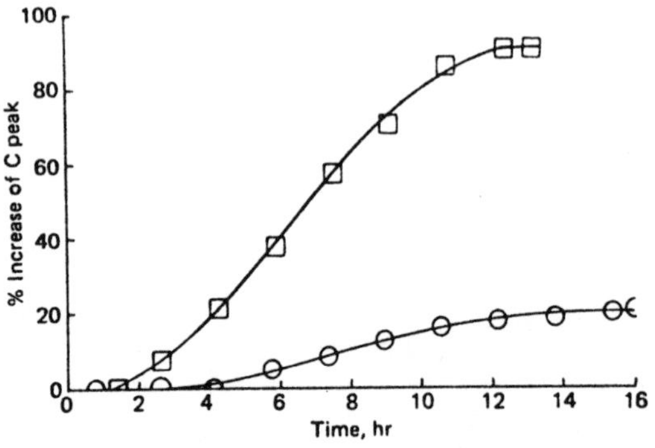

FIGURE 11. Relative change in the ST-EPR C peak (see Figure 10) as a function of time after labeling. (O), Control; (•), DMD. (From Wilkerson, L. S., Perkins, R. C., Roelofs, R., Swift, L., Dalton, L. R., and Park, J. H., *Proc. Natl. Acad. Sci. U.S.A.*, 75, 838, 1978. With permission.)

from Duchenne-MD patients the fatty acid spin label cannot redistribute itself to an equilibrium state as rapidly as normal.

C. Lateral Diffusion in Membranes

Most of the methods for measuring lateral diffusion coefficients in phospholipid bilayer membranes are restricted to relatively fast motion at high temperatures. However, a newly developed technique[10] combining photobleaching and spin labeled phospholipids allows simple analysis of lateral diffusion over a wide range of motional rates. In principle, the method depends on photochemically produced carboxymethyl radicals, which quench the signal of nitroxide-labeled phospholipids in the membrane. The carboxymethyl radicals are generated from the alkylcobalt complex $[Co(CN)_5CH_2COO^-]^{4-}$ by exposure to a laser beam of certain segments of the membrane. Other segments of the membrane are "covered" by metal disks preventing radical generation in those regions. Between light pulses, spin-labeled phospholipids can diffuse into "unprotected" areas of the membrane, and thus are susceptible to a second pulse of light. From the sequential decrease in ESR signal for the spin label, the lateral diffusion of the labeled phospholipid can be determined. The technique of course depends on the absence of diffusion of the photogenerated alkyl radical into the "protected" areas. When photolysis is carried out at temperatures where spin label diffusion is negligible, the ESR signal decays to a constant value expected from the exposed area, confirming that no radical diffusion takes place. In trial studies with dimyristoyl phosphatidylcholine and dipalmitoyl phosphatidylcholine in the fluid phase using a phospholipid with a head group spin label;

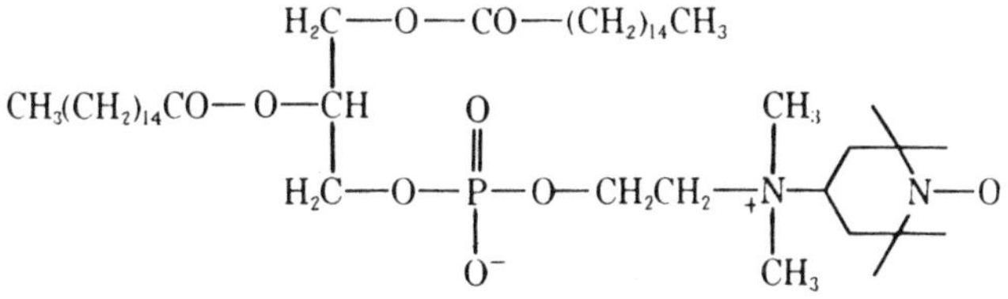

I

excellent agreement was found using the above technique and NMR measurements.

This method is particularly suited to the study of diffusion in biological systems, since it is suited for precise temperature control, and, compared to the photobleach recovery technique[11] introduces far less heat into the system because of the high quantum yield of alkylcobalt photolysis. While the technique has so far been used to measure translational diffusion, using ESR, it is also possible to measure rotational diffusion by utilizing ST-ESR in conjunction with the photolysis.

D. Limitations of ESR in Membrane Studies

As we have seen, electron spin resonance studies can give information both about the mobility of spin-labeled lipid constituents of the membrane, and about the mobility of membrane proteins, and lipids associated with these proteins. Although classical ESR is limited in terms of the time resolution of motion that it can sense (correlation times must be less than 10^{-6} sec), the advent of saturation transfer ESR has largely eliminated this limitation of the method. With ST-ESR, correlation times as short as 10^{-3} sec can be determined.

All ESR measurements have been, and are, of continuing concern that introduction of a spin label into, for example, a fatty acid or a steroid molecule may significantly alter the normal environment around that molecule. As a result, the spin probe would in fact report on its own local environment rather than the environment of the bulk phase into which it has been introduced.

IV. NUCLEAR MAGNETIC RESONANCE

A. Advantages of NMR over ESR

Nuclear magnetic resonance studies have been increasingly used to study membrane systems over the last few years. As theory is developed, and the level of instrumental power and sophistication increased, NMR studies have been able, and will continue to be able to give extremely detailed information about the conformation and mobility of lipids in synthetic bilayers, and in native membranes. The real advantages that NMR has over ESR, and for that matter techniques such as fluorescence, is that membrane environments can be studied without the introduction of a nitroxide probe, or a bulky fluorescent probe, which as just discussed, may perturb the environment it is trying to reflect. Thus NMR techniques have the advantage of reflecting bulk properties of the membrane directly rather than by inference.

B. Types of Nuclei Used

A variety of nuclei have been used in NMR studies of membrane systems; ^{1}H, ^{2}H, ^{13}C, ^{15}N, ^{19}F, and ^{31}P, and we shall deal with each in turn. With the exceptions of ^{1}H and ^{19}F, studies with these nuclei have been facilitated by the ability to enrich the natural abundance of the appropriate nucleus, resulting in improved signal to noise ratios. Much of the early work in membrane systems involved ^{1}H-NMR, however due to various problems (see below) that are not so relevant with ^{2}H-NMR, deuterium labeling of lipids etc. has virtually replaced ^{1}H-NMR. ^{19}F-NMR is becoming increasingly popular, however it retains the problem of the possibility of perturbing the natural environment of the "probe" by introduction of ^{19}F into whatever compound is being used.

1. ^{1}H - NMR

Retinal rod outer segment (ROS) disk membranes, which as discussed earlier have been extensively studied using ESR and ST-ESR have also been studied using ^{1}H-NMR. ^{1}H-NMR of ROS phospholipids show two distinct terminal CH_3 resonances, with characteristic spin-spin coupling. These resonances have been assigned by spin decoupling as indicated in Figure 12.[12] The downfield, well resolved triplet is from the CH_3 group of fatty acids with a double band in the W position relative to the CH_3, whereas the upfield resonance exhibiting more complex spin-spin coupling is from the CH_3 groups on other fatty acid side chains. Both sets of resonances have similar linewidths and the areas reflect their compositional ratio, indicating that the motional state of the unsaturated fatty acid is similar to that of the saturated fatty acids. ^{1}H-NMR spectra of unsonicated ROS membranes have resonances that are relatively broad and unresolved. On sonication the linewidths are reduced and the resonances are more resolved. Both CH_3 resonances are clearly seen in the membrane or liposome spectra.

When distinct sharp and broad components are observed in ^{1}H-NMR spectra of membranes as is seen for ROS membranes or liposomes (Figure 13), it indicates that the phospholipid protons must be subjected to different degrees of motional averaging of magnetic dipolar interactions. There are three possible causes of such behavior: distribution of vesicle size, clustered domains of phospholipids with different extents of mobility, or a change in mobility from one end of the fatty acid chain to the other. Since ROS membranes appear to form fairly uniformly sized vesicles,[13] and the ^{1}H-NMR spectra are not affected by prolonged sonication, it would appear that the first of these possibilities can be eliminated. The methylene and terminal CH_3 groups of sonicated phosphatidylcholine or dipalmitoyl-phosphatidylcholine show different T_1 relaxation rates, suggesting that coupling of these spin systems does not occur.[14] As a

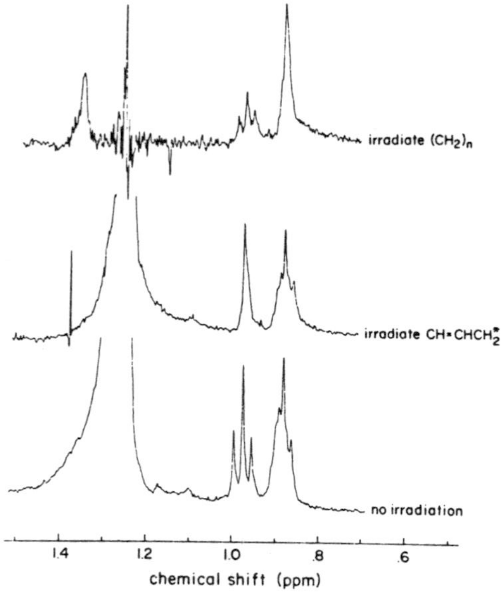

FIGURE 12. Spin-decoupling of the terminal CH_3 resonances from the directly bonded side-chain methylene resonances of total extracted ROS phospholipids in 19:1 $CDCl_3$-CD_3OD, 20°C at 360 MH_z. (From Brown, M. F., Miljanich, G. P., and Dratz, E. A., *Biochemistry*, 16, 2640, 1977. With permission.)

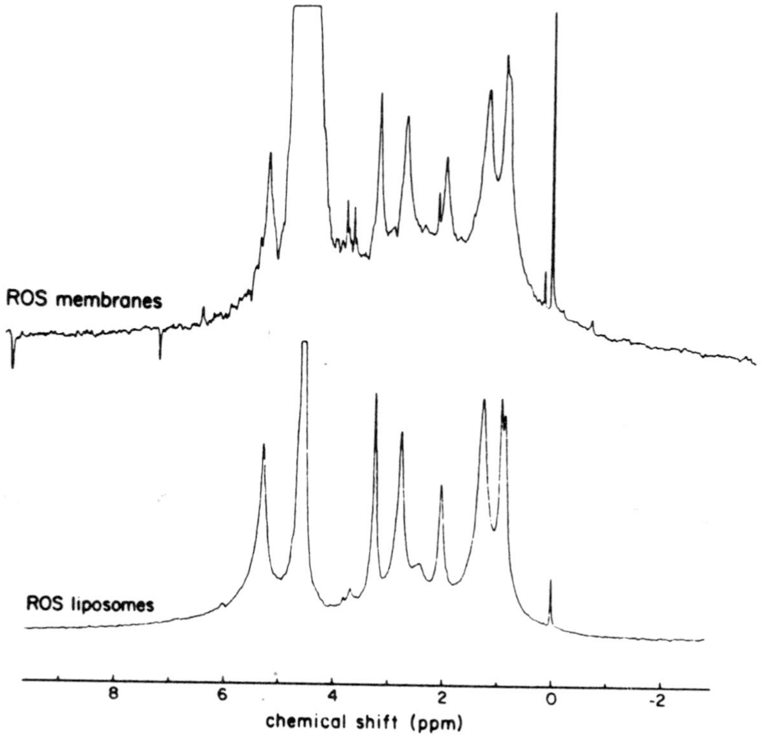

FIGURE 13. 360-MH_z ^1H-NMR spectra of sonicated ROS membranes and liposomes in 0.1*M* borage buffer, 0.1*M* KCl, pH 7 at 40°C. (From Brown, M. F., Miljanich, G. P., and Dratz, E. A., *Biochemistry*, 16, 2640, 1977. With permission.)

result, 100% of the intensity of terminal CH_3 groups might be expected to be resolved, if there was a motional gradient. Since significantly less than 100% of the terminal CH_3 group intensity is resolved in ROS membranes, it is difficult to interpret the ^1H-NMR spectra on the basis of a motional gradient in the fatty acid chains. Thus in the case of ROS membrane vesicles or liposomes, it is suggested that the ^1H-NMR spectra indicate the existence of domains of phospholipids with distinctly different mobilities. The fact that membrane vesicles and liposomes show similar spectra suggests that these domains are inherent in the lipid bilayer, and not associated with proteins bound to the bilayers.

^1H-NMR has also been used to establish that the acyl chains of phospholipids are in similar states of motion in small vesicles and multilamellar membranes.[15] ^1H-NMR spectra obtained from ultrasonically generated vesicles are much sharper than those from unsonicated dispersions. This is to be expected for small vesicles whose rotational Brownian motion causes "motional narrowing" of the NMR lines. Bloom et al.[15] have, however, presented an extensive theoretical treatment (which is beyond the scope of this chapter) using a theory of motional narrowing that takes into account the symmetry properties of bilayer systems, as well as the fact that all of the spins along the hydrocarbon chain are not equivalent. From this theoretical treatment, predicted line shapes and widths for nonterminal carbons, resolved α-CH_2 peaks, and terminal CH_3 peaks were found to agree with experimentally determined shapes and widths, suggesting that the local orientation of phospholipid acyl chains in small vesicles was similar to that in multilamellar membranes.

Lateral diffusion of membrane lipids may be determined using pulsed gradient ^1H-NMR spin echo techniques. Spin echo techniques were not applicable to lipid systems because anisotropic motion of the lipids produced only a partial averaging of the dipolar interactions and a spin echo could not be observed. However, it was found[16] that the dipolar interactions are minimized if the lipid chains are oriented at 54° 44′ to the applied field. Thus if oriented multibilayer samples are used, a spin echo can be observed and diffusion measured. The echo response is governed by the equation,[17]

$$A(on)\ A(off) = e^{-\alpha^2 D g^2 \delta^2 [\Delta - (1/3)\delta]}$$

which assumes an homogeneous static magnetic field H_o. A (on) is the amplitude of the spin echo peak in the ft spectrum with gradient pulses on, A (off) is the amplitude with gradient pulses off, γ is the nuclear gyromagnetic ratio, δ is the width of the gradient pulse, and g is the field gradient. In studies with model membrane — D_2O multilayers of dipalmitoyl, dimyristoyl, dilauryl, and egg yolk lecithins it was found[18] that for saturated lipids the diffusion decreased monotonically as the chain length increased. It was also shown that cholesterol at low molar percentages increased lipid diffusion in dipalmitoyl lecithin multilayers. However, at larger concentrations decreased diffusion was observed. It might be noted that the results obtained using this approach are significantly different from those obtained using classical ESR, but do agree well with values obtained using the photo-spin label technique discussed earlier.

2. *^2H-NMR*

Deuterium Nuclear Magnetic Resonance has a number of advantages both over ^1H-NMR, and most other types of NMR. The natural abundance of ^2H is 0.016%, which means that by specific deuteration it is possible to enrich a site by upwards of 6000-fold compared to its natural abundance, thus considerably reducing the effects of the background. In addition, to this, deuterium has the benefit of not being a large perturbing moiety, which could disrupt its environment compared to ^1H, which it replaces,

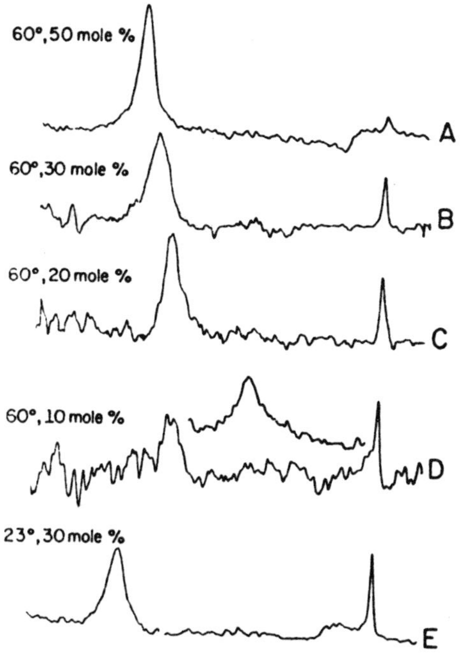

FIGURE 14. Temperature and cholesterol-concentration dependence of the deuterium quadrupole splitting of cholesterol-3α-d$_1$ in a DMP bilayer system. Experimental details as described previously.[20] (From Oldfield, E., Meadows, M., Rice, D., and Jacobs, R., *Biochemistry*, 17, 2727, 1978. With permission.)

is generally nontoxic to living systems, and may be incorporated to very high levels in living systems. The commercial availability of deuterium depleted water means that working in an aqueous medium is no problem. In addition to these benefits of deuterium there are two theoretical advantages of working with ^{2}H-NMR. The ^{2}H nucleus has a quadrupole moment (i.e., spin I = 1) and thus electric quadrupolar interactions perturb the nuclear Zeeman levels, which makes the ^{2}H nucleus a particularly sensitive probe. Secondly, since the relaxation of the ^{2}H nucleus is totally dominated by the quadrupolar mechanism,[19] analysis of relaxation data is relatively simple.

Oldfield et al.[20] have shown that the angle of tilt of the specifically deuterated cholesterol molecule

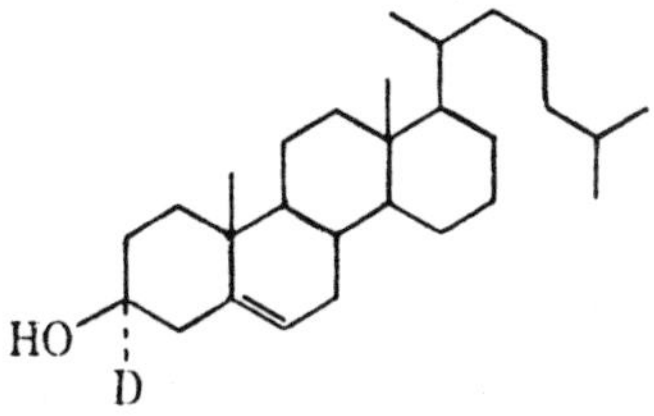

can be determined from the observed quadrupole splitting. Typical spectra for this cholesterol derivative in a bilayer membrane of dimyristoylphosphatidylcholine in water are shown in Figure 14 at two different temperatures and varying mole percentages of the derivative. When the quadrupole splitting is plotted as a function of mole

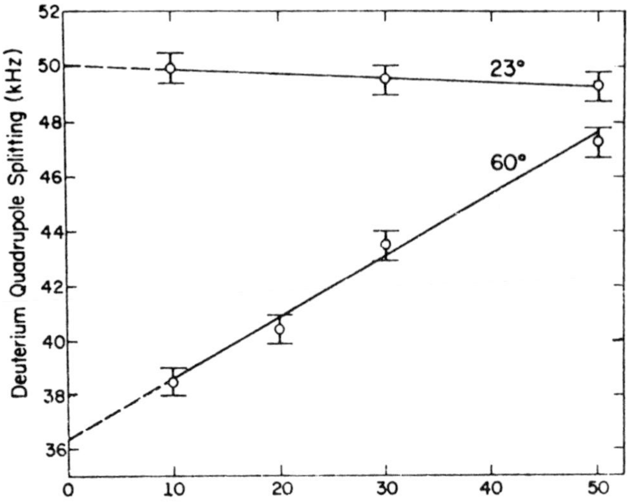

FIGURE 15. Effect of mole percent of cholesterol on quadrupole splitting at 23°C and 60°C. (From Oldfield, E., Meadows, M., Rice, D., and Jacobs, R., *Biochemistry,* 17, 2727, 1978. With permission.)

percent cholesterol, Figure 15, it is seen that the splitting is independent of cholesterol content at the phase transition temperature of 23°C, but is highly dependent on cholesterol content at high temperature. Since the observed splitting is much less than the theoretically predicted value of 63.8 kHz for a molecule undergoing fast rotation about the bilayer normal, it may be deduced that the observed quadrupole splitting must be caused by tilt of the steroid nucleus in the bilayer. It might be noted that the observed independence of the quadrupole splitting on mole percent of cholesterol at 23°C suggests that the steroid nucleus does not perturb its local environment, which would probably change the most probable angle of tilt. The temperature effects observed suggest that the presence of high cholesterol concentrations in the plane of the membrane introduces a temperature independence of the most probable angle of tilt of the steroid molecule. Assuming the molecular tilt is a result of concerted fluctuations in the membrane, it may be that one role of cholesterol in biological membranes is to conserve membrane thickness.

Studies with dimyristoylphosphatidylcholines (DMPC) labeled in various positions with deuterium, where the quadrupole splitting can be related to the chain length of the acyl chain, have been used to study membrane thickness. Spectra of two-chain specifically labeled DMPC-30% cholesterol bilayers are shown in Figure 16. When transmembrane thicknesses at C-2' of the two-chain, C-6' of the two-chain, or C-12' of the two-chain are calculated from the quadrupole splitting are compared with results obtained from neutron diffraction, excellent agreement is seen (Table 1). When the quadrupolar splitting of variously labeled DMPC's were examined at 10°C in the DMPC-Cholesterol system (Figure 17), distinct differences are observed at the 2' position depending on whether the one-chain, or the two-chain is labeled. At 10°C, the two-chain 2' resonances are totally broadened from the spectrum. In Figure 17 A, the resonance arises totally from the one-chain 2' position. This behavior can only be explained if the 2' positions in the one and two chains behave magnetically independently, and can "freeze" separately. This observation may indicate that the two-chain is in a bent configuration at the 2' position.

These studies demonstrate that a considerable amount of information concerning

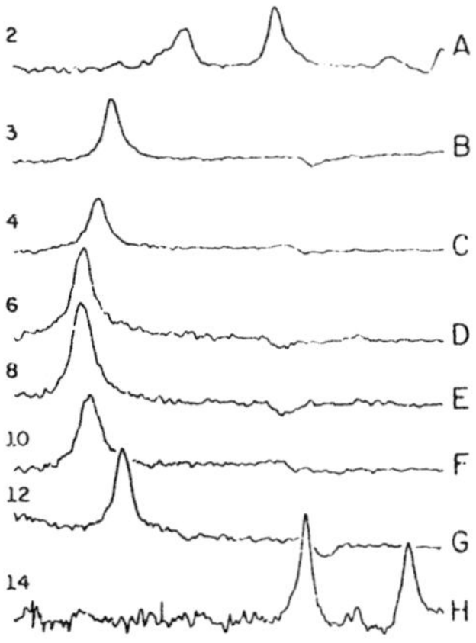

FIGURE 16. Dependence of ²H quadrupole splitting of two-chain specifically labeled DMPC-30 mol. % cholesterol bilayers as a function of position labeled. Experimental details as described previously.[20] (From Oldfield, E., Meadows, M., Rice, D., and Jacobs, R., *Biochemistry*, 17, 2727, 1978. With permission.)

Table 1
COMPARISON OF TRANSMEMBRANE THICKNESS DETERMINED BY NMR MEASUREMENTS OR NEUTRON DIFFRACTION MEASUREMENTS

Position C-η′ of two-chain	Thickness by ²H-NMR	Thickness by neutron diffraction
2′	31.0Å	33.0Å
6′	22.0Å	24.0Å
12′	8.0Å	9.0Å

chain length, packing, and orientation can be obtained from quadrupolar splitting measurements in ²H-NMR.

Such studies have been extended to examine the effects of proteins and peptides on the hydrocarbon chain order of specifically deuterated DMPC bilayers.[21] Using a quadrupole echo pulse sequence[22] the ²H-NMR spectra shown in Figure 18 were obtained. The quadrupolar splitting observed for pure DMPC-d_3 of 3.4 kH_z is very much less than the expected value for such a group in a rigid-lattice crystal powder as a result of motional narrowing. As discussed above, the quadrupolar splitting in such a system is related to the order and mobility of the hydrocarbon chain to which the methyl group is attached. In the case of each of the additions, the observed quadrupole splitting is less than that observed for pure DMPC-d_3 or are collapsed into a single narrower line, as with gramicidin A, bacteriophage fl coat protein, and myelin proteolipid apoprotein. These observations suggest that the association of these proteins and peptides with DMPC-d_3 bilayers results in an increase in the mobility and disorder of the hydrocarbon chains. Similar effects are seen below the gel to liquid crystal phase transition

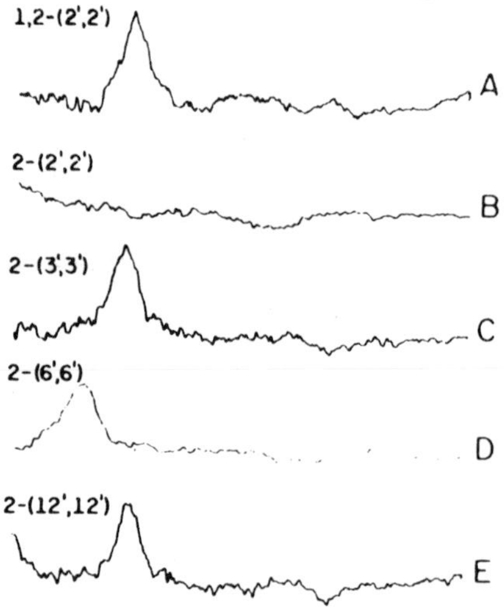

FIGURE 17. Low temperature spectra of deuterium-labeled DMPC's in the DMPC-cholesterol system. (From Oldfield, E., Meadows, M., Rice, D., and Jacobs, R., *Biochemistry*, 17, 2727, 1978. With permission.)

temperature, Figure 19. In this case, it is apparent that the presence of the protein or polypeptide chain prevents the hydrocarbon chains from crystallizing into the hexagonal α-crystalline gel phase.

These results conflict with classical ESR experiments,[23] which suggest that when protein is added to spin-labeled lipid, some of the spin label becomes less mobile. There is, however, a difference in the rates of motion required for motional averaging in these two types of experiments. In ^{2}H-NMR rates of the order of 10^5 sec^{-1} are sufficient, while with classical ESR, the rates must be at least 10^8 sec^{-1}. Thus ^{2}H-NMR can sense slower motions than can ESR.

Quadrupolar echo measurements have also been applied to the mobility of cholesterol esters in membranes.[24] When ^{2}H-NMR spectra of multilamellar liposomes composed of 50% egg yolk lecithin containing variable amounts of cholesterol palmitate-d_{31}, and 50% water, distinctly different spectra are obtained, depending on the cholesterol content, Figure 20. Both spectra can be analyzed in terms of two complex spectra, one consisting of peaks labeled A and B in Figure 20, and the other consisting of peaks C and D. The broad features of the C + D spectrum is most probably due to solid cholesterol palmitate. The intensity of the more mobile signal remains essentially constant when the mole % of cholesterol is reduced from 5 to 1%, however, that of the solid phase signal is reduced fivefold, indicating that once the cholesterol mole % reaches a certain level in egg yolk lecithin liposomes it forms a solid phase.

3. ^{13}C-NMR

As with ^{2}H-NMR studies, the use of ^{13}C-NMR can benefit greatly from the fact that specific enrichment with ^{13}C increases the sensitivity to the point that acceptable signal to noise ratios can be obtained in spectra of millimolar concentrations of phospholipid dispersions in a matter of hours.

Using 1,2-dihexanoyl-3-sn-phosphatidylcholine (DHPC) and 2,3-dipalmitoyl-3-sn-phosphatidylcholine (DPPC) enriched in the carbonyl carbons with ^{13}C, interactions with the carbonyls as DHPC undergoes micelle formation have been studied.[25] This

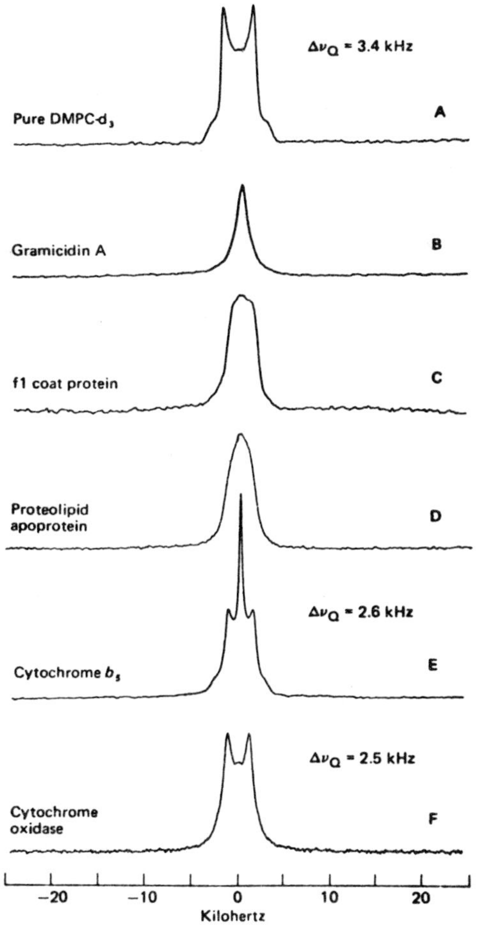

FIGURE 18. ^{2}H-NMR spectra of DMPC-d_3 bilayers containing protein or polypeptide. $\Delta\nu Q$ is the ^{2}H quadrupole splitting. (From Oldfield, E., Gilmore, R., Glaser, M., Gutowsky, H. S., Hshung, J. C., Kang, W. Y., King, T. E., Meadows, M., and Rice, D., *Proc. Natl. Acad. Sci. U.S.A.*, 75, 4657, 1978. With permission.)

region, which can be regarded as an interface between the head group and the hydrophobic chain is thought to be important in determining molecular packing in micelles and bilayers.[26] When ^{13}C-NMR spectra for the carbonyl region of β-DPPC and α, β-DPPC are examined (Figure 21), it is seen that the α carbonyl ^{13}C resonance is upfield of the β resonance. In organic solvents this is reversed. Similar results were seen with DHPC. When the effects of DHPC concentration on the $^{13}C=O$ chemical shifts for α- and β-carbonyls was examined, Figure 22, it was found that the shifts become equal at about 65 n*M*. There are a number of possible explanations for these observations. The fact that there are no changes in the carbonyl ^{13}C shifts of egg phosphatidylcholine in chloroform over the concentration range of 1 to 100 m*M*, and a temperature range of 24 to 50°C[27] where the cmc is 14 m*M* at 24°C, indicates that the carbonyl groups do not participate in micelle formation. The head group, or the head group orientation could affect the carbonyl ^{13}C chemical shifts. Such an effect would be via an electric field effect, which is the effect of a molecular dipole on a nearby polarizable bond.

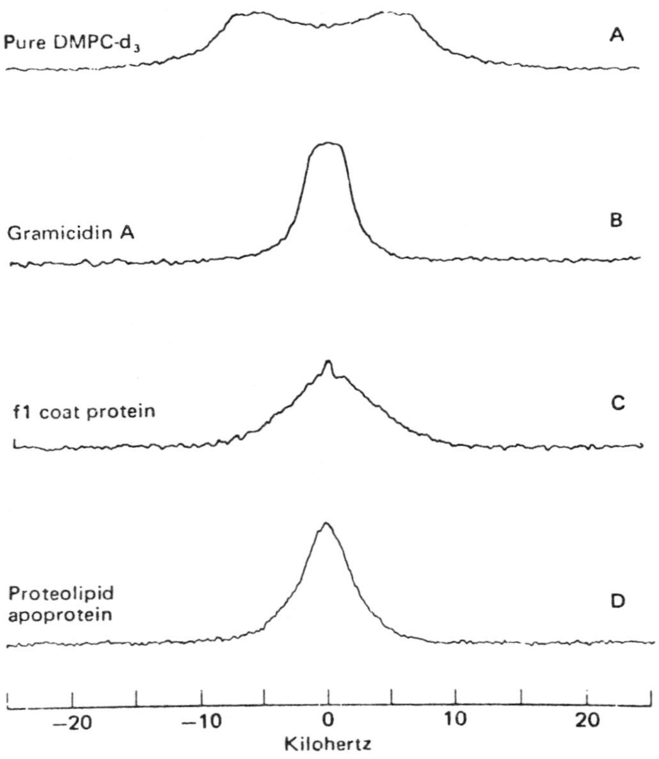

FIGURE 19. ²H-NMR spectra of DMPC-d_3 bilayers below the gel to liquid crystal phase transition temperature. (From Oldfield, E., Gilmore, R., Glaser, M., Gutowsky, H. S., Hshung, J. C., Kang, W. Y., King, T. E., Meadows, M., and Rice, D., *Proc. Natl. Acad. Sci. U.S.A.*, 75, 4657, 1978. With permission.)

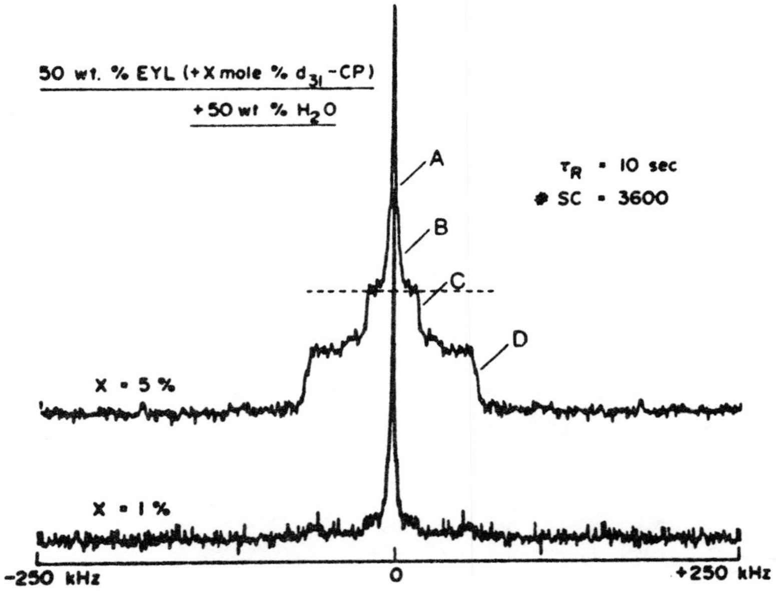

FIGURE 20. ²H-NMR spectra of multilamellar liposomes containing different amounts of cholesterol palmitate-d_{31}. (From Valic, M. I., Gorrisseu, H., Cushley, R. J., and Bloom, M., *Biochemistry*, 18, 854, 1979. With permission.)

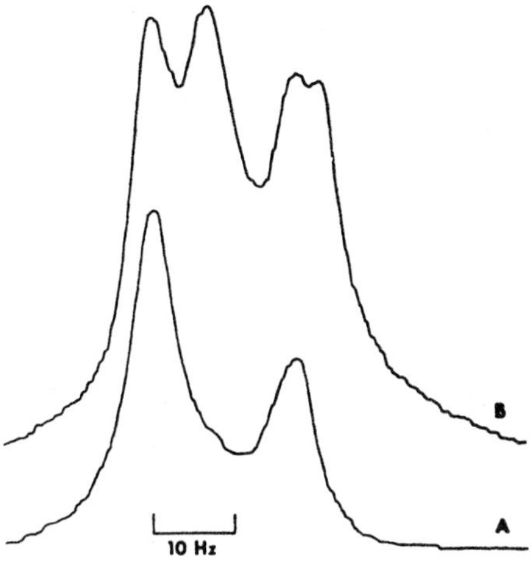

FIGURE 21. Carbonyl ^{13}C-NMR spectra of (A) B-DPPC, and (B) α,β-DPPC for small, single walled vesicles DPPC conc. ∿ 40m*M*, 42°C. (From Schmidt, C. F., Barenholz, Y., Huang, C., and Thompson, T. E., *Biochemistry*, 16, 3948, 1977. With permission.)

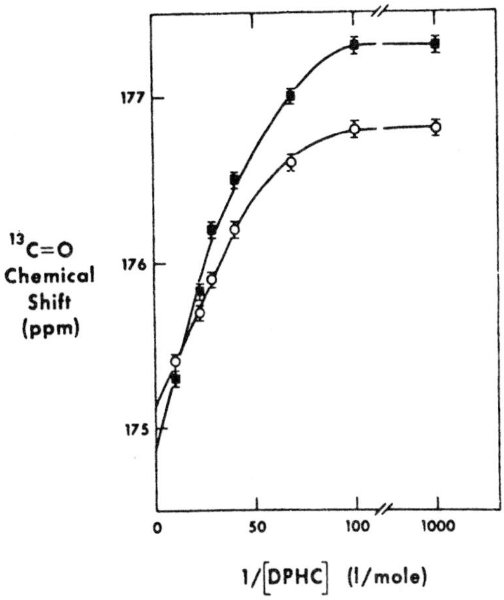

FIGURE 22. Carbonyl ^{13}C chemical shifts of α,β-DHPC as a function of concentration. (■) α carbonyls, (o) β carbonyls. (From Schmidt, C. F., Barenholz, Y., Huang, C., and Thompson, T. E., *Biochemistry*, 16, 3948, 1977. With permission.)

Such an effect could be possible because of the polarizability of the C = O bond and the charge on the head group. However, the chemical shifts of dipalmitoylphosphatidic acid in chloroform are the same as those for DPPC, and not significantly different

(0.1 ppm) from those of 1,2-sn-dipalmitoylglycerol. In addition, electric field effect shifts decrease as dielectric constant increases, and thus would be less expected in aqueous solution. Thus, the observed shifts appear to be caused by solvent-solute interactions. The results with α,β-DHPC, Figure 22, show that these effects are the result of molecular aggregation. There are two general mechanisms for aggregation shifts, solvent effects, and conformational effects. Changes in chain conformation about the C_2-C_3 axis, or the ester O-C bond might affect the carbonyl ^{13}C chemical shifts. Previous work[28,29] has shown that the three bond coupling constant T_{ccch} is sensitive to the C_2-C_3dihedral angle, as well as the two bond coupling constant J_{cch}. The J_{coch} coupling constant will reflect the ester O-C bond. When J_{cch} and J_{ccch} are measured, they are invariant between 10 m*M*, where the solution is all monomer, and 55 m*M*, where more than 70% of the molecules will be in the micellar form. J_{coch} coupling constants could not be determined in this study, but the ester linkage confers rigidity on the C−O bond[30] making conformational effects about this bond improbable as well. The two possible medium effects are hydration, and electric field effects due to interactions between carbonyls on adjacent molecules. Since the latter are sterically highly unlikely, it seems that hydration effects must account for the observed shifts. Taking the carbonyl ^{13}C shift in carbon tetrachloride as a reference for "unhydrated" carbonyls and the shift for DHPC monomers as "hydrated", it is possible to calculate the hydration fraction from the equation

$$f_{H_2O} = \frac{\delta A - \delta ccl_4}{\delta M - \delta ccl_4}$$

where δA is the observed chemical shift in the bilayer or micelle, δccl_4 is the shift in carbon tetrachloride, and δM is the shift in monomers. f_{H_2O} is thus the fraction of time a given carbonyl spends hydrogen bonded. Using this relationship it was shown that the hydration difference between carbonyls of molecules on the inside or the outside of single walled vesicles was 0.05. It was also shown that the carbonyl on the fatty acid on the 2-carbon of glycerol is more accessible to water than the carbonyl on the fatty acid on the 1-carbon of glycerol. This approach obviously has significant application into the question of the extent of water penetration into the bilayer and how this may change when membrane interactions occur.

Natural abundance ^{13}C-NMR has been used to obtain information concerning the mobility of the hydrocarbon chains in sphingomyelin.[31] When the temperature dependence of the ^{13}C-NMR spectra of sphingomyelin is examined, Figure 23, a distinct change with decreasing temperature is observed. The broad peak, next to the sharp N-methyl carbon peak undergoes significant broadening as the temperature is decreased, suggesting that a gelling of the sphingomyelin hydrocarbon chains is occurring.

Chain motion in the pentaglycine bridge of *Staphylococcus aureus* has been studied by ^{13}C-NMR, taking advantage of biological enrichment. The growth medium for the *S. aureus* was supplemented with $^{13}C_2$-glycine in place of $^{12}C_2$-glycine. The proton-decoupled spectra of cell walls, Figure 24, consists of three Lorentzian-shaped lines at 1.1, 43.0, and 43.7 ppm. The T_1, T_2, and NOE values and gated-decoupled relative intensities of these three ^{13}C glycyl resonances are given in Table 2. When the cell walls are subjected to partial autolysis by incubation at pH 9.0 at 37°C, the ^{13}C spectrum, Figure 25, shows a new resonance at 42.0 ppm, and a shift of 0.3 ppm downfield of the 41.1 ppm resonance. The spectral parameters obtained for these spectra are also given in Table 2. The cell wall resonances have been assigned to terminal and non-terminal glycyl groups by pH titration. The 43.0 and 43.7 ppm resonances are insensitive to pH, whereas the 41.1 ppm resonance shifts 3.4 ppm downfield at pH 11 (Figure 26). Such behavior is expected for an N-terminal glycyl residue.

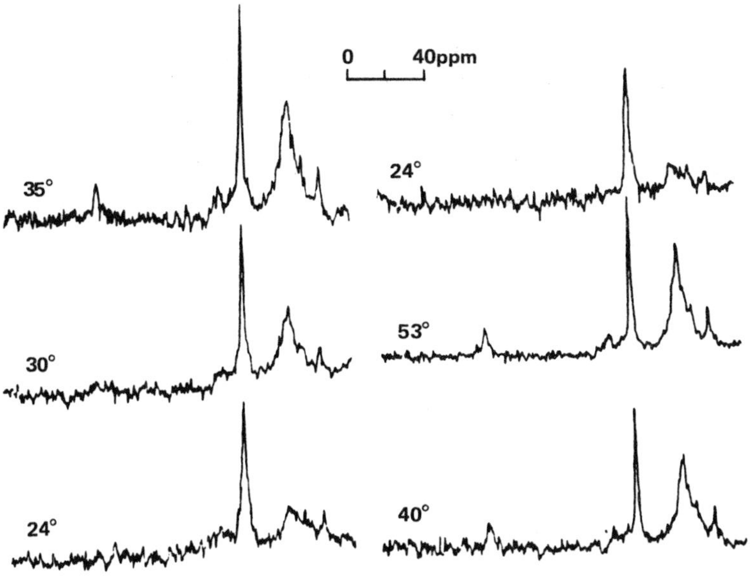

FIGURE 23. Temperature dependence of sphingomyelin ^{13}C-NMR spectra. The upper right hand spectrum is obtained from a return to 24°C from 53°C. (From Yeagle, P. L., Hutton, W. C., and Martin, R. B., *Biochemistry,* 17, 5745, 1978. With permission.)

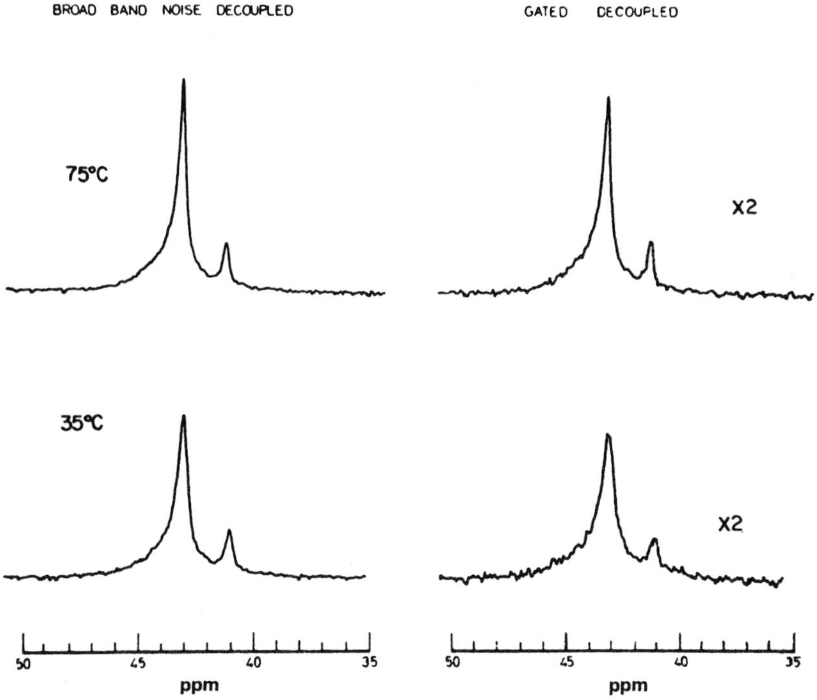

FIGURE 24. Proton broad-band noise-decoupled, and gate-decoupled ^{13}C-NMR spectra of $^{13}C_2$-Gly-labeled *S. aureus* cell walls at 75°C and 35°C. (From Lapidot, A. and Irving, C. S., *Biochemistry,* 18, 1788, 1979. With permission.)

Table 2
OBSERVED RELAXATION PARAMETERS OF PENTAGLYCINE CROSS-BRIDGE ^{13}C RESONANCES IN $^{13}C_2$-GLY-LABELED *S. AUREUS* CELL WALLS

Resonance	T(°K)	Rel. Area	T_1 (msec.)	T_2 (msec.)	NOE
41.1	308	1.0	203 ± 21	10.6	2.05
	328	1.0	446 ± 20	15.9	2.0
	348	1.0	(870)	14.5	1.6
43.0	308	4.5	142 ± 5	7.9	2.2
	328	4.8	202 ± 4	12.7	2.12
	348	4.4	360 ± 10	11.8	2.1
43.7	308	6.1	96 ± 2	2.1	1.25
	328	6.2	101 ± 8	2.1	1.79
	348	7.2	125 ± 6	2.1	1.80
Partially Autolyzed Cell Walls					
41.4	308	1.0	444 ± 43	32	2.4
42.0	308	5.1	2405 ± 196	42	2.5
43.0	308	8.1	174 ± 7	23	2.3
43.7	308	1.0		22	3.0

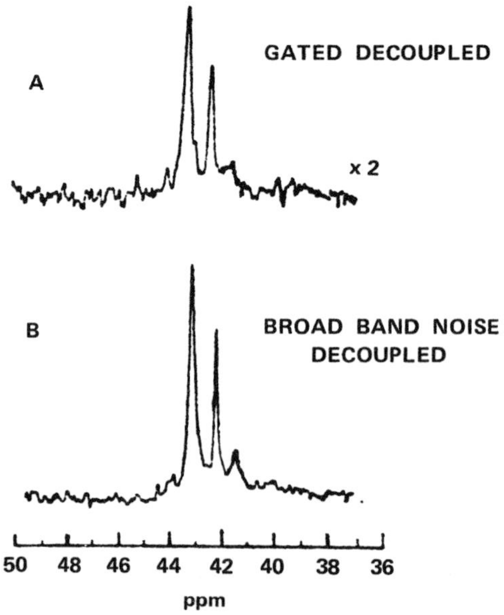

FIGURE 25. Decoupled ^{13}C-NMR spectra of partially autolysed $^{13}C_2$-Gly-labeled *S. aureus* cell walls at 29°C, pH 7.0. (From Lapidot, A. and Irving, C. S., *Biochemistry*, 18, 1788, 1979. With permission.)

The very fact that relatively intense ^{13}C resonances are seen, shows that at least a fraction of the pentaglycine bridges undergo rapid segmental motion. If the pentaglycine chains experienced isotropic overall rotation, a single isotropic reorientation correlation time T_r would be obtained from T_1, T_2, or NOE values of a single resonance at fixed temperature. This is not the case in this instance. The distribution of correla-

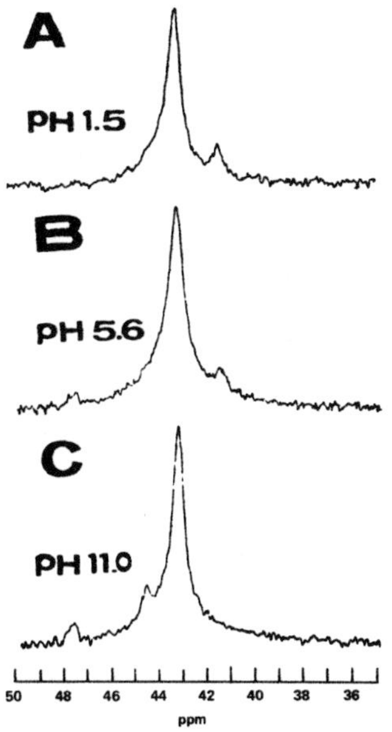

FIGURE 26. Effects of pH on proton-decoupled ^{13}C-NMR spectra at 29°C of normal $^{13}C_2$-Gly-labeled *S. aureus* cell walls. (From Lapidot, A. and Irving, C. S., *Biochemistry,* 18, 1788, 1979. With permission.)

tion times in the backbone motions of polymer chains can be described by the log-χ^2 distribution introduced by Schaefer.[33] In this distibution, a logarithmic time scale is defined by

$$S = \log_b [1 + (b-1)\, T/\tilde{T}]$$

where b is an adjustable parameter, usually taken to be 1000. The distribution function is given by:

$$H(s) = \frac{1}{T(P)} (PS)^{P-1} e^{-PS} p$$

where p is the width parameter and T(p) is a normalization factor equal to the T function of P. The width parameter is a measure of the number of degrees of motional freedom: when p > 100, the function is equivalent to a single correlation time. The spectral density function for the log-χ^2 distribution is

$$J(w) = \int^{\infty} \frac{\tilde{\Gamma}\, H(s)\, (\exp_b S-1)}{(b-1)\, \{1 + W^2 \tilde{\Gamma}^2 [(\exp_b S-1)/(b-1)]^2\}}\, ds$$

If the model applies, three independent values of T_1, T_2, and NOE can be accounted for by the adjustable parameters p and $\tilde{\Gamma}$. Values of T_1, T_2, and NOE at 308°K of 203 msec, 10.6 msec, and 2.05, and the values of 328°K of 388 msec, 16.4 msec, and

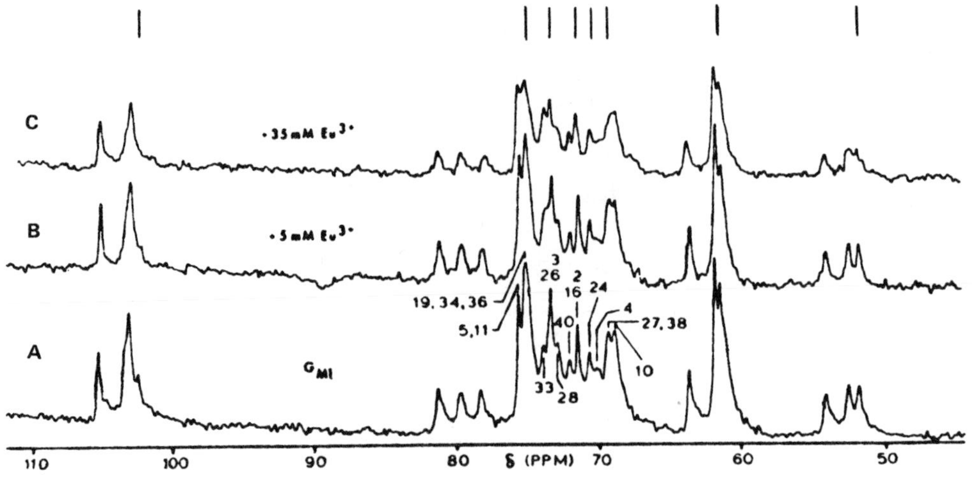

FIGURE 27. Proton noise decoupled ^{13}C-NMR spectra of GM_1 (86 m*M* @ 42°C) in the absence of europium, and in the presence of either 5 m*M* $EuCl_3$ or 35 m*M* $EuCl_3$. Peaks altered by Eu^{3+} are indicated at the top of the figure. (From Sillerud, L. O., Prestegard, J. H., Yu, R. K., Schafer, D. E., and Konigsberg, W. H., *Biochemistry*, 17, 2619, 1978. With permission.)

1.93 calculated using this model for the 41.1 ppm resonance can be compared with the experimental values of Table 2. This suggests that a model involving varying segmental motion may be applicable to this system. When this type of analysis is applied to all three cell wall resonances, it is found that the 43.0 ppm and the 41.1 ppm resonances, which are due to nonterminal glycyl residues and N-terminal glycyl residues, respectively, have similar spectral parameters indicating that they may arise from similar types or regions of the pentaglycine chains. The 43.7 ppm resonance, however, appears to undergo much slower segmental motions (Γ is much longer). Models for the packing of cell wall polymers must be consistent with motional analysis by ^{13}C-NMR. For bacterial cell wall peptidoglycan, tight packing of the cell wall polymers is not consistent with T_1 values for resonances representing significant amounts of the cell wall polymer. It might also be noted that the existence of intramolecular hydrogen bonding is inconsistent with the observation that guanidine hydrochloride does not affect the nature or rate of the bridge chain segmental motions as judged by ^{13}C-NMR.

Solution studies using ^{13}C-NMR have provided information on the linkage patterns of oligosaccharides[34] and the conformation of glycosidic bonds,[35] as well as the nature of metal binding sites.[36] These types of information can also be obtained for glycolipids in micelles, for example. For the maximum amount of information to be obtained from such studies, complete assignments of the resonances are vital. The paramagnetic shift probe europium (III) has been used to investigate GM_1 groups in micelles in terms of the groups involved in metal binding.[37] When europium binds to GM_1 the ^{13}C-NMR spectrum shows many altered resonances, Figure 27. The europium shifts some resonances, and broadens some resonances. Using established assignments of the resonances, it is obvious that europium affects more than just the terminal sialic acid residue, indicating that other sugar residues in the ganglioside make up parts of the metal binding site. A structure that is consistent with the observed effects of europium on the micellar GM_1 ^{13}C-NMR resonances is shown in Figure 28. As is seen, moieties on the β-D-galactopyranoside and 2-acetamido-2-deoxy-β-D-galactopyranoside rings are in the binding site. These additional binding residues probably account for the enhanced affinity of GM_1 over sialic acid.

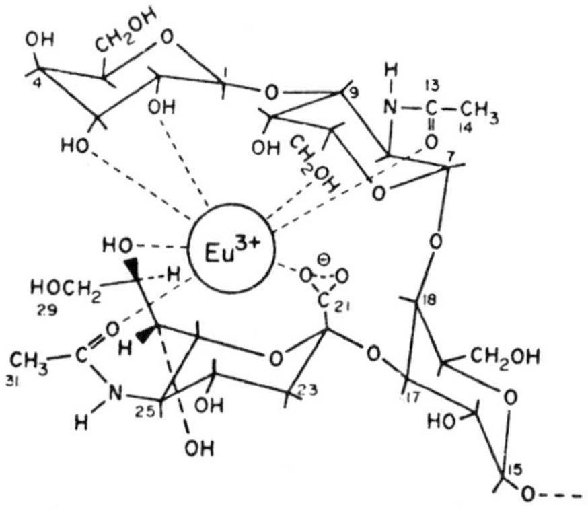

FIGURE 28. Proposed structure of the cation-binding site on GM1 based on the spectral perturbations of europium in the ^{13}C-NMR spectrum. (From Sillerud, L. O., Prestegard, J. H., Yu, R. K., Schafer, D. E., and Konigsberg, W. H., *Biochemistry,* 17, 2619, 1978. With permission.)

4. ^{15}N-NMR

As with the ^{13}C-NMR studies discussed above, ^{15}N-NMR studies are facilitated where enrichment of ^{15}N is possible. This has been particularly easy with bacterial systems, and as a result much work has been done on cell wall structure in bacteria such as *Bacillus subtilis.* There are several technical features of ^{15}N-NMR that must be taken into account, however, in studies on mobility of ^{15}N-containing compounds, and the rationale behind the interpretation of such studies is illustrated by the recent work of Lapidot and Irving.[38] In most instances, when a proton-decoupled resonance is not observed, it is assumed that it has been broadened beyond detection by dipolar interactions that result from low mobility. Narrow resonances can however, in ^{15}N-NMR be lost from proton decoupled spectra due to partial quenching of nuclear Overhauser enhancement by factors such as paramagnetic ions and not mobility. In such cases, these resonances would be observed in gated-decoupled spectra, where proton-decoupling occurs only during spectral accumulation, but not in broad-band, noise-decoupled spectra. Thus, with ^{15}N-NMR, it is necessary to apply both modes of decoupling when examining resonances.

Figure 29 shows a series of spectra obtained for *B. subtilis* intact cells, isolated cell walls, or lysozyme digests of cell walls. With intact cells two main regions of resonances are seen in gated-decoupled spectra, a broad region (250 to 270 ppm) containing the amide resonances, and the 340 to 350 ppm region that contains resonances from techoic acid (335.5 ppm), terminal NH_3^+ groups (337.8 ppm), and protein Lys-N_W (354.6 ppm). Proton broad-band noise decoupling (resulting in nuclear Overhauser effects nulling of many of the resonances in this region) leaves well-resolved resonances at 253.4 ppm and 266.8 ppm. Both resonances also occur in broad-band noise decoupled spectra of isolated cell walls, suggesting that they result from cell wall components. The change in their intensities probably is a result of a change in their mobility, which affects the NOE factor. The 253.4 ppm resonance is easily assigned since it is abolished by specific isotope dilution with *N*-acetylglucosamine derivatives. The 266.8 ppm resonance appears to be from the NH group of amidated D-glucose and meso-diaminopimelic acid, since this resonance is not found in ^{15}N spectra of *E. coli* whose peptidoglycan is identical to that of *B. subtilis,* with the exception of amidated

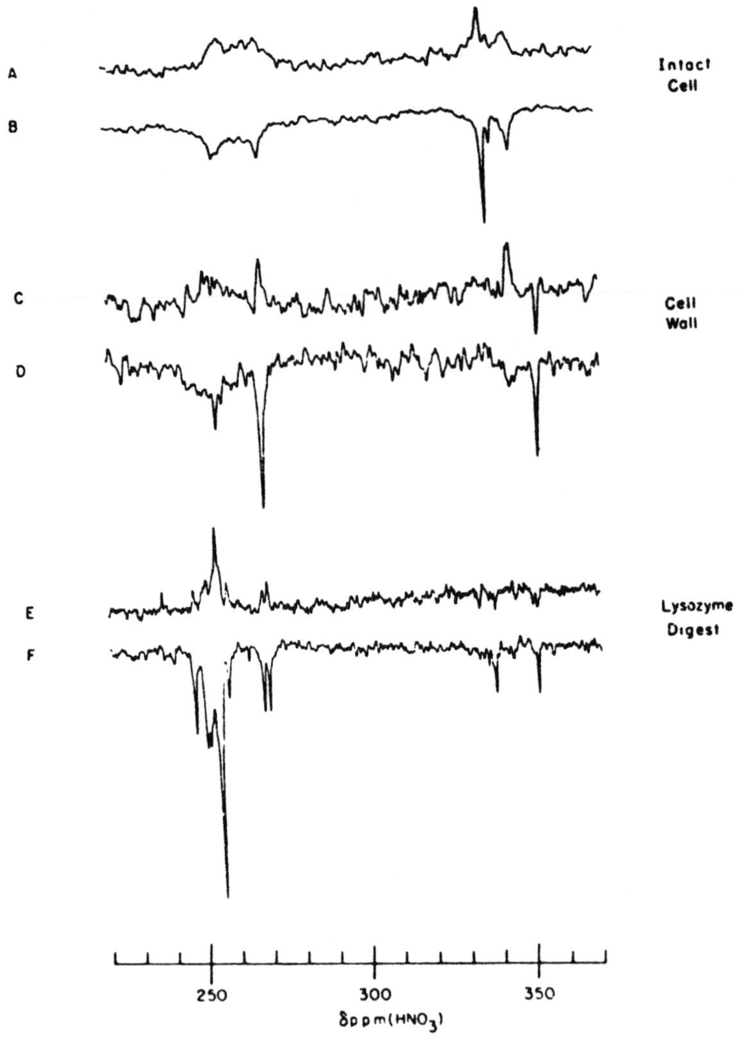

FIGURE 29. Gated-decoupled (A,C,E) and broad-band noise-decoupled (B,C,F) ^{15}N-NMR spectra of *B. subtilis.* (From Lapidot, A. and Irving, C. S., *Biochemistry,* 18, 704, 1979. With permission.)

groups.[39] When the cell wall is subjected to lysozymal digestion,a number of new resonances appear. While some of these resonances were assigned to sugar residue acetamido groups, pH studies were required to identify resonances of peptide groups adjacent to free carboxyl groups. This is facilitated when the acetamido resonances are diluted by growth on ^{14}C-*N*-acetylglucosamine. When the pH is progressively lowered from pH 11 (Figure 30), the 246.1, 249.4, and 250.3 ppm resonances merge into a single resonance at 250 ppm, characteristic of alamine, suggesting that these three resonances arose from different alanine peptide bands in the cell wall structure. The remaining resonances were assigned to sugar residues. With the assignments of the resonances, it is possible to examine the intact cell resonances in terms of mobility. In the intact cell, none of the resonances can be assigned to either the stem region of the peptide, or the glycan strands. All the observable resonances arise from the cross-bar or bridge regions of the peptide. Thus it would appear that the glycan strands and the peptide stems are held rigidly, while there is considerable mobility in the peptide cross-

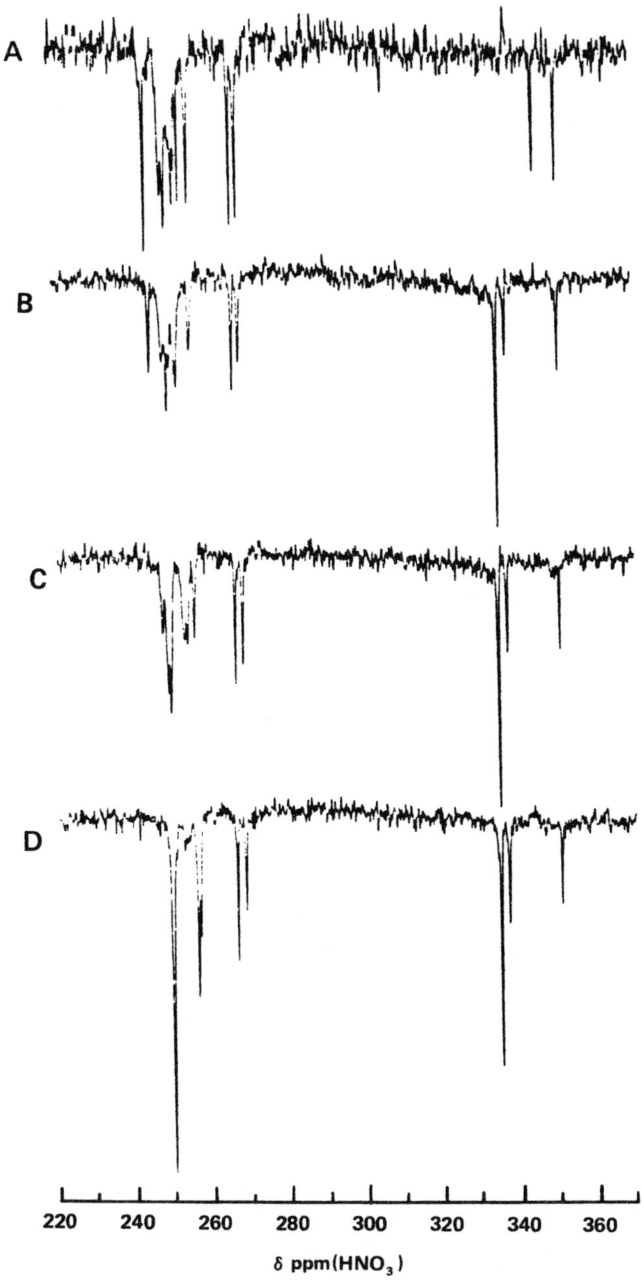

FIGURE 30. Proton broad-band noise-decoupled ^{15}N-NMR spectra of ^{15}N-NMR spectra of ^{15}N, ^{14}N BlcNAc-labeled *B. subtilis* cell wall lysozyme digest. The effects of pH. (From Lapidot, A. and Irving, C. S., *Biochemistry*, 18, 704, 1979. With permission.)

bar and bridge regions. It might be noted that similar conclusions were reached concerning motion of the pentaglycine bridge in *S. aureus* using ^{15}N-NMR,[33] and ^{13}C-NMR discussed earlier. With the assignments of the resonances in such a system, ^{15}N-NMR offers considerable potential in examining the effects of drugs such as vancomycin on the cell walls.

5. ^{19}F-NMR

Fluorine NMR has a number of distinct advantages over the other nuclei discussed in this chapter. While all NMR approaches suffer from inherent low sensitivity, ^{19}F-NMR is better than most other nuclei, showing a sensitivity almost equal to that of protons. While the problem of overlapping resonances due to proteins and other membrane components can be overcome with ^{2}H, and ^{13}C-NMR by specific isotope enrichment, ^{19}F-NMR does not have this complication, as ^{19}F is not normally present in biological systems. ^{19}F can be used to label fatty acids in the form of a difluoromethylene group, which in many instances can be incorporated biosynthetically. Since the CF_2 group is very similar to the methylene group in terms of its size, geometry, and physical properties, perturbation of the membrane by its introduction would be expected to be minimal. ^{19}F-NMR thus offers the advantages of minimal perturbation combined with a single resonance of high sensitivity, which can be incorporated in a specific site into a membrane.

Just as with the quadrupole splitting of ^{2}H nuclei (see earlier) or the nonaveraged chemical shift anisotropy (CSA) of the ^{31}P nucleus (see later) an order parameter, which is related to the linewidth of the resonance, can be defined for the motion of the region of the phospholipid to which ^{19}F is attached. However, with ^{19}F both dipolar interactions and CSA interactions are involved. When these effects can be dissected, even more information can be obtained regarding the dynamics of ^{19}F motion. The orientation of the acyl chain of the phospholipid will influence the width of the resonance if incompletely averaged dipolar interactions significantly contribute to the broadening of the resonance. Non-averaged CSA interactions can in turn cause a shift in resonance frequency that will also depend on the orientation of the chain axis. However, it has been shown theoretically[40] that at a given orientation, the CSA related shift is proportional to the dipolar broadening. It has also been shown theoretically that ^{19}F-NMR line shapes are very different for frozen and fluid lipid regions. In the gel phase where there are less diffusional motions, non-averaged intermolecular interactions contribute to the line shape in addition to the broadening caused by decreased motion. In the fluid state a much narrower resonance is seen. Because of the complexity of interactions causing line broadening in ^{19}F-NMR, maximum information is realized by spectral simulation. Because of the very different shapes of ^{19}F resonances in the gel and fluid states, it is very easy to follow each phase, even when both exist simultaneously.

When ^{19}F-NMR spectra of *E. coli* grown on 8,8-difluoromyristate are studied as a function of temperature (Figure 31) it is apparent that at low temperatures, 18 and 23°C there is a broad, but quite distinct ^{19}F resonance. Between 25 and 35°C, a sharper resonance appears, which above 35°C becomes even sharper. The broad resonance however persists at temperatures below the phase transition temperature. When the temperature effects on the narrow resonance are studied (Figure 32), it is apparent that in whole cells, or isolated *E. coli* membrane vesicles the narrow resonance persists even at low temperatures. This is not seen with extracted lipids or purified phospholipids. When the linewidth of the central narrow resonance is studied as a function of temperature (Figure 33), clear differences between cell membrane and purified phospholipids can be seen. These results can be interpreted in terms of a two-stage melting process, with superimposed effects of membrane proteins on the fluidity of the lipids. Real fluidity of the lipids does not occur till the high-temperature end of the phase transition. At the low temperature end of the transition, the membrane proteins are segregated into the fluid regions that remain, despite the fact that the nominally frozen lipid experiences some motion (the broad resonance that slowly is broadened out of the spectrum). Even as the membrane is cooled further however, a mobile lipid region

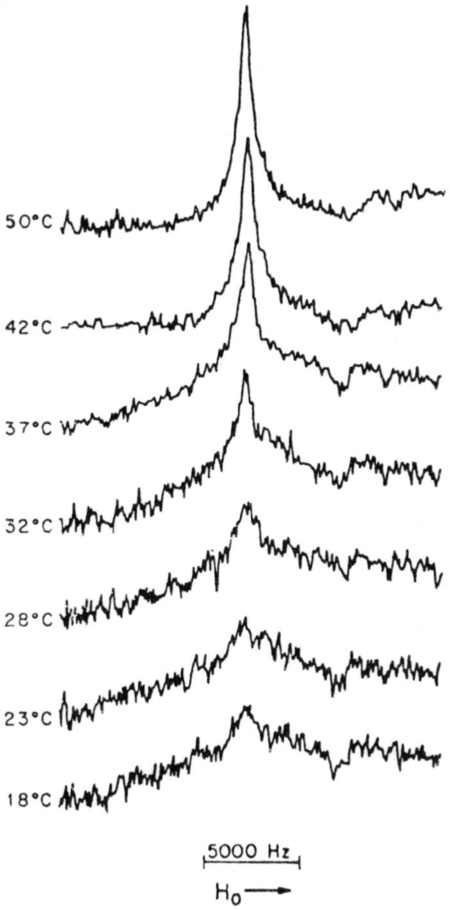

FIGURE 31. ^{19}F-NMR spectra of *E. coli* labeled with 8,8-difluoromyristate as a function of temperature. (From Gent, M. P. N. and Ho, C., *Biochemistry*, 17, 3023, 1978. With permission.)

remains associated with the membrane protein (the anomalous behavior of the narrow peak intensity as temperature is decreased.)

6. ^{31}P-NMR

Since ^{31}P-NMR spectra are dominated by chemical shift anisotropy effects, the complete interpretation of ^{31}P spectra of phospholipids requires a knowledge of the chemical shift tensor, and its orientation. To this end, many studies with ^{31}P-NMR have been directed at establishing values for the elements of the chemical shielding tensors for model compounds, and then applying these values to the interpretation of spectra from lamellar phospholipid dispersions or native membranes. Kohler and Klein[41] have reported values for these shielding tensors determined from ^{31}P-NMR spectra of phospholipid powders. Figure 34 shows the proton-decoupled ^{31}P-NMR spectra of phospholipid powders of phosphatidylcholine (PC), phosphatidylethanolamine (PE), phosphatidylserine (PS), and phosphatidic acid (PA). The three diesters (PC, PE, and PS) all have similar spectra, while the monoester (PA) is significantly different, which would be expected since in PA the difference from the other three molecules arises at

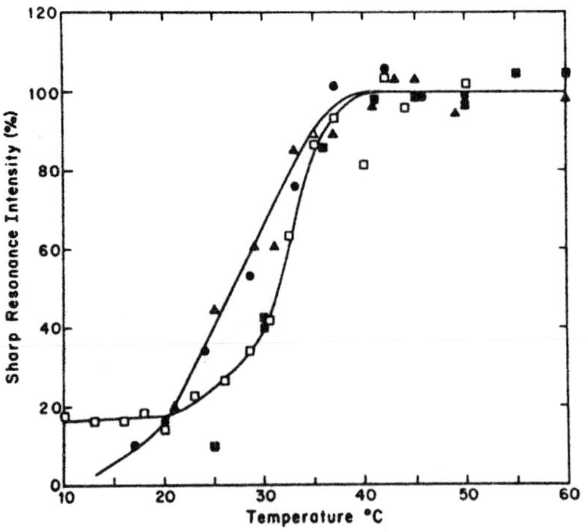

FIGURE 32. Temperature dependence of the narrow ^{19}F-NMR resonance seen in whole cells (□), membrane vesicles (■), total lipids extracted (•), or purified phospholipids (▲). (From Gent, M. P. N. and Ho, C., *Biochemistry*, 17, 3023, 1978. With permission.)

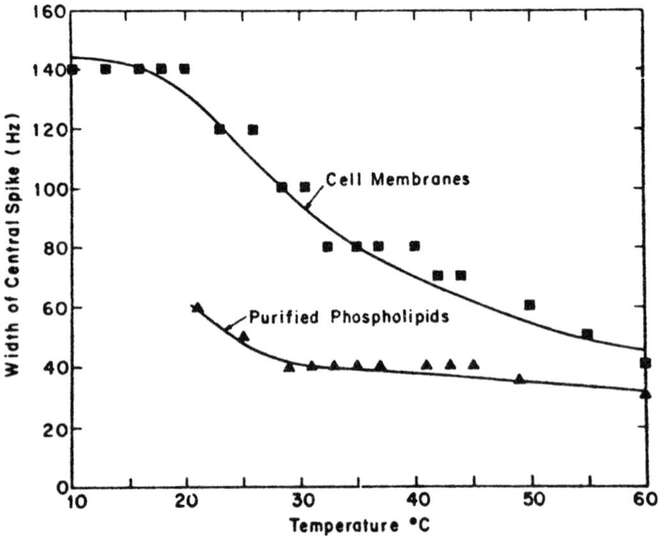

FIGURE 33. Linewidth of the narrow resonance of ^{19}F-NMR spectra of cell membranes (■) or purified lipids (▲) as a function of temperature. (From Gent, M. P. N. and Ho, C., *Biochemistry*, 17, 3023, 1978. With permission.)

the second atom removed from the phosphorus, rather than at the third, which is the case with PC, PE, and PS. Analysis shows that the shielding tensors are all nonaxial in these cases.

When spectra of unsonicated lamellar dispersions of dipalmitoyl-PE, dipalmitoyl-PC, egg-PC, and brain-PS are observed (Figure 35) each of the spectra show a collapse to a pattern characteristic of axial symmetry of the shielding tensors. This effect results from averaging of the shielding tensors due to phosphate head group motion. When

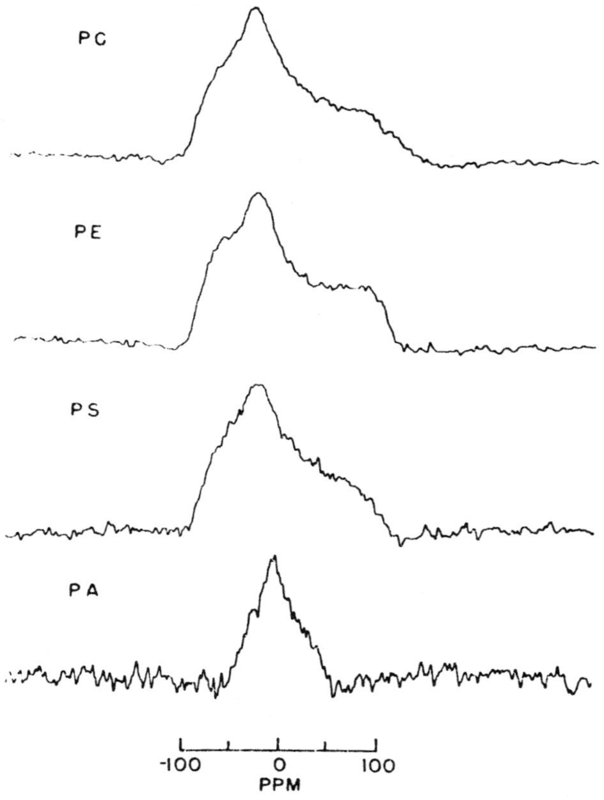

FIGURE 34. ^{31}P-NMR spectra of phospholipid powders obtained using proton decoupling. (From Kohler, S. J. and Klein, M. P., *Biochemistry*, 16, 519, 1977. With permission.)

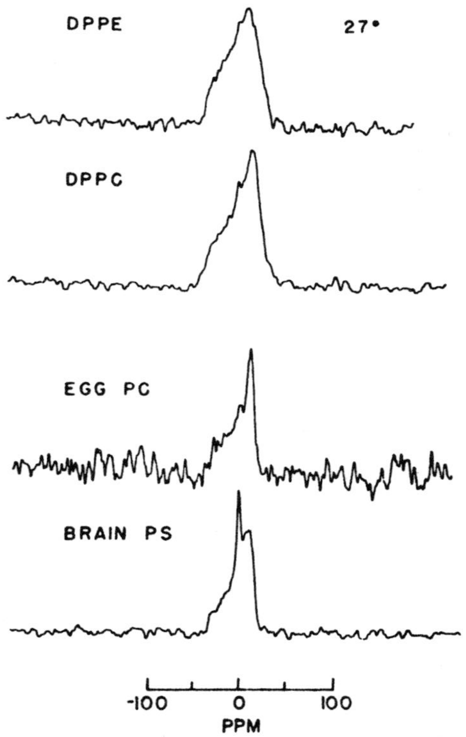

FIGURE 35. ^{31}P-NMR spectra of lamellar dispersions of phospholipids obtained using proton decoupling. (From Kohler, S. J. and Klein, M. P., *Biochemistry*, 16, 519, 1977. With permission.)

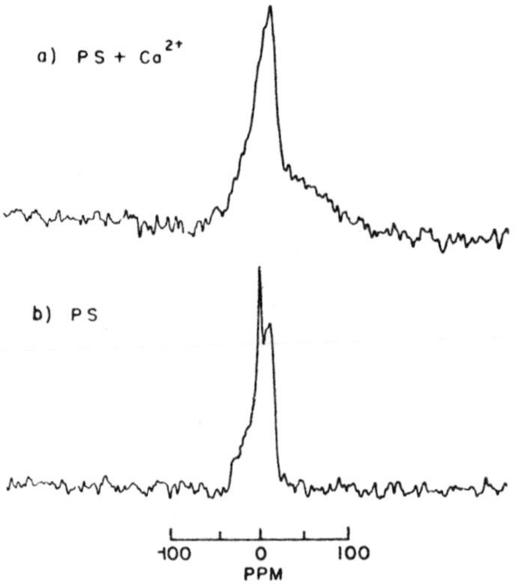

FIGURE 36. ^{31}P-NMR spectra of phosphatidylserine in the presence and absence of Ca.$^{2+}$ (From Kohler, S. J. and Klein, M. P., *Biochemistry*, 16, 519, 1977. With permission.)

the residual anisotropies for DPPC and DPPE are studied as a function of temperature, it is found that there is a sharp discontinuity, towards decreased anisotropy, as the sample passes through its transition temperature. Although the residual anisotropies represent head group motion, it is apparent that they undergo transitions at temperatures which are usually associated with melting of the hydrocarbon tails. Egg-PC and brain-PS also undergo temperature transitions detected in terms of the residual anisotropy, however, these transitions occur over much wider temperature ranges, reflecting the fact that these "natural" phospholipids contain a mixture of saturated and unsaturated chains. When calcium is added to PS (Figure 36) the resonance resembles a combination of the typical lamella resonance (b) superimposed on a much broader resonance. Since the peak width of the broad resonance is similar to that of PS in a rigid lattice, the spectrum (a) suggests that in the presence of calcium there is a significant proportion of immobilized phospholipid head groups. Spectra obtained for chick embryo fibroblasts, Figure 37, can be interpreted in terms of a narrow resonance at 0 ppm, relative to H_3PO_4 and a broader resonance. The narrow resonance is presumed to be due to intracellular phosphate while the broader resonance represents, presumably, membrane phospholipid.

When the spectra are analyzed in terms of the shielding tensors, using the orientation of phosphate in crystalline dilauroyl-PE[42] to relate the rotation axes to the shielding tensor reference frame, it is found that the spectra are consistent with a model that includes rotations about the P−O (glycerol) bond and the C1−C2 bond, as well as the ability of the molecule to "wobble" about the bilayer normal. Increased temperature causes the "wobble" to increase with resultant narrowing of the resonances.

Similar studies with dipalmitoyl-PC bilayers[43] came to similar conclusions regarding the shielding tensors, but extended the study to include the effects of orienting the bilayers with respect to H_o on the chemical shifts. When proton-decoupled ^{31}P-NMR spectra are observed with the bilayer normal parallel to H_o (Figure 38a), a single broad resonance is observed. When however, the bilayer normal is tilted at an angle of 50° with respect to H_o (Figure 38b), two distinct resonances are seen. Barium diethyl phosphate (Figure 38c) is also included as a mobility reference. From previous studies of

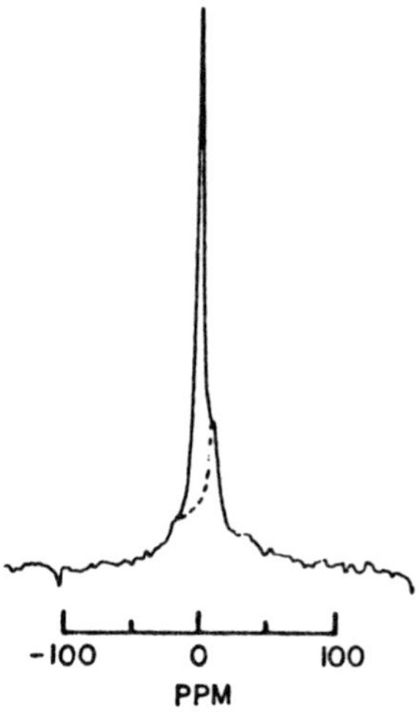

FIGURE 37. ^{31}P-NMR spectrum of chick embryo fibroblasts. The narrow resonance centered at 0 ppm is probably due to intracellular phosphate. The broader resonance (shown by the dashed line) is probably due to membrane phospholipids. (From Kohler, S. J. and Klein, M. P., *Biochemistry,* 16, 519, 1977. With permission.)

powdered samples it was determined[44] that the principle values of the chemical shift tensors $_{11}$ (which is approximately perpendicular to the O-P-O plane) and σ_{22} (which bisects the O-P-O angle) and σ_{33} are −81, −25, and +110 ppm, respectively. Hence, chemical shifts of less than −81 ppm or greater than 110 ppm are not allowed. In addition, if during orientation of the bilayer with respect to H_o a shift of −81 ppm or +110 ppm is observed, then $_{11}$ or $_{33}$ is along H_o. The observation in Figure 38b that when the bilayer is tilted at 50° with respect to H_o a shift of −80 is observed indicates that at this point a significant fraction of the molecules have $_{11}$ parallel to H_o. Since this tensor element is approximately perpendicular to the O-P-O plane (where O's are the non-esterified oxygens in the phosphodiester), it is apparent that this plane is tilted at approximately 50° with respect to the bilayer normal. This orientation of the phosphate probably results in the choline moiety being extended parallel to the bilayer plane. In conjunction with model studies as descibed here, ^{31}P-NMR is of great potential in assessing membrane structure.

Although it does not strictly fall into the category of membrane studies, it is worth mentioning in this section that ^{31}P-NMR has become a powerful tool in the study of intracellular phosphate-containing metabolites. Since this approach was first introduced,[45] high resolution ^{31}NMR has been used to study metabolic processes in a variety of cell tissues, cellular organelles, and even perfused organs. Once resonances can be assigned, concentrations of metabolites may be measured from peak intensities, as well as the state of ionization states of various phosphate compounds determined from chemical shift data. In this way, the internal pH can be determined. A recent study[46] has demonstrated that internal and external phosphates in rat mitochondria can be distinguished due to chemical shifts induced by Mg^{2+}. Figure 39 shows oxidative phosphorylation of external ADP by mitochondria at 0°C. The difference spectrum (B) is

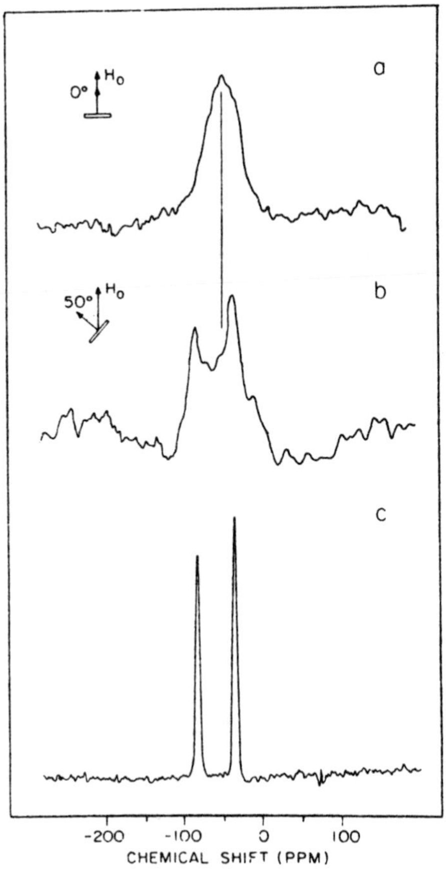

FIGURE 38. Proton-decoupled ^{31}P-NMR spectra illustrating the angular dependence of the ^{31}P chemical shifts. (From Griffin, R. G., Powers, L., and Pershan, P. S., *Biochemistry,* 17, 2718, 1978. With permission.)

generated from the spectrum after oxygenation with 5 m*M* succinate and H_2O_2 (Spectrum C) and the spectrum before oxygenation (Spectrum A). Internal phosphorylation in aerobic mitochondria has also been studied Figure 40. From these studies internal and external phosphate potentials can be estimated.

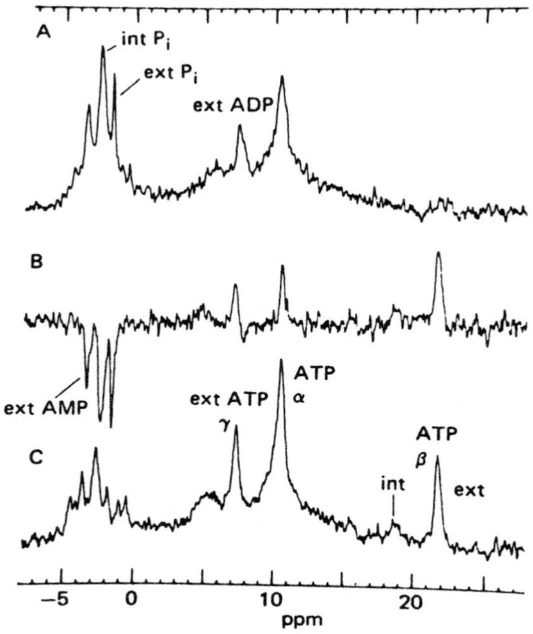

FIGURE 39. Oxidative phosphorylation of external ADP by mitochondria at 0°C. Mitochondria are suspended at 120 mg/m*ℓ* protein in 0.25 *M* Sucrose containing 2.4 m*M* adenine dinucleotide phosphate. (From Ogawa, S., Rottenberg, H., Brown, T. R., Shulman, R. G., Castillo, C. L., and Glynn, P., *Proc. Natl. Acad. Sci. U.S.A.*, 75, 1796, 1978. With permission.)

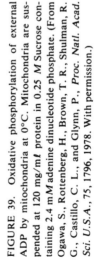

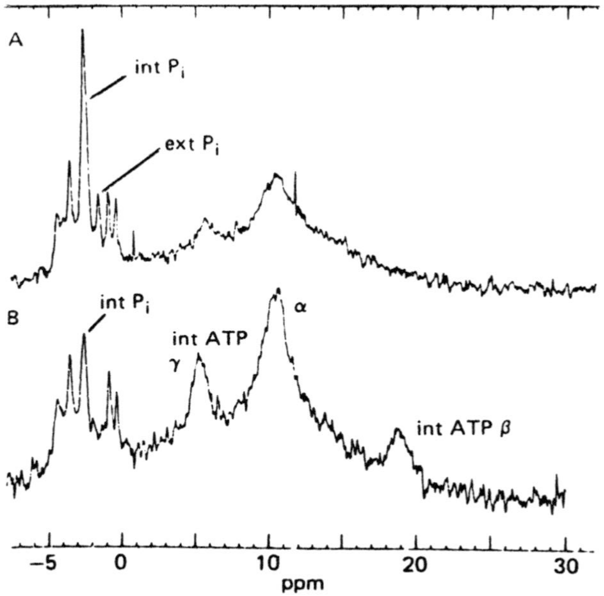

FIGURE 40. Internal phosphorylation in aerobic mitochondria at 0°C. A, Anaerobic state, B, Aerobic state. (From Ogawa, S., Rottenberg, H., Brown, T. R., Shulman, R. G., Castillo, C. L., and Glynn, P., *Proc. Natl. Acad. Sci. U.S.A.*, 75, 1796, 1978. With permission.)

REFERENCES

1. **Marsh, D. and Barrantes, F. J.**, Immobilized lipid in acetylcholine receptor-rich membranes from Torpedo marmorata, *Proc. Natl. Acad. Sci. U.S.A.*, 75, 4329, 1978.
2. **Knowles, P. F., Marsh, D., and Rattle, H. W. E.**, *Magnetic Resonance of Biomolecules*, John Wiley & Sons, New York, 1976, chap. 6.
3. **Dahl, C. E. and Levine, R. P.**, Electron spin resonance studies on the interaction of complement proteins with erythrocyte membranes, *Proc. Natl. Acad. Sci. U.S.A.*, 75, 4930, 1978.
4. **Butterfield, D. A.**, Electron spin resonance studies of erythrocyte membranes in muscular dystrophy, *Acc. Chem. Res.*, 10, 111, 1977.
5. **Butterfield, D. A., Chesnut, D. B., Appel, S. H., Roses, A. D.**, Spin label study of erythrocyte membrane fluidity in myotonic and Duchenne muscular dystrophy and congenital myotonia, *Nature (London)*, 263, 159, 1976.
6. **Hyde, J. S. and Dalton, L.**, Very slowly tumbling spin labels: adiabatic rapid passage, *Chem. Phys. Lett.*, 16, 568, 1972.
7. **Favre, E., Baroin, A., Bienvenue, A., and Devaux, P. F.**, Spin-label studies of lipid-protein interactions in retinal rod outer segment membranes. Fluidity of the boundary layer, *Biochemistry*, 18, 1156, 1979.
8. **Baroin, A., Bienvenue, A., and Devaux, P. F.**, Spin-label studies of protein-protein interactions in retinal rod outer segment membranes. Saturation transfer electron paramagnetic resonance spectroscopy, *Biochemistry*, 18, 1151, 1979.
9. **Wilkerson, L. S., Perkins, R. C., Roelofs, R., Swift, L., Dalton, L. R., and Park, J. H.**, Erythrocyte membrane abnormalities in Duchenne muscular dystrophy monitored by saturation transfer electron paramagnetic resonance spectroscopy, *Proc. Natl. Acad. Sci. U.S.A.*, 75, 838, 1978.
10. **Sheats, J. R. and McConnell, H. M.**, A photochemical technique for measuring lateral diffusion of spin-labeled phospholipids in membranes, *Proc. Natl. Acad. Sci. U.S.A.*, 75, 4661, 1978.
11. **Smith, B. A. and McConnell, H. M.**, Determination of molecular motion in membranes using periodic pattern photobleaching, *Proc. Natl. Acad. Sci. U.S.A.*, 75, 2759, 1978.
12. **Brown, M. F., Miljanich, G. P., and Dratz, E. A.**, Interpretation of 100- and 360-MHz proton magnetic resonance spectra of retinal rod outer segment disk membranes, *Biochemistry*, 16, 2640, 1977.
13. **Brown, M. F., Miljanich, G. P., Franklin, L. K., and Dratz, E. A.**, H-NMR studies of protein-lipid interactions in retinal rod outer segment disc membranes, *FEBS Lett.*, 70, 56, 1976.
14. **McLaughlin, A. C., Podo, F., and Blasie, J. K.**, Temperature and frequency dependence of longitudinal proton relaxation times in sonicated lecithin dispersions, *Biochim. Biophys. Acta*, 330, 109, 1973.
15. **Bloom, M., Burnell, E. E., Mackay, A. L., Nichol, C. P., Valic, M. I., and Weeks, G.**, Fatty acyl chain order in lecithin model membranes determined from proton magnetic resonance, *Biochemistry*, 17, 5750, 1978.
16. **Wennerstrom, H. and Lindblom, G. A.**, Biological and model membranes studied by nuclear magnetic resonance of spin one half nuclei, *Q. Rev. Biophys.*, 10, 67, 1977.
17. **Stejskal, E. O. and Tanner, J. E.**, Spin diffusion measurements: spin echoes in the presence of a time-dependent field gradient, *J. Chem. Phys.*, 42, 288, 1965.
18. **Kuo, A.-L. and Wade, C. G.**, Lipid lateral diffusion by pulsed nuclear magnetic resonance, *Biochemistry*, 18, 2300, 1979.
19. **Abragam, A.**, *The Principles of Nuclear Magnetism*, Clarendon Press, Oxford, 1961.
20. **Oldfield, E., Meadows, M., Rice, D., and Jacobs, R.**, Spectroscopic studies of specifically deuterium labeled membrane systems. (NMR investigation of the effects of cholesterol in model systems, *Biochemistry*, 17, 2727, 1978.
21. **Oldfield, E., Gilmore, R., Glaser, M., Gutowsky, H. S., Hsung, J. C., Kang, S. Y., King, T. E., Meadows, M., and Rice, D.**, Deuterium nuclear magnetic resonance investigation of the effects of proteins and polypeptides on hydrocarbon chain order in model membrane systems, *Proc. Natl. Acad. Sci. U.S.A.*, 75, 4657, 1978.
22. **Davis, J. H., Jeffrey, K. R., Bloom, M., Valic, M. I., and Higgs, T. P.**, Quadrupolar echo deuteron magnetic resonance spectroscopy in ordered hydrocarbon chains, *Chem. Phys. Lett.*, 42, 390, 1976.
23. **Jost, P. C., Griffith, O. H., Capaldi, R. A., and Vanderkooi, G.**, Evidence for boundary lipid in membrane, *Proc. Natl. Acad. Sci. U.S.A.*, 70, 480, 1973.
24. **Valic, M. I., Gorrissen, H., Cushley, R. J., and Bloom, M.**, Deuterium magnetic resonance study of cholesterol esters in membranes, *Biochemistry*, 18, 854, 1979.
25. **Schmidt, C. F., Barenholz, Y., Huang, C., and Thompson, T. E.**, Phosphatidylcholine ^{13}C-labeled carbonyls as a probe of bilayer structure, *Biochemistry*, 16, 3948, 1977.

26. **Tanford, C.,** *The Hydrophobic Effect,* John Wiley & Sons, New York, 1973.

27. **Haque, R., Tinsley, I. J., and Schmedding, D.,** Nuclear magnetic resonance studies of phospholipid micelles, *J. Biol. Chem.,* 247, 157, 1972.

28. **Karabatsos, G. J., Orzech, C. E., and Hsi, N.,** Proton-carbon-13, spin-spin coupling. VII, the relative magnitudes of trans and gauche J13cccH, *J. Am. Chem. Soc.,* 88, 1817, 1966.

29. **Schwarcz, J. A. and Perlin, A. S.,** Orientational dependance of vicinal and geminal ^{13}C-^{1}H coupling, *Can. J. Chem.,* 50, 3667, 1972.

30. **Pauling, P.,** in *Structural Chemistry and Molecular Biology,* Rich, A. and Davidson, N., Eds., W. H. Freeman, San Francisco, 1968, 593.

31. **Yeagle, P. L., Hutton, W. C., and Martin, R. B.,** Sphingomyelin multiple phase behavior as revealed by multinuclear magnetic resonance spectroscopy, *Biochemistry,* 17, 5745, 1978.

32. **Lapidot, A. and Irving, C. S.,** ^{15}N and ^{13}C dynamic NMR study of chain segmental motion of the peptidoglycan pentaglycine chain of ^{15}N and $^{13}C_2$-Gly-labeled *Staphylococcus aureus* cells and isolated cell walls, *Biochemistry,* 18, 1788, 1979.

33. **Schaefer, J.,** Distribution of correlation times and the carbon-13 nuclear magnetic resonance spectra of polymers, *Macromolecules,* 6, 882, 1973.

34. **Jennings, H. J. and Smith, I. C. P.,** Determination of the composition and sequence of a glucan containing mixed linkages by carbon-13 nuclear magnetic resonance, *J. Am. Chem. Soc.,* 95, 606, 1973.

35. **Perlin, A. S., Casu, B., and Koch, H. J.,** Configurational and conformational influences on the carbon-13 chemical shifts of some carbohydrates, *Can. J. Chem.,* 48, 2596, 1970.

36. **Czarniecki, M. F. and Thornton, E. R.,** ^{13}C NMR chemical shift titration of metal ion-carbohydrate complexes. An unexpected dichotomy for Ca^{2+} binding between anomeric derivatives of N-acetylneuramic acid, *Biochem. Biophys. Res. Commun.,* 74, 553, 1977.

37. **Sillerud, L. O., Prestegard, J. H., Yu, R. K., Schafer, D. E., and Konigsberg, W. H.,** Assignment of the ^{13}C nuclear magnetic resonance spectrum of aqueous ganglioside GM_1 micelles, *Biochemistry,* 17, 2619, 1978.

38. **Lapidot, A. and Irving, C. S.,** Comparative in vivo nitrogen — 15 nuclear magnetic resonance study of the cell wall components of five gram-positive bacteria, *Biochemistry,* 18, 704, 1979.

39. **Irving, C. S. and Lapidot, A.,** The dynamic structure of the *Escherichia Coli* cell envelope as probed by ^{15}N nuclear magnetic resonance spectroscopy, *Biochim. Biophys. Acta,* 470, 251, 1977.

40. **Gent, M. P. N. and Ho, C.,** Fluorine-19 NMR studies of lipid phase transitions in model and biological membranes, *Biochemistry,* 17, 3023, 1978.

41. **Kohler, S. J. and Klein, M. P.,** Orientation and dynamics of phospholipid head groups in bilayers and membranes determined from ^{31}P nuclear magnetic resonance chemical shielding tensors, *Biochemistry,* 16, 519, 1977.

42. **Hitchcock, P. B., Mason, R., Thomas, M., and Shipley, G. G.,** Structural chemistry of 1,2 dilauroyl-DL-phosphatidylethanolamine; molecular conformation and intermolecular packing of phospholipids, *Proc. Natl. Acad. Sci. U.S.A.,* 71, 3036, 1974.

43. **Griffin, R. G., Powers, L., and Pershan, P. S.,** Head-group conformation in phospholipids: a ^{31}P-NMR study of oriented monodomain dipalmitoylphosphatidylcholine bilayers, *Biochemistry,* 17, 2718, 1978.

44. **Herzfeld, J., Griffin, R. G., and Haberkorn, R. A.,** ^{31}P chemical-shift tensors in barium diethyl phosphate and urea-phosphoric acid: model compounds for phospholipid head-group studies, *Biochemistry,* 17, 2711, 1978.

45. **Houlf, D. I., Busby, S. J. N., Gadian, D. G., Radda, G. K., Richards, R. E., and Seeley, P. J.,** Observation of tissue metabolites using ^{31}P nuclear magnetic resonance, *Nature (London),* 252, 285, 1974.

46. **Ogawa, S., Rottenberg, H., Brown, T. R., Shulman, R. G., Castillo, C. L., and Glynn, P.,** High-resolution ^{31}P NMR study of rat liver mitochondria, *Proc. Natl. Acad. Sci. U.S.A.,* 75, 1796, 1978.

Chapter 5

LASER LIGHT SCATTERING

Charles S. Johnson, Jr. and Don A. Gabriel

TABLE OF CONTENTS

I. INTRODUCTION

Light scattering has provided an important method for characterizing macromolecules for at least three decades. However, the replacement of conventional light sources by lasers in recent years has qualitatively changed the field and has sparked renewed interest. Through the use of intense, coherent laser light and efficient spectrum analyzers and autocorrelators, experiments in the frequency and time domains can now be used to study molecular motion, e.g., diffusion and flow, and other dynamic processes, as well as the equilibrium properties of solutions. The technology for clarifying samples has also significantly improved. Recirculation through submicron filters in closed loop systems reduces the effects of dust and other contaminants, and the new time domain techniques provide built-in tests for the presence of such particles. These advances make laser light scattering a powerful form of spectroscopy for use in biochemistry.

Classical light scattering studies are concerned with the measurement of the intensity of scattered light as a function of the scattering angle. In addition to this kind of study, laser light sources now permit spectral information to be obtained from the scattered light. The latter type of experiment is often called quasi-elastic light scattering (QLS), and the various forms of the experiment are known as light beating spectroscopy (LBS), intensity fluctuation spectroscopy (IFS), and photon correlation spectroscopy (PCS). Related experiments in laser doppler velocimetry (LDV) now permit very low rates of uniform motion to be measured. A special case of LDV is electrophoretic light scattering (ELS) where mobilities are determined. The aim of this chapter is to provide an introduction to both the classical and quasi-elastic forms of laser light scattering, which can serve as an introductory text for students and a reference for research workers. The same purpose has been served for prelaser classical light scattering since 1960 by Chapter 5 of Tanford's excellent book, *Physical Chemistry of Macromolecules*.[1] As in that chapter, we emphasize concepts and the kinds of information that can be obtained from the various experiments rather than either presenting a comprehensive survey of the literature or discussing experimental techniques in great detail. For the most part we stick to well-developed techniques such as the measurement of translational and rotational diffusion coefficients. A few specialized applications (such as the study of motility) are also discussed; however, we have not included scattering from internal modes in polymers or gels. Interesting work is going on in the latter areas, but the theory is still in flux.

We attempt to keep the objectives in sight at all times and to avoid distracting complications as much as possible. Accordingly, certain definitions and derivations have been relegated to a set of appendixes. In general, we have opted for simplicity rather than elegance, or even rigor in certain cases, in the hope that a wider audience can be reached. For those who desire more extensive treatments there is an abundance of sources. For classical scattering, one of the latest reviews is that by Timasheff and Townsend, which is concerned with proteins.[2] Also Huglin[3] has edited an extensive volume that emphasizes experimental methods and data handling in the study of polymer solutions. The book by Fabelinskii[4] is a comprehensive, but difficult treatment mainly of prelaser light scattering without special attention to macromolecules. The classical physics of light scattering has been treated in detail by Kerker.[5] The book by Long,[6] while emphasizing Raman scattering, provides a good introduction to Rayleigh scattering, especially polarization and electronic resonance effects. Quasi-elastic scattering, also called dynamic light scattering, has received considerable attention in the literature, and the recent review papers are too numerous to list in full. An encyclopedic review of dynamic light scattering from biopolymers and biocolloids has been provided by Schurr.[7] Pusey and Vaughan have given a much briefer review of the principles of intensity fluctuation spectroscopy,[8] and Carlson[9] has covered applications in molecular biology. Laser velocimetry has been reviewed by Ware.[10,11] An introduction to the theoretical foundations of dynamic light scattering is presented in the book by Berne and Pecora,[12] which is an excellent text for students of physical chemistry. The book by Chu[13] is a good reference for many experimental and theoretical details. For those attempting to understand photon statistics and photon correlation spectroscopy, the monograph edited by Cummins and Pike is invaluable.[14]

II. CLASSICAL LIGHT SCATTERING

A. Scattering Intensity

1. Scattering by an Isolated Dipole and by Gases

Lasers produce collimated, quasi-monochromatic radiation having high intensity. In all but the least expensive lasers, the output is highly polarized. For discussion of the scattering experiment we adopt the coordinate system shown in Figure 1, where the incident laser light propagates in the +y-direction, and the x and y axes define the scattering plane. Only the electric field of the incident light is of interest here, and we assume that the incident light is polarized so that the electric field is in the z-direction. We express the electric field of the incident light as

$$E_z = E_{zo}\cos(k_o y - \omega_o t) \qquad (1)$$

where $k_o = 2\pi/\lambda_o$, λ_o is the wavelength in vacuum, and $\omega_o = 2\pi\nu_o$ is the laser frequency. The electric field E_z interacts with electrons in an atom or molecule to induce an electric dipole moment, which oscillates at the angular frequency ω_o. The usual expression for the induced dipole is

$$\underset{\sim}{p} = \underset{\approx}{\alpha} \cdot \underset{\sim}{E} \qquad (2)$$

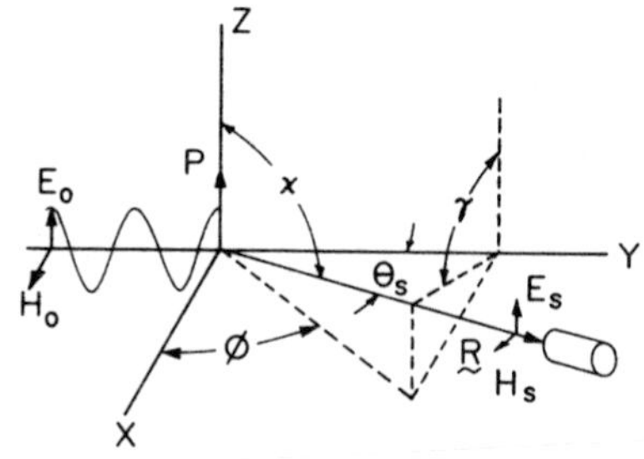

FIGURE 1. Scattering geometry.

where in general $\underset{\sim}{p}$ is a vector not necessarily in the same direction as $\underset{\sim}{E}$, and $\underset{\approx}{\alpha}$ is the *polarizability* tensor that relates the two vectors. A discussion of the polarizability tensor is given in Appendix A. We shall mainly be concerned with isotropic scatterers for which $\underset{\approx}{\alpha}$ is independent of orientation and Equation 2 becomes

$$p = p_z = \alpha E_z = \alpha E_{oz} \cos(k_o y - \omega_o t) \tag{3}$$

where α is a constant.

It is well known in electromagnetic theory that an oscillating dipole produces radiation — as for example with the oscillating currents in antennas of radio transmitters. The oscillating electric moment p_z thus provides the source of the scattered light. This is equivalent to the statement that an accelerating charge generates electromagnetic radiation. The solution of Maxwell's equations for an oscillating dipole shows that at large distances, i.e., where $R >> \lambda_o$, the electric field E_s of the scattered light is proportional to $d^2p/dt^2 = -\omega_o^2 p$ and inversely proportional to distance R as required by the conservation of energy. The complete expression in SI units for the field E_s at R resulting from dipole at the origin in Figure 1 is

$$E_s = \frac{-\omega_o^2 p \sin \chi}{(4\pi\epsilon_o)\, c^2 R} \tag{4}$$

where χ is the angle between the induced moment p and $\underset{\sim}{R}$, ε_o is the permittivity of free space, and c is the speed of light. When the detector is in the xy plane, the factor $\sin\chi$ is of course unity.

Detectors, e.g., photomultiplier tubes, respond to the intensity rather than the electric field of incident light. The instantaneous intensity, which is defined as the rate of passage of energy through unit area perpendicular to the direction of propagation, is given by

$$I = c\epsilon_o E^2 \tag{5}$$

Since all measurements require times that are much larger than the oscillation period of the radiation field, we are usually concerned with the cycle average of I. Thus for the incident radiation

$$\overline{I_i} = c\epsilon_o\overline{E^2} = c\epsilon_o E_{oz}^2 \frac{\omega_o}{2\pi}\left[\int_o^{2\pi/\omega_o} \cos^2 \omega_o t\, dt\right] \tag{6}$$

$$= \frac{c\epsilon_o}{2} E_{oz}^2$$

In calculations involving electric fields it is often convenient to use complex variables, e.g.,

$$E = E_{oz}\, e^{i(k_o y - \omega_o t)}$$

rather than $E = E_{oz} \cos(k_o y - \omega_o t)$. When the complex variables are handled properly, derivations are simplified, but the final results are not changed. For example, if the electric field of the incident radiation is written in complex form, the cycle average of the intensity can be written as[15]

$$\overline{I_i} = c\epsilon_o \overline{(\mathrm{Re}\, E)^2} = \frac{c\epsilon_o}{2} E^* E = \frac{c\epsilon_o}{2} |E|^2 \tag{7}$$

where Re means "the real part of" and E* is the complex conjugate of E.

Using Equation 6 the intensity of the scattered light is given by $\overline{I_s} = c\varepsilon_o\overline{E^2}_s$ and Equation 4 yields

$$\overline{I_s} = c\epsilon_o \left[\frac{\omega_o^4 \sin^2 \chi}{(4\pi\epsilon_o)^2\, c^4 R^2}\right] \overline{p^2} \tag{8}$$

From Equations 3 and 6 we see that $c\varepsilon_o\overline{p^2} = \alpha^2\, \overline{I_i}$, and Equation 8 gives

$$\frac{\overline{I_s}}{\overline{I_i}} = \frac{\omega_o^4 \alpha^2 \sin^2 \chi}{(4\pi\epsilon_o)^2\, c^4 R^2} = \frac{16\, \pi^4\, \alpha^2 \sin^2 \chi}{(4\pi\epsilon_o)^2\, \lambda_o^{\,4} R^2} \tag{9}$$

The inverse fourth power dependence on λ_o was predicted by Lord Rayleigh on the basis of simple dimensional arguments.[16] It accounts for the blue color of the sky, since molecules in the atmosphere tend to scatter the blue part of the solar spectrum with greater intensity than the longer wavelength red components. At optical frequencies almost all of the scattering results from electrons, and the number of electrons increases with molecular volume. Therefore, the polarizability is roughly proportional to molecular volume, especially for larger molecules, and the intensity accordingly depends on the square of the volume. In Equation 9 and in other parts of this chapter we have used SI units, which are based on the MKS system, except when established definitions were preserved in cgs units. The equations can be converted to the cgs system simply by replacing $4\pi\varepsilon_o$ with unity.

The intensity of scattering per unit volume for a gas at low pressure can be obtained by multiplying Equation 9 by N, the number of scatterers per unit volume. It is conventional, however, to rewrite this equation in terms of the change of refractive index with concentration, which is the experimentally determined quantity, rather than the molecular polarizability. If C is the mass per unit volume of the gas, then at low pressures the refractive index n can be expanded in a Taylor's series

$$n = 1 + \left(\frac{\partial n}{\partial C}\right)_o C \qquad (10)$$

so that

$$n^2 \simeq 1 + 2\left(\frac{\partial n}{\partial C}\right)_o C \qquad (11)$$

However, as discussed in Appendix B, n^2 is also given by

$$n^2 = 1 + N\alpha/\epsilon_o \qquad (12)$$

The combination of Equations 11 and 12 permits us to write

$$\alpha = \frac{2\epsilon_o}{N}\left(\frac{\partial n}{\partial C}\right)_o C = 2\epsilon_o m\left(\frac{\partial n}{\partial C}\right)_o \qquad (13)$$

where m is the mass per scattering particle. To obtain the intensity of scattering per unit volume, we substitute Equation 13 into Equation 9 and multiply by $N = N_A C/M$ where N_A is Avogadro's number and M is the molecular weight. Thus

$$\frac{\bar{I}_S}{I_i} = \frac{4\pi^2 \sin^2\chi\, CM}{\lambda_o^4 R^2 N_A}\left(\frac{\partial n}{\partial C}\right)_o^2 \qquad (14)$$

Scattering experiments are usually reported in terms of the Rayleigh ratio $\mathcal{R}_\theta$ defined by

$$\mathcal{R}_\theta = \left(\frac{I_\theta}{V}\right)\frac{R^2}{I_i} = \frac{I_S}{I_i} R^2 \qquad (15)$$

where I_θ is the measured intensity at θ_s, V is the volume of the scattering region, and the bars on I_s and I_i have been dropped, as they will be henceforth, for convenience. Equation 15 is not very useful in practice, since the radiant power P rather than the intensity I is usually measured by photomultiplier tubes and the scattering volume V is not known with accuracy. A more practical expression for $\mathcal{R}_\theta$ is shown in Equation 16.[17]

$$\mathcal{R}_\theta = \left(\frac{P_\theta}{P_i}\right)\frac{1}{(\Delta\Omega)\ell} \qquad (16)$$

Here P_θ is the radiant power of the light collected at the scattering angle θ, P_i is the radiant power of the incident beam, $\Delta\Omega$ is the solid angle of the scattered light that is collected, and ℓ is the length of the scattering volume. The conversion of Equation 15 into Equation 16 proceeds by using the cross-sectional areas A_i and $A_\theta = R^2(\Delta\Omega)$ of the incident and scattered beam, respectively, where R is the distance from the scattering volume to the detector, with the definitions $I_\theta = P_\theta/A_\theta$, $I_i = P_i/A_i$, and $V = \ell A_i$.

One additional comment about Equation 14 is in order before we proceed to the consideration of condensed phases. We have assumed that the incident light is polarized (along the z-direction) and that the scattering particles have dimensions much smaller than the wavelength of the incident light. As a consequence, the intensity is independent of the scattering angle θ_s. However, if either (1) the incident light is unpolarized, or (2) all of the scattered light in the cone θ_s to $\theta_s + d\theta_s$ is collected regardless of the angle χ, then the factor $\sin\chi$ must be replaced by $(1 + \cos^2\theta_s)/2$.[1] This factor is obtained by averaging $\sin^2\chi$ over the angle γ. By inspection of Figure 1 it is seen that $\cos\chi = \sin\theta_s\cos\gamma$. Therefore, $\sin^2\chi = 1-\sin^2\theta_s\cos^2\gamma$ and

$$<\sin^2\chi> = \frac{1}{2\pi}\int_0^{2\pi}\sin^2\chi\, d\gamma = \frac{1}{2}(1 + \cos^2\theta_s) \qquad (17)$$

In practice we find that either χ is set at 90° or that light is collected at all values of χ. In the following χ is assumed to be 90° unless otherwise stated.

2. Scattering by Macromolecules in Solution

In contrast to light scattered from gases, the intensity of scattered light from condensed phases is less than that predicted by Equations 9 and 14. The reduced intensity is the result of destructive interference. In fact, for perfect crystals irradiated with light, the wavelength of which is much greater than the separation of the lattice planes, no light is scattered. This follows from the fact that it is always possible to pair two scattering planes so that destructive interference occurs. Scattering from crystals is possible at certain angles, however, when the wavelength of the incident radiation is roughly equal to the distance d separating the scattering planes. The condition for scattering is the well-known Bragg relation, $\sin(\theta_s/2) = \lambda/d$ where θ_s is defined in Figure 1.

On the other hand, the scattering centers in liquids are not stationary, but undergo Brownian movement, which produces transient optical inhomogenities in the solution.

It is the presence of these inhomogenities that allows a small fraction of the scattered radiation to escape destructive interference and be observed as scattered intensity outside the sample. Our task is to formulate a theory that relates the intensity of scattered light to these transient optical inhomogenities. The basic ideas in the fluctuation theory of light scattering are attributed to Smoluchowski[18] and Einstein.[19] The problem can be approached by considering the solution of scatters as being composed of N small volume elements per unit volume. The volume elements δV are assumed to be small relative to the wavelength of the incident radiation. The light scattered from the independent volume elements largely cancels, but at any instant there will be a deviation from the time average number of particles in any volume element, and the cancellation will be incomplete. The connection with the scattering theory developed in Section II.A.1 is made by realizing that fluctuations in concentration or density lead to fluctuations in the polarizability. The fluctuation in the polarizability of one volume element is defined as $\delta\alpha_v = \alpha_v - \overline{\alpha}_v$ where α_v is the instantaneous polarizability and $\overline{\alpha}_v$ is the time average of α_v. A fluctuation can obviously be either positive or negative, and from the definition the time average of $\delta\alpha_v$ is zero. In Equation 9 the intensity of the scattered radiation is shown to be proportional to the square of the polarizability. To obtain the contribution from an average volume element, we square the quantity, $\alpha_v = \overline{\alpha}_v + \delta\alpha_v$, and take the time average. Thus,

$$\overline{(\overline{\alpha}_V + \delta\alpha_V)^2} = (\overline{\alpha}_V)^2 + \overline{(\delta\alpha_V)^2} \qquad (18)$$

The contribution from $(\overline{\alpha}_v)^2$ cancels exactly as in perfect crystals, and the net scattering intensity depends on $\overline{(\delta\alpha_v)^2}$. Substituting this result into Equation 9 and multiplying by $N = 1/\delta V$ gives

$$\frac{I_S}{I_i} = \frac{16\pi^4\,\overline{(\delta\alpha_V)^2}}{(4\pi\epsilon_0)^2\,\lambda_0^4\,R^2\,\delta V} \qquad (19)$$

for the scattering intensity per unit volume. To obtain $\overline{(\delta\alpha_v)^2}$ in terms of more readily measurable experimental quantities, we note that the mean square fluctuation in the polarization for a given volume element is related to the mean square fluctuation in concentration by

$$(\delta\alpha_V)^2 = \left(\frac{\partial\alpha}{\partial C}\right)^2_{T,V}\overline{(\delta C)^2} \qquad (20)$$

The smaller contributions resulting from temperature and volume fluctuations are neglected in this derivation, since these contributions are expected to be the same for the solution as for the solvent. It is expected that the experimental measurements can be corrected so that only the "excess" scattering by the solute is obtained.

In Appendix B it is shown that the polarizability is related to the refractive index of the solution by

$$n^2 - n_o^2 = \frac{\alpha}{(\delta V)\epsilon_o} \tag{21}$$

This equation can be differentiated with respect to the concentration to obtain an expression for $(\partial \alpha / \partial C)_{T,V}$ in terms of the measurable quantity $(\partial n / \partial C)_{T,V}$. Thus

$$2n \left(\frac{\partial n}{\partial C}\right)_{T,V} = \frac{1}{(\delta V)\epsilon_o} \left(\frac{\partial \alpha}{\partial C}\right)_{T,V} \tag{22}$$

The relationship between the mean-square fluctuation in the polarizability and the mean-square fluctuation in the concentration is obtained by combining Equations 21 and 22.

$$\overline{(\delta \alpha_V)^2} = [2n(\delta V)\epsilon_o]^2 \left(\frac{\partial n}{\partial C}\right)^2_{T,V} \overline{(\delta C)^2} \tag{23}$$

Equation 23 can now be used with Equation 19 to obtain the intensity of the scattered light in terms of the mean-square concentration fluctuations.

$$\frac{I_s}{I_i} = \frac{4\pi^2\, n^2 (\delta V)\, (\partial n / \partial C)^2_{T,V}\, \overline{(\delta C)^2}}{\lambda_o^4\, R^2} \tag{24}$$

B. Concentration Dependence

The magnitude of the average concentration fluctuation will clearly depend on the energy required to produce the fluctuation. A simple calculation that is exhibited below shows that $\overline{\delta C^2}$ depends on $(\partial^2 A / \partial C^2)_{T,V}$ where A is the Helmholtz free energy. Again we imagine that the sample is divided into small volume elements δV, which exchange solute particles and are in thermal equilibrium. The problem then becomes one of determining the probability of fluctuations in the composition of a given volume element, and determining the mean-square concentration fluctuation. By analogy with chemical kinetics the probability of a fluctuation $(\delta C)^2$ is given by $e^{-\delta A / k_B T}$ where δA is the associated increment in the Helmholtz free energy for the volume element. Since the fluctuations are small, δA can be expanded in a Taylor's series

$$\delta A = \left(\frac{\partial A}{\partial C}\right)_{T,V} \delta C + \frac{1}{2!} \left(\frac{\partial^2 A}{\partial C^2}\right)_{T,V} (\delta C)^2 + \ldots \tag{25}$$

The first term on the right vanishes, since the system is at equilibrium. Therefore, the probability of a fluctuation becomes

$$\exp(-\delta A / k_B T) = \exp\left[-\left(\frac{\partial^2 A}{\partial C^2}\right)_{T,V} (\delta C)^2 / 2k_B T\right] \tag{26}$$

This weighting factor can now be used to calculate $\overline{(\delta C)^2}$ as follows:

$$\overline{(\delta C)^2} = \frac{\int_0^\infty (\delta C)^2 \, e^{-\delta A/k_B T} \, d(\delta C)}{\int_0^\infty e^{-\delta A/k_B T} \, d(\delta C)} \tag{27}$$

The above integrals have the standard form $\int_0^\infty x^n e^{-ax^2} dx$ and can be found in tables. Therefore, the evaluation of Equation 27 immediately gives

$$\overline{(\delta C)^2} = k_B T/(\partial^2 A/\partial C^2)_{T,V} \tag{28}$$

The quantity $(\partial^2 A/\partial C^2)_{T,V}$ depends on the concentration, and thus Equation 28 in conjunction with Equation 24 gives the concentration dependence of the scattered intensity. A series expansion of $(\partial^2 A/\partial C^2)_{T,V}$ in terms of the concentration could be introduced at this point. However, it is conventional to relate $(\partial^2 A/\partial C^2)_{T,V}$ to $(\partial \mu_1/\partial C)_{T,V}$ so that the concentration dependence is given in terms of the familiar virial coefficients. The derivation of this relationship is given in Appendix C with the important result that

$$\left(\frac{\partial^2 A}{\partial C^2}\right)_{T,V} = -\frac{\delta V}{C\overline{V}_1} \left(\frac{\partial \mu_1}{\partial C}\right)_{T,V} \tag{29}$$

where $\overline{V}_1$ and μ_1 are the partial molar volume and the chemical potential of the solvent, respectively. The virial expansion of $(\partial \mu_1/\partial C)_{T,V}$, which is also discussed in Appendix C, is given by

$$-\frac{1}{k_B T \overline{V}_1} \left(\frac{\partial \mu_1}{\partial C}\right)_{V,T} = N_A \left[\frac{1}{M} + 2B_2 C + 3B_3 C^2 + \ldots\right] \tag{30}$$

where M is the molecular weight of the solute and the B_n are virial coefficients. Equations 28 to 30 can then be combined with Equation 24 to yield

$$\frac{I_s}{I_i} = \frac{4\pi^2 n^2 C(\partial n/\partial C)_o^2}{\lambda_o^4 R^2 N_A [M^{-1} + 2B_2 C + 3B_3 C^2 + \ldots]} \tag{31}$$

Equation 31 describes the intensity of the "excess scattered light" per unit volume from a dilute solution of scatterers. The term "excess scattering" is used to describe the scattering resulting from the solute alone. In fact scattering from the solvent, which is usually a weak aqueous salt solution, is subtracted from the total scattered intensity in the data analysis as a blank correction. As discussed in Section III.A, Equation 31 has been derived for polarized incident light. Therefore, the $\sin^2\chi$ factor must be replaced by $(1 + \cos^2\theta_s)/2$ for instruments using unpolarized light, or when all the scattered light in the angular range θ_s to $\theta_s + d\theta_s$ is collected.

Equations 15 and 31 can be combined to obtain

$$\frac{\mathcal{K}C}{\mathcal{R}_\theta} = \frac{1}{M} + 2B_2C + 3B_3C^2 + \ldots \tag{32}$$

where $\mathcal{K}$, the optical constant, is defined by

$$\mathcal{K} = \frac{4\pi^2 n^2 (\partial n/\partial C)_0^2}{\lambda_0^4 N_A} \tag{33}$$

It should be noted that the optical constant defined in Equation 33 is twice as large as that used by Tanford.[1] This equation has been derived for small particles that conform to the rule that their major dimension is less than $\lambda/10$. It is obvious that the intercept of a plot of $\mathcal{K}\ C/\mathcal{R}_\theta$ vs. C gives the inverse molecular weight and the initial slope is $2B_2$, where B_2 is the second virial coefficient. As discussed in Appendix D, the molecular weight obtained when there is polydispersity is the weight average molecular weight. In the limit of $C = 0$ Equation 31 becomes equivalent to Equation 14. The purpose of the simple derivation outlined here is to obtain the concentration dependence of the scattered intensity for real solution of macromolecules.

Light scattering intensities are sometimes reported in terms of the turbidity, i.e., the attenuation coefficient τ for a beam passing through the sample. Attenuation, of course, results from scattering in any direction except $\theta_s = O$. For small particles, equations relating R_θ and τ can easily be derived with the help of Equations 15 and 17. For light passing through the volume $V = A_i\,\Delta\ell$, the average intensity scattered into the cone defined by θ_s and $\theta_s + d\theta_s$ is

$$I_\theta = \frac{\mathcal{K}CM}{2R^2}\,[1 + \cos^2\theta_s]\; I_i\, A_i\, \Delta\ell \tag{34}$$

The power scattered into this angular range is $dP_\theta = I_\theta\, 2\pi\, R^2 \sin\theta_s\, d\theta_s$ and the total power lost from the incident beam in the path length $\Delta\ell$ is obtained by integrating over θ_s. Thus

$$\begin{aligned} \Delta P &= -\int dP_\theta \\ &= -\mathcal{K}CM\, P_i\, \Delta\ell\, \pi \int_0^\pi [1 + \cos^2\theta_s] \sin\theta_s\, d\theta_s \\ &= -\mathcal{K}CM\, P_i \left(\frac{8\pi}{3}\right) \Delta\ell \end{aligned} \tag{35}$$

For an infinitesimal path length Equation 35 has the form $dP/P = -\tau d\ell$, which integrates to give $P = P_i \exp(-\tau\ell)$. Equation 35 shows that the attenuation coefficient is just $\tau = (8\pi/3)\mathcal{K}$ CM. Another optical constant $H = (8\pi/3)\mathcal{K}$ is often introduced so that

$$\frac{HC}{\tau} = \left(\frac{\mathcal{K}C}{\mathcal{R}_\theta}\right)_{C = O} \tag{36}$$

This equation becomes equivalent to Equation 32 when the concentration dependence is introduced.

C. Size Dependence

1. Structure Factors

It is evident from Figure 2 that when the size of the scattering particle increases to the point where the largest dimension is greater than $\lambda/10$, the path difference for light scattered from two points in the particle may become large enough to produce a significant phase difference and hence interference in the scattered light. The interference will obviously attenuate the intensity of the scattered light. Initially, the decrease in intensity may seem a nuisance, but it turns out to be a unique source of information pertaining to the size and shape of the particle. The dependence of the scattered intensity on the particle size and shape is contained in the structure factor $S(\underset{\sim}{K})$, which is defined by

$$S(\underset{\sim}{K}) = \frac{\text{scattered intensity observed (at } \theta_S)}{\text{scattered intensity without interference (at } \theta_S = O)} \tag{37}$$

In our development of the expression for $S(\underset{\sim}{K})$ we assume randomly oriented, nonabsorbing particles, composed of optically isotropic material. The refractive index within the particle is assumed to be homogeneous, and the refractive indexes of the particle and solvent are assumed to be only slightly different so that the Rayleigh-Debye condition, $4\pi L(\Delta n)/\lambda << 1$, is satisfied[5] where L is the largest dimension of the particle and Δn is the difference between the refractive indexes of the particle and the solvent.

Figure 2 shows a macromolecule with a scattering segment S at position $\underset{\sim}{\rho}_j$ relative to the center of mass. Parallel incident light with the wave vector $\underset{\sim}{k}_i$ passes through reference plane OO′. Similarly the scattered light passes through reference plane PP′. The path difference between light scattered from segment S and that passing through the origin at O is given by

$$\text{path difference} = O'\,S\,P' - OP$$

The phase difference ϕ for the two rays is just $2\pi n/\lambda_o$ times the path difference. Therefore, we can write

$$\phi = \underset{\sim}{\rho}_j \cdot \underset{\sim}{k}_i - \underset{\sim}{\rho}_j \cdot \underset{\sim}{k}_s = \underset{\sim}{\rho}_j \cdot \underset{\sim}{K} \tag{38}$$

where $\underset{\sim}{K} = \underset{\sim}{k}_i - \underset{\sim}{k}_s$ is called the scattering vector. Simple trigonometry shows that

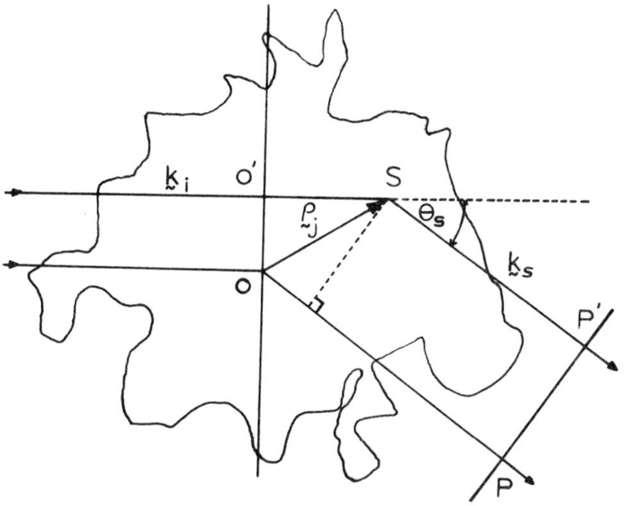

FIGURE 2. Diagram of the path difference for scattering from a segment in a macromolecule.

$$|\underset{\sim}{K}| = \frac{4\pi n}{\lambda_o} \sin\left(\frac{\theta_s}{2}\right) \tag{39}$$

where we have used the fact that $|\underset{\sim}{k}_s| = |\underset{\sim}{k}_i| = (2\pi n/\lambda_o)$. Accordingly, the electric field of the light scattered by the jth segment can be written as

$$E_j = a_j e^{i\underset{\sim}{k} \cdot \underset{\sim}{\rho}_j} E_o e^{-i\omega_o t} \tag{40}$$

where a_j contains the factors such as the polarizability, the distance R to the detector, etc. To calculate the intensity of the scattered light from the total particle for a given scattering angle, the electric fields of the scattered waves from the P segments within the particle must be summed as follows:

$$\sum_{j=1}^{P} E_j = E_o e^{-i\omega_o t} \sum_{j=1}^{P} a_j e^{i\underset{\sim}{K} \cdot \underset{\sim}{\rho}_j} \tag{41}$$

According to Equation 7 the intensity of the scattered light is

$$I_s = \frac{c\epsilon_o}{2} \left| \sum_{j=1}^{P} a_j e^{i\underset{\sim}{K} \cdot \underset{\sim}{\rho}_j} \right|^2 E_o^2 \tag{42}$$

Equation 42 gives the scattering for one particle with a fixed orientation. For an ensemble of randomly oriented particles the intensity is given by

$$I_s = N \frac{c\epsilon_0}{2} < \left| \sum_{j=1}^{P} a_j e^{i\underset{\sim}{K} \cdot \underset{\sim}{\rho}_j} \right|^2 >_{angle} E_0^2 \qquad (43)$$

where N is the number of particles per unit volume. The calculation of the structure factor also requires the intensity in the absence of interference. This condition is met at $\theta_s = O$, where the path distance for all scattering segments becomes identical; therefore, the intensity without interference is given by

$$<I_s>_{\theta=O} = \frac{c\epsilon_0}{2} |a|^2 P^2 E_0^2 N \qquad (44)$$

where all of the a_j have been assumed to be equal, i.e., $a_j = a$ for all j. The expression for the structure factor then follows from Equation 37.

$$S(\underset{\sim}{K}) = < \left| \frac{1}{P} \sum_{j=1}^{P} e^{i\underset{\sim}{K} \cdot \underset{\sim}{\rho}_j} \right|^2 >_{angle} \qquad (45)$$

To illustrate this procedure,we consider two simple examples:

a. *Long Thin Rod*

The rod is considered as a linear array of uniformly polarizable segments, and the cross section is assumed to be sufficiently small that interference can only arise for scattering from different positions along the major axis. For a uniform rod the summation in Equation 45 can be replaced with an integral over the length L

$$\lim_{P\to\infty} \sum_{j=1}^{P} \frac{1}{P} e^{i\underset{\sim}{K} \cdot \underset{\sim}{\rho}_j} = \frac{1}{L}\int_{-L/2}^{L/2} e^{iK\rho \cos\alpha} \, d\rho = \frac{\sin w}{w} \qquad (46)$$

where $w = (KL/2) \cos \alpha$ and α is the angle formed by the scattering vector $\underset{\sim}{K}$ and the position vector $\underset{\sim}{\rho}_j$ as shown in Figure 3. This result can now be substituted into Equation 45 and the averaging carried out over all orientations. We proceed by considering the probability of finding an orientation in the solid angle between Ω and $\Omega + d\Omega$. The integration over the solid angle Ω is carried out using the relations given in Figure 3.

$$S(\underset{\sim}{K}) = \frac{1}{4\pi} \int_{\phi=O}^{2\pi} \int_{\alpha=O}^{\pi} \left| \frac{\sin w}{w} \right|^2 \sin\alpha \, d\alpha d\phi \qquad (47)$$

Equation 47 is then evaluated by standard methods utilizing the trigonometric identity $2 \sin^2 \theta = (1 - \cos 2\theta)$ and integration by parts. The result is[12]

$$S(\underset{\sim}{K}) = \frac{2}{KL}\int_{0}^{KL} \frac{\sin x}{x}\, dx - \left[\frac{\sin(KL/2)}{(KL/2)}\right]^2 \tag{48}$$

This equation was first given by Neugebauer[20] in 1943. The first term is usually evaluated by series expansion when KL is small. For large values of KL, restrictions imposed by the Rayleigh-Debye condition may be a limiting factor in some experimental situations.

b. Uniform Sphere

In order to calculate the structure factor for a sphere of radius R, we again return to Equation 45. This time the summation is replaced by an integration over the volume of the sphere. In spherical polar coordinates this gives

$$\begin{aligned} S(\underset{\sim}{K}) &= \left| 2\pi \int_{\rho=0}^{R}\int_{\alpha=0}^{\pi} e^{iK\rho\cos\alpha} \sin\alpha d\alpha\, \rho^2 d\rho \Big/ 4\pi \int_{\rho=0}^{R} \rho^2 d\rho \right|^2 \\ &= \left\{ \frac{3}{R^3} \int_{\rho=0}^{R} \frac{\sin(K\rho)}{K\rho}\, \rho^2 d\rho \right\}^2 \end{aligned} \tag{49}$$

Integration by parts can be applied in the last step to obtain

$$S(\underset{\sim}{K}) = \left\{ \frac{3}{X^3} (\sin X - X\cos X) \right\}^2 \tag{50}$$

where $X = KR$. It is important to note that integration over all orientations of the particle is unnecessary in this case because of the spherical symmetry.

The preceding two examples serve to illustrate the principles involved in the derivation of the structure factor. In practice it is often convenient to carry out the average over orientations in Equation 45 before applying the equation to a particular particle shape. The procedure is as follows. We first rewrite Equation 45 as

$$S(\underset{\sim}{K}) = \frac{1}{P^2} < \sum_{i=1}^{P}\sum_{j=1}^{P} e^{i\underset{\sim}{K}\cdot\underset{\sim}{r}_{ij}} >_{angle} \tag{51}$$

where $r_{ij} = \underset{\sim}{\rho}_j - \underset{\sim}{\rho}_j$ Using $\underset{\sim}{K} \cdot \underset{\sim}{r}_{ij} = K\, r_{ij} \cos\alpha$ and the relations in Figure 3 we find that

$$\begin{aligned} < e^{iKr_{ij}\cos\alpha} >_{angle} &= \frac{1}{2}\int_{0}^{\pi} e^{iKr_{ij}\cos\alpha} \sin\alpha d\alpha \\ &= \frac{\sin(Kr_{ij})}{Kr_{ij}} \end{aligned} \tag{52}$$

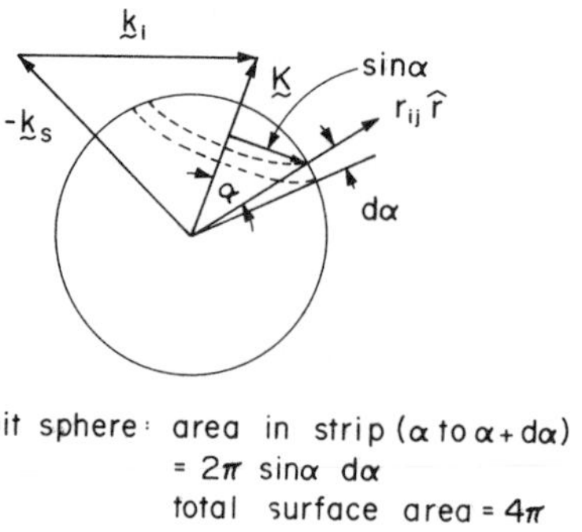

FIGURE 3. The possible orientations of $\underset{\sim}{r}_{ij}$ relative to the scattering vector $\underset{\sim}{K}$. Unit sphere: area in strip (α to $\alpha + d\alpha$) = $2\pi \sin\alpha d\alpha$. Total surface area = 4π.

Equations 51 and 52 combine to give the standard form of the structure factor equation

$$S(\underset{\sim}{K}) = \frac{1}{P^2} \sum_{i=1}^{P} \sum_{j=1}^{P} \frac{\sin(Kr_{ij})}{Kr_{ij}} \tag{53}$$

This is a very useful equation for deriving expressions relating the angular dependence of the scattered intensity to the shapes of particles. The application of Equation 53 to specific shapes has been discussed in many reviews.[1-5,12] The results for some standard particles are listed in Table 1.

In Figure 4 plots of the inverse structure factor $S^{-1}(\underset{\sim}{K})$ vs. $R^2_G \sin^2(\theta_s/2)$ for several common macromolecular shapes are given. One should be extremely cautious, however, in applying these plots to experimental data since the morphology of the experimental curves for a given shape may be distorted by many factors, e.g., polydispersity and optical anisotropy.[21] The resulting ambiguities seriously reduce the usefulness of Equation 53 for many real systems.

2. Radius of Gyration

Most experimental determinations of molecular dimensions require some preliminary assumption about the shape of the molecule. In contrast, light scattering measurements permit the radius of gyration to be determined from the structure factor $S(\underset{\sim}{K})$ without any assumptions about the shape, provided the measurements can be made at sufficiently low angles.[22] To see how this property of the structure factor comes about, we expand $S(\underset{\sim}{K})$ in a Taylor's series so that

$$S(\underset{\sim}{K}) = \frac{1}{P^2} \sum_{i}^{P} \sum_{j}^{P} \left(1 - \frac{K^2 r^2_{ij}}{3!} + \frac{K^4 r^4_{ij}}{5!} + \ldots\right) \tag{54}$$

For small values of K only the first two terms in the expansion are significant, and since

Table 1

STRUCTURE FACTORS $S(\underset{\sim}{K})$ AND RADII OF GYRATION FOR VARIOUS PARTICLES

Particle	$S(\underset{\sim}{K})$	R_G^2
Thin rod	$\frac{2}{x}\int_0^x \frac{\sin u}{u}\,du - \left[\frac{\sin(x/2)}{(x/2)}\right]^2$; $x = KL$; L = length	$\frac{L^2}{12}$
Uniform sphere	$\left[\frac{3}{x^3}(\sin x - x\cos x)\right]^2$; $x = KR$; R = radius	$\frac{3R^2}{5}$
Gaussian coil[a]	$\frac{2}{x^2}(e^{-x} - 1 + x)$; $x = K^2\frac{\langle h^2\rangle}{6}$ $\langle h^2\rangle$ = mean square end-to-end distance	$\frac{\langle h^2\rangle}{6}$

[a]Reference 12, p. 168

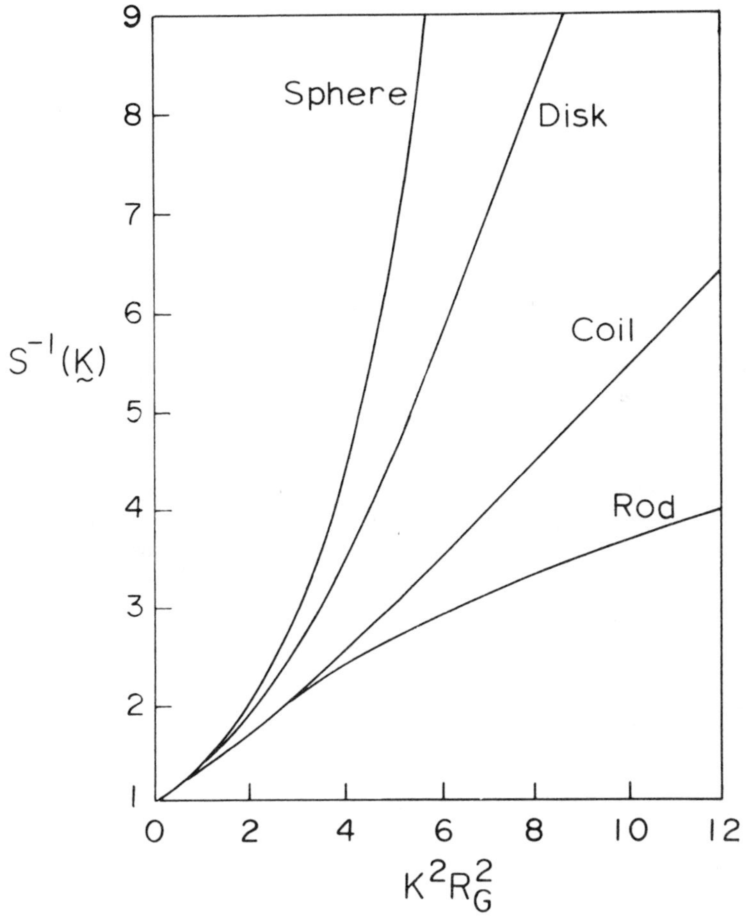

FIGURE 4. The inverse structure factor $S^{-1}(\underset{\sim}{K})$ vs. $K^2R_G^2$ for various shapes (see text).

$$\sum_{i}^{P} \sum_{j}^{P} 1 = P^2$$

we obtain

$$S(\underset{\sim}{K}) \cong 1 - \frac{K^2}{3!P^2} \sum_{i}^{P} \sum_{j}^{P} r_{ij}^2 \tag{55}$$

We can introduce the radius of gyration R_G at this point since one of the definitions of this quantity is[23]

$$R_G^2 = \frac{1}{2P^2} \sum_{i}^{P} \sum_{j}^{P} r_{ij}^2 \tag{56}$$

The use of this definition with Equations 55 and 38 yields the more familiar equation

$$S(\underset{\sim}{K}) = 1 - \frac{16\pi^2 n^2}{3\lambda_o^2} (R_G^2) \sin^2 \left(\frac{\theta_s}{2}\right) + \ldots \tag{57}$$

An expansion for the inverse structure factor $S^{-1}(\underset{\sim}{K})$ can easily be obtained from Equation 57 by noting that for small values of x, $(1 - x)^{-1} \simeq 1 + x$. With this factor, Equation 32 can be corrected for interference effects in large particles to give in the limit of zero concentration

$$\left(\frac{\mathcal{K}C}{\mathcal{R}_\theta}\right)_{C=O} = \frac{1}{M_w} \left[1 + \frac{16\pi^2 n^2}{3\lambda_o^2} (R_G^2) \sin^2 \left(\frac{\theta_s}{2}\right) + \ldots\right] \tag{58}$$

3. Zimm Plots

In a classic article, Zimm[24] reported a graphical technique for simultaneously extrapolating light scattering data to both zero angle and zero concentrations. This is achieved by plotting $\mathcal{K} C/\mathcal{R}_\theta$ vs. $\sin^2(\theta_s/2)$ + (constant) C as shown in Figure 5 where the arbitrary constant is chosen to give a convenient spacing of the data points on the graph. It is seen that the data points fall on a grid, producing two families of curves, one corresponding to constant concentration and the other to constant angle. The data points on the grid corresponding to a given angle are then extrapolated to zero concentration, and similarly the points at a given concentration are extrapolated to zero angle. This double extrapolation then produces two new sets of data points, one at zero angle and the other at zero concentration from which the information about the particle shape and weight is obtained. The inverse of the weight average molecular weight is obtained from the intercept of the θ = O curve, according to Equation 32, and the second virial coefficient, B_2, is obtained from the slope ($2B_2$ = slope). The intercept of the C = O curve again gives the inverse of the molecular weight, and the initial slope is proportional to the radius of gyration as described by Equation 58. These quantities can be combined to yield

$$R_G^2 = \frac{3\lambda_o^2}{16\pi^2 n^2}\left(\frac{\text{slope}}{\text{intercept}}\right) \qquad (59)$$

The radius of gyration is obtained as a z-average (see Appendix D).

There are limitations on the size of the radius of gyration that can be measured by this method. For small values of R_G, the derivation of the structure factor from unity may be so small that an accurate slope cannot be obtained, even when the scattering angle is large. On the other hand, for large values of R_G, it may be impossible to obtain the limiting condition required for Equation 58 to hold. Rough estimates for the limits of R_G using 5° and 160° as the extremes for conveniently obtainable scattering angles and λ_o = 633 nm as the wavelength give 300 Å and 5000 Å. It is possible to extend these limits by changing the wavelength of the incident radiation. However, with most biological polymers the ultraviolet region is unsuitable because of strong absorption bands in that region, and also in some cases because of photo-induced chemical modification of the polymer. At the opposite extreme where the dimension of the molecule exceeds the wavelength of the incident light, the above theory breaks down and the scattered intensity becomes a complicated oscillatory function of angle. Kerker has given a detailed discussion of this problem.[5] An alternative for large particles is to increase the wavelength of the incident radiation. With the advent of infrared laser sources, e.g., Nd-Yag lasers, dye lasers, and infrared diodes (see Section III.H), as well as improved infrared detectors, large particles can be studied using Rayleigh-Debye theory and Zimm plots. Recently, Morris[25] et al. reported an infrared Zimm plot of *Serratia marcescens,* a bacterium approximately 1 μm in length. CO_2 lasers may offer the possibility of studying particles having diameters of the order of 10 μm.

D. Polydispersity

Inhomogeneity in either the size or shape of the particles leads to ambiguity in the interpretation of the angular dependence of the scattered intensity. Since the measured structure factor is an average over the structure factors of the scattering particles, it is often impossible to characterize polydisperse samples with the limited amount of information available. However, if the shapes of the particles are uniform and known, certain features of the distribution can be assessed. For example a plot of $(K\ C/\mathcal{R}_\theta)$ vs. $\sin^2(\theta_s/2)$ can yield M_w, M_n, $\langle R^2_G\rangle_z$ and $\langle R^2_G\rangle_n$ for a collection of Gaussian coils. The basic ideas were presented by Zimm[24] who showed that the scattered intensity is proportional to the z-average of the structure factor so that

$$\left(\frac{\mathcal{R}_\theta}{Kc}\right)_{C=0} = M_w \langle S(\underset{\sim}{K})\rangle_z \qquad (60)$$

The consequences of this result are discussed in Appendix D.

If the particles are composed of subunits, Equation 60 can be written as

$$\left(\frac{\mathcal{R}_\theta}{Kc}\right)_{C=0} = M_w \frac{\int S(\underset{\sim}{K})_N\, Nf(N)\, dN}{\int Nf(N)\, dN} \qquad (61)$$

where $S(\underset{\sim}{K})_N$ is the structure factor for a particle containing N subunits, N is the degree of polymerization, and f(N)dN is the fractional mass in the range N to N + dN. The following definitions are useful:

$$1 = \int f(N)\, dN \tag{62}$$

$$N_n = 1 \bigg/ \int \frac{f(N)}{N}\, dN \tag{63}$$

$$N_w = \langle N \rangle = \int N f(N)\, dN \tag{64}$$

$$\langle (\Delta N)^2 \rangle = \int (N - \langle N \rangle)^2 f(N)\, dN = \langle N^2 \rangle - \langle N \rangle^2 \tag{65}$$

The quantity $\langle(\Delta N)^2\rangle$ is known as the dispersion of the distribution.

In the limit of small scattering angles the series expansion for the structure factor from Equation 57 can be obtained with Equation 61 to give

$$\left(\frac{\mathcal{R}_\theta}{Kc}\right)_{C=0} = M_w \left[1 - \frac{16\pi^2 n^2}{3\lambda_o^2 N_w} \int N f(N)\, R_G^2\, dN \sin^2\left(\frac{\theta_s}{2}\right)\right] \tag{66}$$

To proceed further, a particular particle shape must be assumed. For example, for thin rods $R^2{}_G = L^2/12$, and we assume that L = Nb where b is the length of a subunit. Therefore

$$\int N f(N)\, R_G^2\, dN = \frac{b^2}{12} \int N^3 f(N)\, dN = \frac{b^2}{12} \langle N^3 \rangle \tag{67}$$

In the case of linear Gaussian coils, Table 1 gives $R^2{}_G = \langle h^2\rangle/6$ and $\langle h^2\rangle$ can be written as Nb^2. Therefore, the required integral becomes

$$\int N f(N)\, R_G^2\, dN = \frac{b^2}{6} \int N^2 f(N)\, dN = \frac{b^2}{6} \langle N^2 \rangle \tag{68}$$

The following discussion considers the case of the Guassian coil. By combining Equations 66 and 68 and taking the inverse we obtain

$$\left(\frac{\mathcal{K}C}{\mathcal{R}_\theta}\right)_{C=0} = \frac{1}{M_w}\left[1 + \frac{8\pi^2 n^2 b^2}{9\lambda_o^2 N_w} <N^2> \sin^2\left(\frac{\theta_s}{2}\right)\right] \tag{69}$$

This equation can be put in standard form by using Equation 65 to give

$$\left(\frac{\mathcal{K}C}{\mathcal{R}_\theta}\right)_{C=0} = \frac{1}{M_w}\left\{1 + N_w\left[1 + \frac{<(\Delta N)^2>}{N_w^2}\right]\frac{8\pi^2 n^2 b^2}{9\lambda_o^2}\sin^2\left(\frac{\theta_s}{2}\right)\right\} \tag{70}$$

Equation 66 and subsequent equations all show that the weight average molecular weight and the z-average radius of gyration can be obtained from a plot of $(\mathcal{K}\, C/\mathcal{R}_\theta)_{C=0}$ vs. $\sin^2(\theta_s/2)$. Equation 70 is important because it relates the slope of the plot to the dispersion of the distribution.

Additional information can be obtained from the slope and intercept of the asymptote of the curve $(\mathcal{K}\, C/\mathcal{R}_\theta)_{C=0}$ vs. $\sin^2(\theta_s/2)$.[26-28] For example, with the Guassian coil the structure factor listed in Table 1 clearly approaches the asymptotic limit:

$$\lim_{x\to\infty} S(\underset{\sim}{K}) = \frac{2}{x} - \frac{2}{x^2} \tag{71}$$

where $x = K^2Nb^2/6$. For a collection of polydisperse Gaussian coils the expression for the z-average of $S(\underset{\sim}{K})$ from Equation 61 an be used to obtain the z-average of Equation 71. Thus

$$\begin{aligned} <S(\underset{\sim}{K})>_z &= \frac{2}{N_w u}\int f(N)\, dN - \frac{2}{N_w u^2}\int \frac{f(N)}{N}\, dN \\ &= \frac{2}{N_w u}\left[1 - \frac{1}{N_n u}\right] \end{aligned} \tag{72}$$

where $u = K^2b^2/6$, and in the last step we have used Equation 63 to introduce the number average of the degree of polymerization N_n. Using the fact that $N_n u$ is large, the inverse of Equation 72 can be written as

$$<S(\underset{\sim}{K})>_z^{-1} = \frac{N_w}{2N_n} + \frac{N_w u}{2} \tag{73}$$

Equation 73 describes a straight line when $<S(\underset{\sim}{K})>_z^{-1}$ is plotted vs. $\sin^2(\theta_s/z)$. This expression can now be combined with Equation 60 to obtain the asymptotic equation

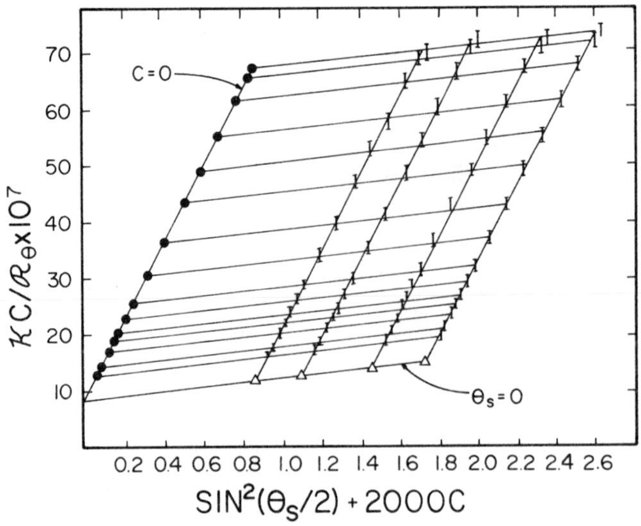

FIGURE 5. Zimm plot for cellulose nitrate. (Reprinted with permission from Benoit, H., Holtzer, A. M., and Doty, P., *J. Phys. Chem.*, 58, 635, 1954. Copyright by the American Chemical Society.)

$$\left(\frac{Kc}{\mathcal{R}_\theta}\right)_{C=0} = \frac{N_w}{2M_wN_n} + \frac{N_w}{2M_w}u = \frac{1}{2M_n} + \frac{N_n}{2M_n}u$$

$$= \frac{1}{2M_n} + \frac{(N_nb^2)}{12M_n}\left(\frac{4\pi n}{\lambda_o}\right)^2 \sin^2(\theta_s/2) \qquad (74)$$

If M_o is the molecular weight of a subunit, then $M_w = M_oN_w$ and $M_n = M_oN_n$. These relations were used in deriving Equation 74. The conclusion is that the intercept of the asymptote gives $1/(2M_n)$ and the slope is proportional to N_nb^2, which is equal to the number average of the end-to-end distance. The features of this type of plot are illustrated in Figure 6.

A similar analysis has been presented by Holtzer[29] and Goldstein[30] for rigid rods. Also, Rice[31] and Benoit[26] have considered polydispersity of shape. In spite of the intriguing possibilities suggested by these analyses, it should be realized that the asymptote may not be attainable in many experimental situations.[32] Carpenter[33] has shown that multiple wavelengths of incident light may be used in some situations in order to achieve the asymptotic condition.

III. DYNAMIC LIGHT SCATTERING

A. Time and Frequency Dependence

1. Background

In Section II we were concerned with the average intensity of the scattered light and its dependence on both the concentration of scattering particles and the scattering angle. It was concluded that in favorable cases both the weight average molecular weight and the z-average radius of gyration could be determined from such experiments. In this section we consider the use of light scattering to determine the rates of motion of macromolecules. Motional effects are evident in many kinds of spectroscopies, one of the best known examples being the Doppler broadening, which is encountered in gas

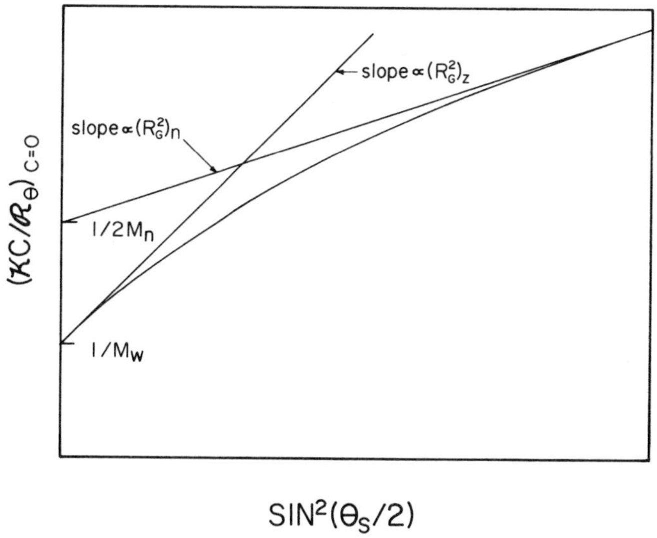

FIGURE 6. Plot showing limiting behavior for a polydisperse system of Gaussian coils. (Reprinted with permission from Benoit, H., Holtzer, A. M., and Doty, P., *J. Phys. Chem.*, 58, 635, 1954. Copyright by the American Chemical Society.)

Table 2
PERFORMANCE FIGURES FOR SPECTROMETERS FOR INCIDENT LIGHT HAVING THE FREQUENCY 5×10^{-14} Hz

Instrument	Resolution	Minimum line width (Hz)	Time scale
Grating spectrograph	10^5	3×10^9	300—0.1 psec
Fabry-Perot interferometer	10^8	5×10^6	200 nsec—30 psec
Lightbeating spectrometer	$\sim10^{14}$	~1	1 sec—1 μsec

phase absorption spectroscopy. A consideration of the velocities of macromolecules, both in diffusive motion and in directed flow, such as that produced in electrophoresis, shows that linewidths and shifts in the range from 1 to 10^5 Hz are expected. Somewhat higher frequencies are found if rotational motions are considered.

To put the frequency shifts in context, one should recall the capabilities of the available types of spectroscopic instruments. In Table 2 we list typical performance figures for grating spectrographs, interferometers, and lightbeating spectrometers for a spectral range in the vicinity of $5 \times 10^{+14}$ Hz, i.e., in the middle of the visible spectrum. Roughly speaking, the resolution is defined as the spectral frequency of a line divided by the contribution to the linewidth, which results from instrumental limitations alone. It is clear from the figures given that the measurement of linewidths in the range 1 to 10^4 Hz requires several orders of magnitude higher resolution than can be obtained by standard optical techniques. In fact, this frequency range was totally inaccessible prior to the development of the light beating experiment in 1964.[34] This remarkable improvement in resolution results from an analysis of the fluctuations in the intensity of the scattered light rather than an attempt to disperse the light according to wavelength. As we shall show, by dealing with the time dependence of the intensity we effectively

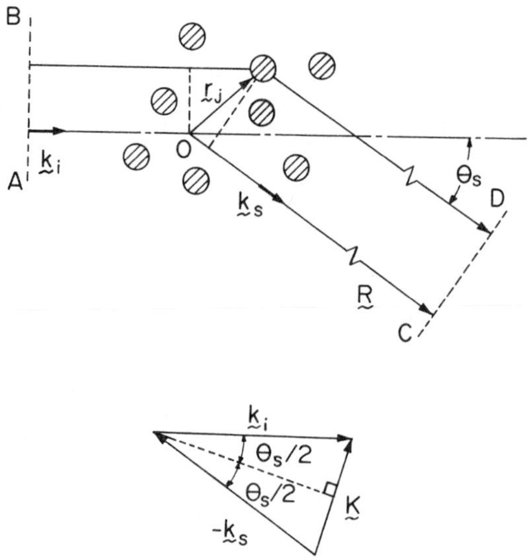

FIGURE 7. The scattering wave vector $\underset{\sim}{K}$ for a solution of small particles.

shift the spectral line down to zero frequency where electronic means can be used for the spectral analysis.

2. The Intensity

In considering the time and frequency dependence of the intensity, we choose to express the electric field in complex form. This is done only for convenience in handling the mathematics. The relationship between the cycle average intensity and the complex electric field is given in Equation 7 as $I = (c\varepsilon_o/2)|E|^2$. Since we are concerned only with the relative intensity, we suppress the factor of $c\varepsilon_o/2$ and write

$$I = EE^* = |E|^2 \tag{75}$$

In spite of the cycle average, I_s will still have time dependence resulting from fluctuations on a longer time scale. To see how the fluctuations in I_s might arise, consider the situation in Figure 7 where incident light having the wave vector $\underset{\sim}{k}_i$ is scattered at the angle θ_s by small particles in suspension. Neglecting the scattering by solvent molecules, the electric field of the scattered light is a superposition of contributions from the N solute particles in the scattering volume.

$$E_s = \sum_{i=1}^{N} E_i \tag{76}$$

The line AB in Figure 7 represents a plane of constant phase for the incident light, and all of the scattering particles are assumed to be coherently illuminated, i.e., to lie within one coherence length along the beam as specified in Section III.H.1. Also, the structure factor $S(\underset{\sim}{K})$ for each particle is assumed to be unity. Interference will still occur at the plane CD in the far field limit because of the different path lengths traversed by light scattered from different particles. Relative to the ray path that passes through the reference point O, the field E_j at CD associated with the jth particle will have the phase

$$\phi_j = \underset{\sim}{k}_i \cdot \underset{\sim}{r}_j - \underset{\sim}{k}_s \cdot \underset{\sim}{r}_j$$
$$= \underset{\sim}{K} \cdot \underset{\sim}{r}_j$$

where we have defined $\underset{\sim}{K} = \underset{\sim}{k}_i - \underset{\sim}{k}_s$, the scattering vector of the sample. Here $\underset{\sim}{r}_j$ specifies the position of the center of mass of the jth particle. By selecting the scattering angle θ_s, we have chosen a particular scattering vector for study. It is appropriate to view this experiment as Bragg scattering from the component of density fluctuation, which is propagating in the direction of $\underset{\sim}{K}$ and which has the wavelength $\Lambda_f = 2\pi/|\underset{\sim}{K}|$. The crests, which represent maximum particle density, are analogous to the scattering planes of atoms in X-ray crystallography. In order to obtain an expression for the magnitude of $\underset{\sim}{K}$, we note that in quasi-elastic light scattering the wavelength of the scattered light is almost unchanged, i.e., $|\underset{\sim}{k}_i| = |\underset{\sim}{k}_s| = 2\pi/\lambda$. The use of the construction in Figure 7 with elementary trigonometry yields the magnitude f $\underset{\sim}{K}$.

$$|\underset{\sim}{K}| = \frac{4\pi n}{\lambda_o} \sin(\theta_s/2) \tag{38'}$$

The angular dependence in light beating spectroscopy enters through this relation.

The total scattered field can now be written as

$$E_s = \sum_{j=1}^{N} A_j e^{i\underset{\sim}{K}\cdot\underset{\sim}{r}_j} E_o e^{-i\omega_o t} \tag{77}$$

where A_j contains the amplitude factors for the jth particle, which were discussed in Section II.A. In the notation of Section II.C, A_j is equal to aP for a molecule containing P segments. The intensity is given by

$$I_s = E_s E_s^* = \sum_{\ell,m=1}^{N} A_\ell A_m^* e^{i\underset{\sim}{K}\cdot(\underset{\sim}{r}_\ell - \underset{\sim}{r}_m)} E_o^2 \tag{78}$$

If the scattering particles are in motion, each $\underset{\sim}{r}_j$ is time dependent; and I_s fluctuates in time. From Equation 78 the time average of $I_s(t)$ is

$$\langle I_s(t)\rangle = \left[\langle \sum_{j=1}^{N} |A_j|^2 \rangle + \sum_{\ell \neq m}^{N} A_\ell A_m^* \langle e^{i\underset{\sim}{K}\cdot(\underset{\sim}{r}_\ell - \underset{\sim}{r}_m)} \rangle \right] E_o^2 \tag{79}$$

For particles that are identical, but independent, the second term on the right-hand side (rhs) of Equation 79 vanishes since the exponent is randomly distributed; and the equation becomes

$$<I_s(t)> = N<|A_j|^2> E_o^2 \tag{80}$$

3. Correlation Functions

The scattered field E_s is a random variable because of the time dependence implicit in Equation 77. For example, the position vectors $\underset{\sim}{r}_j$ depend on time. Also, if the scatterers are not isotropic, A_j depends on the orientation; and rotational motions produce fluctuations in E_s. Finally, the number of particles N in the scattering volume fluctuates about its mean value <N>. The latter effect is only important at very low concentrations and can usually be neglected for molecular solutes. It should be noted that N refers to the number of particles in the small volume defined by the laser beam and the imaging optics, while in Section II, N referred to the number of particles per unit volume. The most useful way to characterize the time dependence of $E_s(t)$ is through the first-order field autocorrelation function $G^{(1)}(\tau)$ defined by

$$G^{(1)}(\tau) = <E_s^*(t)E_s(t+\tau)> = <E_s^*(O)E_s(\tau)> \tag{81}$$

The rhs. of Equation 81 represents the time average of the product of the complex conjugate of the field at one time with the field at a time τ later. It is permissible to replace t in Equation 81 with zero, since $E_s(t)$ is a stationary random variable, and its properties are independent of the time origin. For τ = O, $G^{(1)}(\tau) = <I_s>$, while for large values of τ, $E_s(t)$ and$E_s(t + \tau)$ are uncorrelated and $G^{(1)}(\tau)$ approaches $<E_s(t)>2$, which is equal to zero.

For the scattered field E_s described in Equation 77 with N = <N>, Equation 82 gives:

$$G^{(1)}(\tau) = \sum_{\ell,m=1}^{N} <A_\ell^*(O)A_m(\tau)e^{i\underset{\sim}{K}\cdot[\underset{\sim}{r}_m(\tau) - \underset{\sim}{r}_\ell(O)]}> E_o^2 e^{-i\omega_o\tau} \tag{82}$$

If the motions of the different particles are uncorrelated and the translational and rotational motions are independent, Equation 82 reduces to:

$$G^{(1)}(\tau) = N<A_\ell^*(O)A_\ell(\tau)><e^{i\underset{\sim}{K}\cdot[\underset{\sim}{r}_\ell(\tau) - \underset{\sim}{r}_\ell(O)]}> E_o^2 e^{-i\omega_o\tau} \tag{83}$$

The summation has been dropped and the factor N inserted, since each particle has the same statistical behavior. The subscript ℓ is superfluous and can also be dropped. Thus, the field correlation function $G^{(1)}(\tau)$ is found to be proportional to the product of single particle correlation functions for rotation and translation.

The scattered intensity $I_s(t)$ is usually detected with a photomultiplier tube (PMT), and fluctuations in $I_s(t)$ appear either as fluctuations in the photocurrent i(t) or as fluctuations in the photocount rate. Thus with an ideal detection system

$$I_s(t) \propto i(t) \propto n(t,\Delta T)$$

Here $n(t,\Delta T)$ means the number of photons detected in the time interval from t to t + ΔT. The measured quantity is always $I_s(t)$ rather than $E_s(t)$, and the directly measured autocorrelation function is the second order function $G^{(2)}(\tau)$ defined by

$$G^{(2)}(\tau) = \langle I_s(O) I_s(\tau) \rangle \tag{84}$$

In general $G^{(2)}(\tau)$ is not simply related to $G^{(1)}(\tau)$; however, in special cases useful relations exist. For example, if $E_s(t)$ is a Gaussian random variable the Siegert relation gives

$$G^{(2)}(\tau) = \langle I_s \rangle^2 + |G^{(1)}(\tau)|^2 \tag{85}$$

Practically speaking this equation holds for scattering from solutions at room temperature, except at such low concentrations that number fluctuations become important (see Section III.F). The derivation of Equation 85 is not trivial, but elementary discussions are available.[35] It is conventional to define the reduced first and second order correlation functions $g^{(1)}(\tau)$ and $g^{(2)}(\tau)$, respectively, through the equations

$$g^{(1)}(\tau) = \langle E_s^*(O) E_s(\tau) \rangle / \langle I \rangle \tag{86}$$

$$g^{(2)}(\tau) = \langle E_s^*(O) E_s^*(O) E_s^*(\tau) E_s(\tau) \rangle / \langle I \rangle^2 \tag{87}$$

To see how an approximation to $G^{(2)}(\tau)$ can be derived from experimental data, consider Figure 8. The intensity is measured at time intervals Δt, and $G^{(2)}(\tau)$ is calculated through the relation

$$G^{(2)}(\tau) = \lim_{n\to\infty} \frac{1}{n} \sum_{i=O}^{n-1} I_s(t_i) I_s(t_{i+j}) \tag{88}$$

where $\tau = t_{i+j} - t_i = j\Delta t$. Commercial digital autocorrelators are available, which simultaneously calculate the products for all required values of τ and average these products to obtain an accurate representation of $G^{(2)}(\tau)$ in a few seconds if the intensity is sufficiently large.

While the function $G^{(2)}(\tau)$ is not familiar to most biochemists, its major features are easy to grasp. First consider the limits of short and long times.

$$\lim_{\tau\to O} \langle I_s(O) I_s(\tau) \rangle = \langle I_s^2(O) \rangle \tag{89}$$

$$\lim_{\tau\to\infty} \langle I_s(O) I_s(\tau) \rangle = \langle I_s(O) \rangle^2 \tag{90}$$

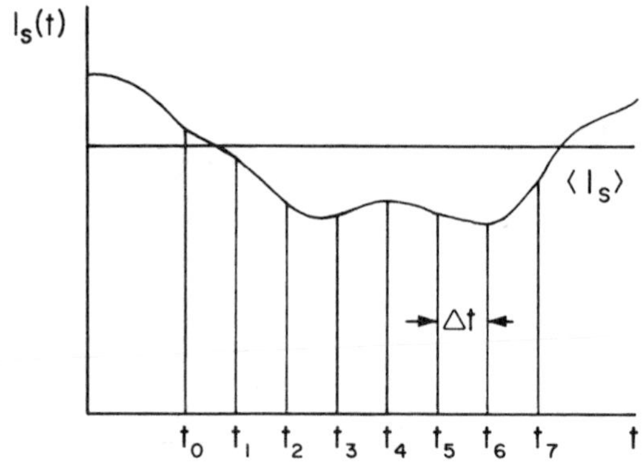

FIGURE 8. The scattered intensity $I_s(t)$ vs. time.

At the short time limit we obtain simply the average of $I^2_s(O)$. This is the largest value that $G^{(2)}(\tau)$ can attain. As τ increases the correlation between $I_s(t)$ and $I_s(t+\tau)$ decreases until the two values are completely independent. The average of their product becomes equal to the product of their averages. This behavior is illustrated in Figure 9 which shows a typical plot of $G^{(2)}(\tau)$ vs. τ. The monotonically decreasing function of τ is, for example, encountered for scattering particles that are undergoing Brownian motion. This particular case is discussed in detail in Section III.B. In other types of experiments oscillations are sometimes found in $G^{(2)}(\tau)$, but always the condition

$$G^{(2)}(\tau) \leqq \langle I_s^2(O)\rangle \tag{91}$$

is satisfied.

4. The Frequency Spectrum

The rate at which $G^{(2)}(\tau)$ decreases indicates how rapidly $I_s(t)$ fluctuates. Rapid fluctuations in turn indicate that high frequency components are present, i.e., a rapid decay in the time domain corresponds to a broad line in the frequency domain. The quantitative expression of these ideas is given in Appendix E, which relates the power spectrum of I_s to its correlation function.

$$I_s^{(2)}(\omega) = \frac{\text{Re}}{\pi}\int_0^\infty G^{(2)}(\tau)e^{i\omega\tau}d\tau \tag{92}$$

In experimental situations, $I_s^{(2)}(\omega)$ can be obtained directly by analysis of the PMT photocurrent using a spectrum analyzer. It should be realized that $I_s^{(2)}(\omega)$ is not equivalent to the usual optical spectrum $I_s^{(1)}(\omega)$, which is obtained by filter type optical

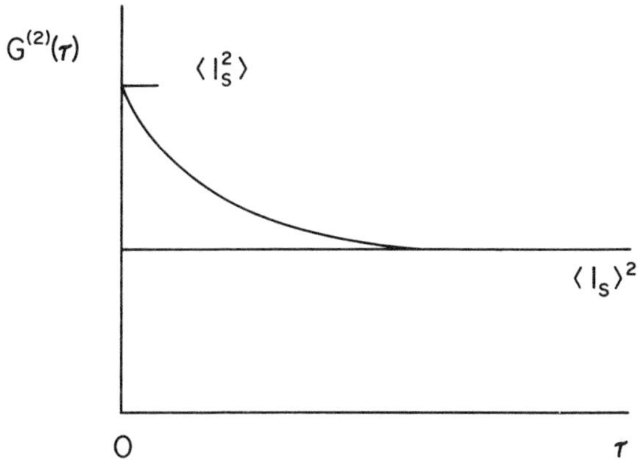

FIGURE 9. Typical behavior of the second-order correlation function $G^{(2)}(\tau)$ vs. τ.

instruments such as grating monochrometers. However, $I_s^{(1)}(\omega)$ is related to $G^{(1)}(\tau)$ and can in principle be calculated using the results of Appendix E.

$$I_s^{(1)}(\omega) = \frac{\text{Re}}{\pi} \int_0^\infty G^{(1)}(\tau) e^{i\omega\tau} d\tau \tag{93}$$

A very important feature of $G^{(2)}(\tau)$ is that E_s appears only in products with its complex conjugate. Factors involving the angular frequency ω_o of the incident light cancel, and only fluctuations in the audio range remain. Accordingly, the spectrum $I^{(2)}(\omega)$ is referred to zero frequency. In contrast to this $G^{(1)}(\tau)$ still contains ω_o, and the frequencies in $I_s^{(1)}(\omega)$ are referenced to ω_o.

B. Translational Diffusion

1. The Diffusion Equation

Probably the most important application of photon correlation spectroscopy is the study of translational diffusion for macromolecules in solution.[36] This technique often permits the z-average diffusion coefficient D_T to be measured quickly and accurately. D_T directly gives information about molecular diameters and when combined with other results, e.g., the Svedberg constant, permits the molecular weights to be determined. The equations that relate D_T to the measured intensity correlation function are based on Equation 83 for $G^{(1)}(\tau)$. For small isotropic scatters Equations 80 and 83 can be combined to obtain the reduced correlation function $g^{(1)}(\tau)$ as defined in Equation 86.

$$g^{(1)}(\tau) = \langle e^{i\underset{\sim}{K}\cdot[\underset{\sim}{r}_\ell(\tau) - \underset{\sim}{r}_\ell(O)]} \rangle e^{-i\omega_o\tau} \tag{94}$$

The heart of the problem is the evaluation of the function $F_s(K,t) = \langle e^{iK\cdot r} \rangle$ where for simplicity we have introduced $r = r_\ell(t) - r_\ell(O)$. For translational diffusion the appropriate average can be calculated using the weighting factor $P(O|r,t)$, which is the conditional probability that a particle will be found in the volume element d^3r at the position r at time t if it were located at $r = O$ initially. Thus

$$F_s(K,t) = \int_0^\infty P(O|r,t)e^{iK\cdot r} d^3r \tag{95}$$

It should be recognized that $F_s(K,t)$ is just the spatial Fourier transform of $P(O|r,t)$.

According to Fick's first law of diffusion, the particle flux J, i.e., the rate of flow of mass at r, is proportional to the gradient of the concentration.

$$J(r,t) = -D_T \nabla C(r,t) \tag{96}$$

The continuity equation, which assures the conservation of mass, is

$$\frac{\partial C(r,t)}{\partial t} = -\nabla \cdot J(r,t) \tag{97}$$

and this equation can be combined with the definition of J to give Fick's second law of diffusion

$$\frac{\partial C(r,t)}{\partial t} = D_T \nabla^2 C(r,t) \tag{98}$$

In this derivation it is assumed that the translational diffusion coefficient D_T is independent of concentration. This is not strictly true, and Equation 98 is only expected to be accurate in the limit of low concentrations. Equations for D_T and its concentration dependence are discussed in Appendix H. At low concentrations it is reasonable to assume that $P(O|r,t)$ also obeys the diffusion equation so that

$$\frac{\partial P(O|r,t)}{\partial t} = D_T \nabla^2 P(O|r,t) \tag{99}$$

The brute force procedure is to solve Equation 99 for $P(O|r,t)$ with the initial condition $P(O|r,O) = \delta(r)$ where $\delta(r)$ is the Dirac delta function. $P(O|r,t)$ is then substituted into Equation 95, and the integral is evaluated to obtain $F_s(K,t)$. It is much simpler to use Fourier transform methods to obtain $F(K,t)$ directly from Equation 99. We first multiply both sides of Equation 99 by $e^{iK\cdot r}$ and then integrate over volume to obtain

$$\int_0^{\infty} e^{i\underset{\sim}{K}\cdot\underset{\sim}{r}} \frac{\partial P(O|\underset{\sim}{r},t)}{\partial t} d^3r = D_T \int_0^{\infty} e^{i\underset{\sim}{K}\cdot\underset{\sim}{r}} \nabla^2 P(O|\underset{\sim}{r},t)d^3r$$

$$\frac{\partial F_s(\underset{\sim}{K},t)}{\partial t} = -D_T K^2 F_s(\underset{\sim}{K},t) \qquad (100)$$

On the left-hand side (lhs) we have simply factored $\partial/\partial t$ from the integral and used Equation 95 for $F_s(\underset{\sim}{K},t)$. The rhs of Equation 100 follows from a property of Fourier transforms, namely,

$$\int_{-\infty}^{+\infty} e^{iKy} \frac{\partial^n}{\partial y^n} P(y)dy = (-iK)^n \int_{-\infty}^{+\infty} e^{iKy} P(y)dy \qquad (101)$$

which can be proved using integration by parts.

Equation 100 is a first order differential equation that can easily be integrated to give

$$F_S(\underset{\sim}{K},t) = F_S(\underset{\sim}{K},O)\, e^{-D_T K^2 t} \qquad (102)$$

The appropriate initial condition is $F_s(\underset{\sim}{K},O) = 1$, which can be derived either by referring to the definition of $F_s(\underset{\sim}{K},t)$ in Equation 94 or by substituting the initial condition $P(O|\underset{\sim}{r},0) = \delta(\underset{\sim}{r})$ into Equation 95. Now having Equation 102 for $F_s(\underset{\sim}{K},t)$ we return to Equation 94 to obtain the desired expression

$$g^{(1)}(\tau) = \frac{G^{(1)}(\tau)}{<I_s>} = e^{-D_T K^2 \tau}\, e^{-i\omega_o \tau} \qquad (103)$$

The optical spectrum associated with $G^{(1)}(\tau)$ can be derived using Equation 93. Thus,

$$I^{(1)}(\omega) = \frac{Re}{\pi} \int_0^{\infty} \left[<I_S> e^{-D_T K^2 \tau}\, e^{-i\omega_o \tau}\right] e^{i\omega\tau} d\tau$$
$$= \frac{<I_S>}{\pi} \left[\frac{D_T K^2}{(D_T K^2)^2 + (\omega_0 - \omega)^2}\right] \qquad (104)$$

This equation describes a Lorentzian curve centered at the laser frequency ω_o and having a half-width at half-height (HWHH) of $D_T K^2$ (see Figure 10a). However, as previously pointed out, ω_o is at least 10^{10} tmes larger than the linewidth if the scattering particles are macromolecules; and it is impossible with optical techniques to measure the linewidth. Experiments which do permit D_T to be determined are based on $G^{(2)}(\tau)$.

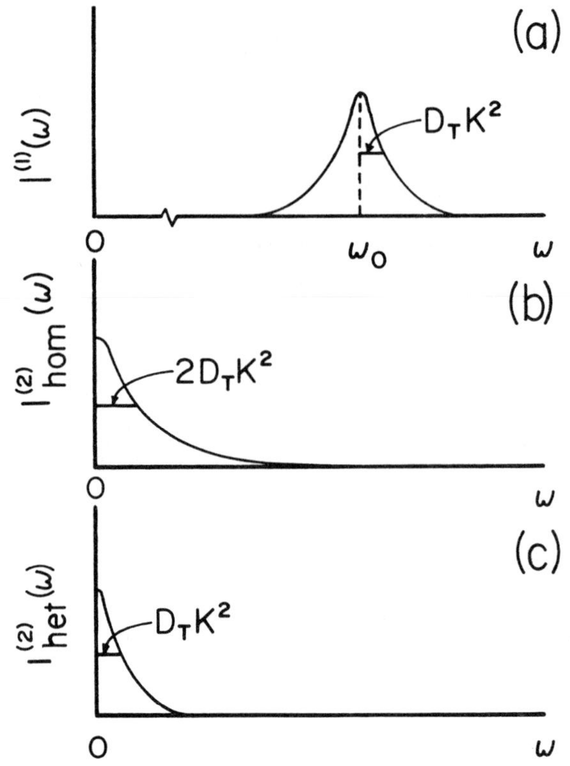

FIGURE 10. Calculated spectra for light scattered from diffusing particles in three cases: (a) the optical spectrum, (b) the power spectrum in a homodyne experiment, and (c) the power spectrum in a heterodyne experiment.

2. Homodyne Experiment

In this experiment the scattered light is detected with a PMT and the photon count rate or the photocurrent is used to generate an approximation to $G^{(2)}(\tau)$. The Siegert relation, Equation 85, is then used in the modified form:

$$\frac{G^{(2)}(\tau)}{\langle I_S \rangle^2} = g^{(2)}(\tau) = 1 + \beta |g^{(1)}(\tau)|^2 \qquad (105)$$

where β is a factor of order unity, which accounts for the temporal and spatial integrations which are unavoidable in real experiments.[37] Approximations to β can be derived, but it is usually just taken to be an instrumental parameter which is maximized, if possible, and then measured in each experiment. The output of an autocorrelator provides sufficient information for the calculation of $g^{(2)}(\tau)$. This reduced correlation function is then analyzed using Equation 105 in the form:

$$Y(\tau) = \ln \left[g^{(2)}(\tau) - 1 \right]^{1/2} = \frac{1}{2} \ln \beta + \ln |g^{(1)}(\tau)| \qquad (106)$$

For translational diffusion in a monodisperse solution Equation 103 is introduced for $g^{(1)}(\tau)$ to give

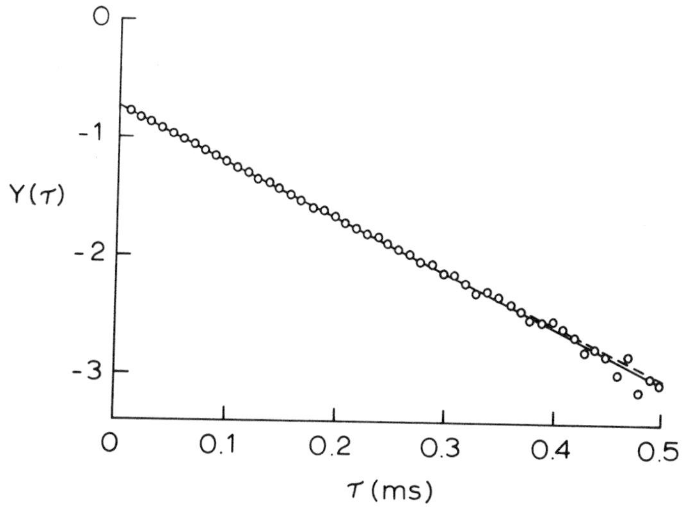

FIGURE 11. Experimental values of Y(τ) plotted vs. τ for a sample of oxyhemoglobin A at a concentration of 16.3 g/dl and T = 20°C. The curves represent the best linear (solid) and best quadratic (dashed) fits to the data (see Equations 107 and 115). (From Jones, C. R., Ph.D. thesis, University of North Carolina, Chapel Hill, 1977.)

$$Y(\tau) = \frac{1}{2} \ell n \, \beta - D_T K^2 \tau \tag{107}$$

A plot of Y(τ) vs. τ then gives a straight line having the slope $-D_T K^2$ and the intercept $(\ell n\beta)/2$. An example of such a plot for oxy-hemoglobin A is shown in Figure 11.[38] The small deviation from linearity indicates polydispersity and suggests that Equation 107 should have additional terms on the rhs which depend on higher powers of τ.

The power spectrum associated with $G^{(2)}(\tau)$ can be obtained using Equations 92, 103, and 105.

$$\begin{aligned} I_{hom}^{(2)}(\omega) &= \frac{Re}{\pi} \int_0^{\infty} \langle I_s \rangle^2 \left[1 + \beta e^{-2D_T K^2 \tau} \right] e^{i\omega\tau} \, d\tau \\ &= \langle I_S \rangle^2 \delta(\omega) + \frac{\beta \langle I_S \rangle^2}{\pi} \left[\frac{2D_T K^2}{(2D_T K^2)^2 + \omega^2} \right] \end{aligned} \tag{108}$$

This equation represents a spike (delta function) at $\omega = O$ and a superimposed Lorentzian curve having a HWHH of $2D_T k^2$ centered at $\omega = O$ (see Figure 10B). The power spectrum lies in the audio range and can be obtained directly from the fluctuating photocurrent by means of an electronic spectrum analyzer. In fact, light-beating experiments were first done this way before correlators were available. An example of an experimental power spectrum is shown in Figure 12 for a 1% solution of RNase A at $\theta_S = 90°$ and T = 24°C.[39] This experiment in principle gives the same information as that obtained from $G^{(2)}(\tau)$; however, in practice the power spectrum sometimes gives less warning of polydispersity. For example the data points in Figure 12 fit the calculated Lorentzian curve within experimental error even though the sample contains dimers and tetramers as well as monomers. The measured linewidth (HWHH) for this sample is 7.63 kHz, while that obtained for a solution containing only monomers is

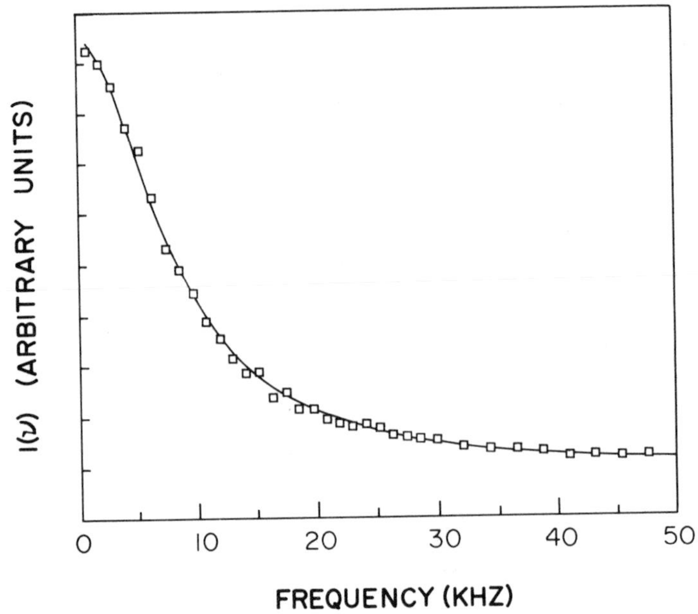

FIGURE 12. The power spectrum $I^{(2)}(\omega)$ of a 1% solution of RNase A measured at $\theta_s = 90°$ and T = 24°C. The solid curve is the best single Lorentzian fit to the experimental data (every fourth data point is displayed). The HWHH value is 7.63 KHz. (From Wilson, W. W., Ph.D. thesis, University of North Carolina, Chapel Hill, 1973.)

12.9 kHz corresponding to $D_T = 11.7 \times 10^{-7}$ cm²/sec at 20°C. In this case the determination of the fraction of monomers required the use of chromatography.

3. Heterodyne Experiment

In this experiment a reference beam is obtained either by deflecting a small fraction of the intensity of the laser light before it reaches the sample or by scattering part of the incident light from a solid object in the scattering volume, e.g., a needle. In either case the reference beam is mixed with the light scattered from the sample on the PMT. The reference beam is called the local oscillator, and its electric field and intensity are denoted by $E_L(t)$ and $I_L(t)$, respectively. The total electric field $E_{det}(t)$ of the light at the PMT detector is the sum of contributions from the local oscillator and light scattered from the sample. Thus

$$E_{det}(t) = E_S(t) + E_L(t) \tag{109}$$

and the associated intensity correlation function is

$$G^{(2)}(\tau) = \langle E^*_{det}(t)\, E_{det}(t)\, E^*_{det}(t+\tau)\, E_{det}(t+\tau)\rangle \tag{110}$$

This function is considered in Appendix F for the special case that $I_L \gg \langle I_S \rangle$, and it is shown that

$$G^{(2)}(\tau) = I_L^2 + 2\,I_L\,\mathrm{Re}\left[G_S^{(1)}(\tau)\,e^{i\omega_o\tau}\right] \tag{111}$$

where $G_s^{(1)}(\tau) = \langle E^*_s(O)\,E_s(\tau)\rangle$. When the intensity fluctuations result from translational diffusion, $G_s^{(1)}(\tau)$ is given by Equation 103, and 111 becomes

$$G^{(2)}(\tau) = I_L^2 + 2\,I_L \langle I_S\rangle\, e^{-D_T K^2 \tau} \tag{112}$$

In contrast to the $G^{(2)}(\tau)$ function which was derived for the homodyne experiment, here the background (base line) depends on the intensity of the reference beam, and the time dependent term contains $\mathrm{Re}[G_s^{(1)}(\tau)]$ rather than $|G_s^{(1)}(\tau)|^2$.

The power spectrum associated with Equation 112 is obtained by using Equation 92,

$$\begin{aligned} I_{het}^{(2)}(\tau) &= \frac{\mathrm{Re}}{\pi}\int_0^\infty \left[I_L^2 + 2\,I_L\langle I_S\rangle\,\mathrm{Re}\left(e^{-D_T K^2\tau}\right)\right] e^{i\omega\tau}\,d\tau \\ &= I_L^2\,\delta(\omega) + \frac{2}{\pi}\,I_L\langle I_S\rangle\,\mathrm{Re}\int_0^\infty e^{-D_T K^2\tau}\,e^{i\omega\tau}\,d\tau \\ &= I_L^2\,\delta(\omega) + \frac{2}{\pi}\,I_L\langle I_S\rangle\left[\frac{D_T K^2}{(D_T K^2)^2 + \omega^2}\right] \end{aligned} \tag{113}$$

The result is a delta function at zero frequency and a Lorentzian curve having the same width as the optical spectrum but also centered at zero frequency rather than at the frequency of the incident light (see Figure 10c). The linewidth is thus a factor of two, smaller in the heterodyne experiment than in the homodyne experiment. There is usually no particular advantage to using the heterodyne arrangement when translational diffusion is being studied, but there is the danger that an unknown amount of the heterodyne component may be present in what is thought to be a homodyne experiment because of stray light or because of dust particles in solution. However, when the scattering particles are undergoing uniform motion, e.g., in flowing liquids or in electrophoresis experiments, the Doppler effect associated with the velocity of the particles relative to the source and the detector can be measured in the heterodyne experiment, but not in the homodyne experiment. This forms the basis of laser velocimetry, some applications of which are discussed in Section III.C.

4. Data Analysis and Experimental Results

Quasi-elastic scattering has become the standard method for the measurement of diffusion coefficients for macromolecules. Any of the arrangements discussed above can be used to measure D_T; however, the determination of $G^{(2)}(\tau)$ by means of the homodyne experiment is the method of choice. Major reasons for this are experimental convenience and ease of data analysis. Since all samples are polydisperse to some extent, the function $|g^{(1)}(\tau)|$, required in Equation 106, must in general be written as

$$|g^{(1)}(\tau)| = \int_0^\infty G(\Gamma)\, e^{-\Gamma\tau} d\Gamma \qquad (114)$$

where G(Γ) is the normalized distribution function for the decay constants and $\Gamma = D_T K^2$. The consequence of the distribution is that Equation 107 must be replaced with the cumulant expansion[40]

$$Y(\tau) = \frac{1}{2} \ln \beta + \sum_{m=1}^{\infty} K_m \frac{(-\tau)^m}{m!} \qquad (115)$$

It is shown in Appendix G that the first four cumulants K_m are

$$K_1 = <\Gamma>, K_2 = \mu_2, K_3 = \mu_3, K_4 = \mu_4 - 3(\mu_2)^2 \qquad (116)$$

where the mth moment of the distribution is defined by

$$\mu_m = \int_0^\infty (\Gamma - <\Gamma>)^m G(\Gamma) d\Gamma \qquad (117)$$

and

$$<\Gamma> = \int_0^\infty \Gamma G(\Gamma) d\Gamma \qquad (118)$$

The z-average diffusion coefficient is given by $<D_T> = <\Gamma>/K^2$, the standard deviation of the distribution is specified by $\sqrt{\mu_2}$, the skewness by $\mu_3/\mu_2^{3/2}$, etc. The normalized variance $\mu_2/<\Gamma>^2$ is often quoted as a measure of polydispersity.

Several schemes have been proposed for determining the moments from experimental data. For example, since

$$(-1)^n \frac{d^n}{d\tau^n} Y(\tau) \bigg|_{\tau = 0} = K_n \qquad (119)$$

a plot of the second derivative of Y(τ) with respect to τ vs. τ can be extrapolated to τ = O to obtain μ_2. For samples which are supposed to be monodisperse it is also reasonable to fit Y(τ) to a quadratic function in τ. According to Equation 115 the linear term contains $<D_T>$ and the coefficient of τ^2 can be used to calculate the variance. If the variance is greater than 0.02, an artifact such as dust in the sample or stray light is probably present. In principle the variance should be useful in the determination of equilibrium constants for dissociation reactions. However, a simple calculation for monomer-dimer equilibria shows that the variance never exceeds ~0.014.

Experience shows that $\langle D_T \rangle$ can be measured in a few minutes with an error of only about 1% for nonabsorbing samples. Variances greater than 0.1 can be measured with fair accuracy, but the higher cumulants have very large uncertainties. Further progress can be made only if the functional form of $G(\Gamma)$ is known or assumed.[41] Additional information about polydispersity can be obtained from a plot of $\langle \Gamma \rangle$ vs. K^2; therefore, a determination of the angular dependence of $\langle \Gamma \rangle$ is always recommended.

The translational diffusion coefficient D_T was introduced in Equation 96. For a sphere in a viscous medium, D_T is given by

$$D_T = \frac{k_B T}{f_T} = \frac{k_B T}{6\pi\eta a_h} \tag{120}$$

where f_T is the <u>translational friction coefficient</u>, η is the coefficient of viscosity of the solvent, and a_h is the hydrodynamic radius of the sphere. The derivation of this equation, which was first presented by Einstein,[42] is discussed in Appendix H. Equation 120 permits the calculation of molecular radii; however, it should be kept in mind that the hydrodynamic radius may include a hydration layer. Also, it is clear from experimental results that D_T depends on the average concentration $C(\underset{\sim}{r},t) \equiv c$. The concentration dependence enters the theory in two ways. First, the derivation of Equation 120 requires that the gradient of the chemical potential μ_2, which is the driving force of the diffusion, be related to the gradient of the concentration of the solute. At finite concentrations this introduces a virial expansion in the numerator of Equation 120. And second, the friction coefficient f depends on the concentration and it can be expanded in terms of the concentration. The result is that D_T can be written as (see Appendix H)[43]

$$D_T(C) = \frac{k_B T}{f_o} \frac{(1 + 2B_2\, MC + 3B_3\, MC^2 + \ldots)}{(1 + B'\, C + \ldots)} \tag{121}$$

where the B_i are the virial coefficients which occur in the expansion of the chemical potential μ_1 of the <u>solvent</u> in terms of C. For small concentrations Equation 121 has the form

$$D_T(C) = D_o\,(1 + K_D\, C + \ldots) \tag{122}$$

where $K_D = (2MB_2 - B')$. The partial cancellation which occurs here probably accounts for the small concentration dependence which is often found for diffusion coefficients. Numerous theoretical treatments have been presented which yield equations for D_T similar to Equation 122, but with different values of K_D. There is still no consensus on the proper way to include hydrodynamic interactions in the calculation of K_D.

Table 3
TRANSLATIONAL DIFFUSION COEFFICIENTS FROM QUASI-ELASTIC LIGHT SCATTERING FOR SELECTED SYSTEMS

Macromolecule	Conditions	$D_T(10^{-7}\ cm^2/s)$	Ref.
Myosin (monomer)		1.24	46
(dimer)		0.84	46
Fibrinogen		2.04 ± 0.09	47
γ-Globulin		3.8	43
Bovine glutamate dehydrogenase (GDG) monomer, pH 7		4.51[a]	48
Oxy-hemoglobin (oxy-Hb)	pH 6.9 (HbA); pH 7.1 (HbS); 0.1 *M* KCl	6.9[b]	49
Bovine serum albumin (BSA)	pH 6.91; no salt	10.2[c]	50
	pH 6.8; 0.5 *M* KCl	6.7	50
Lysozyme	pH 4.2; sodium acetate-acetic acid buffer	10.6[d]	51
Ribonuclease A (RNase)	pH 8.1; 0.5 *M* KCl; 24°C	12.6	52
Avian myeloblastosis virus (AMV)		0.268	53
Vesicular stomatitis virus (VSV)		0.29	54
Rous sarcoma virus (RSV)		0.291	53
Tobacco mosaic virus (TMV)		0.39	55
Infectious pancreatic nucrosis virus (IPNV)		0.67	56
Turnip yellow mosaic virus (TYMV)		1.44	56
R17		1.534 ± 0.015	57
Qβ		1.423 ± 0.014	57
BSV		1.246 ± 0.013	57
PM2		0.650 ± 0.007	57
T2		0.644 ± 0.007	57
Ribosomes 70S		1.81	58
50S		2.1	58
30S		2.2	58

[a] Corrected to water at 25°C.
[b] Corrected to water at 20°C.
[c] Corrected to water at 25°C.
[d] Corrected to water at 20°C.

We emphasize that quasi-elastic light scattering detects mutual diffusion, which is driven by a concentration gradient of solute molecules. This is to be contrasted with tracer diffusion where the solute concentration is uniform, but the migration of "labeled" solute molecules can be followed. The classical diaphragm method in which diffusion through a millipore membrane is measured, while satisfactory for the measurement of tracer diffusion coefficients, is apparently invalid for mutual diffusion measurements on protein solutions because of the osmotic back flow of solvent across the membrane.[44,116] This explains the marked discrepancies between the concentration dependences of D_T obtained by quasi-elastic light scattering and by the membrane method.[45,49] For charged macromolecules the situation is even more complicated. One important consequence of long range interactions is a $\underset{\sim}{K}$ dependence of the apparent diffusion coefficient. Recent work in this area is reviewed in Reference 7.

The number of diffusion coefficients determined by quasi-elastic light scattering is now very large, and we only list a few representative examples in Table 3. In reporting diffusion coefficients, it is common to correct the measured values of D_T so that they represent the hypothetical case of the same hydrodynamic particle moving in pure water at 20°C. This is accomplished using Equation 120 and the fact that $f = \eta(T)G$

where $\eta(T_1)$ is the viscosity of the solvent and G is a form factor, which is characteristic of the molecule.[59] For macromolecules in salt solutions this treatment assumes that there are only two components, i.e., the solvent is taken to be the salt solution without the macromolecule and $\eta(T)$ refers to the viscosity of this salt solution. Thus, if D_T is the measured quantity at the temperature T, the correction gives

$$D_T(20°\text{C, water}) = D_T(\text{T, salt sol}) \left[\frac{293.2}{T(°K)}\right] \left[\frac{\eta(\text{T, salt sol})}{\eta(20°\text{C, water})}\right] \tag{123}$$

where $\eta(20°, \text{water}) = 1.008$ cP.

It should be noted that accurate diffusion constants can be obtained for large particles such as viruses and ribosomes. This is important in the determination of weights of such particles since their diffusion coefficients are not easily obtained by other methods. The procedure is to combine the measured z-average diffusion coefficient D_T with the sedimentation coefficient S_o in the Svedberg equation to obtain the weight average molecular weight in the limit of infinite dilution.[60]

$$M_w = \frac{RT\,S_0}{D_T(1 - \bar{v}\rho)} \tag{124}$$

Here $\bar{v}$ is the partial specific volume of the solute and ρ is the density of the solvent. For example with reovirus, where $S_o = 734 \times 10^{-13}$, $D_T = 0.44 \times 10^{-7}$ cm²/s, and $\bar{v} = 0.690$ mℓ/g, Equation 124 gives $M_w = 12.5 \times 10^6$.[56]

C. Directed Flow

1. Constant Velocities and Laser Velocimetry

Frequency shifts for light scattered from particles in uniform motion are conceptually simpler than for diffusing particles. The reason for this is that the concept of the Doppler shift can be applied more directly for constant velocity. Consider for example the situation in Figure 13. The angular frequency of the light emitted from the laser is ω_o; however, the apparent frequency seen by the scattering particle is Doppler shifted by the amount $\Delta\omega = -(nv_d \cdot \hat{e}/c)\omega_o$ where $\hat{e}$ is a unit vector in the direction of $\underset{\sim}{k_i}$ and $\underset{\sim}{v_d}$ is the velocity of the scattering particle. A second Doppler shift enters the measurement since the scattering particle is also moving relative to the detector. The situation can be summarized as follows using $n\omega_o/c = |\underset{\sim}{k_i}| = |\underset{\sim}{k_s}|$.

1.	Source frequency:	ω_o
2.	Frequency seen by moving particle	$\omega_o - \underset{\sim}{k_i} \cdot \underset{\sim}{v_d}$
3.	Frequency seen by detector	$(\omega_o - \underset{\sim}{k_i} \cdot \underset{\sim}{v_d}) + \underset{\sim}{k_s} \cdot \underset{\sim}{v_d}$

Thus the detected frequency is shifted relative to the source frequency by the amount $\Delta\omega = -\underset{\sim}{K} \cdot \underset{\sim}{v_d}$. This result is, in fact, already contained in Equation 83 for $G^{(1)}(\tau)$ since for uniform motion $\underset{\sim}{r_\ell}(\tau) - \underset{\sim}{r_\ell}(O) = \underset{\sim}{v_d}\tau$. The correlation function is then

$$G^{(1)}(\tau) = N < |A_\ell|^2 > e^{i\underset{\sim}{K} \cdot \underset{\sim}{v_d}\tau} E_0^2\, e^{-i\omega_o\tau} \tag{125}$$

which is equivalent to replacing the source frequency of $\omega_o - \underset{\sim}{K} \cdot \underset{\sim}{v_d}$.

As previously discussed, the magnitude of $\underset{\sim}{v}_d$ expected for a macromolecule corresponds to such a small frequency shift that a light-beating experiment must be used to measure $\Delta\omega$. Suppose that a homodyne experiment is employed. In this case Equations 85 and 125 combine to show that $G^{(2)}_{hom}(\tau) = 2 <I_s>^2$, i.e., the intensity correlation function contains no information at all about $\underset{\sim}{v}_d$! The reason for this is that in a homodyne experiment the phase of the scattered wave does not affect the measured intensity. In order to observe the effect of $\underset{\sim}{v}_d$, a fixed reference frequency must be introduced, so that interference terms will alter the intensity, and this is just what is done in the heterodyne experiment. The information content of the heterodyne experiment can be seen by combining Equation 125 for $G^{(1)}(\tau)$ with Equation 111 to obtain

$$\begin{aligned} G^{(2)}_{het}(\tau) &= I_L^2 + 2 I_L \, \mathrm{Re}[N<|A_\ell|^2> e^{i\underset{\sim}{K}\cdot\underset{\sim}{v}_d\tau}] E_o^2 \\ &= I_L^2 + 2 I_L <I_S> \cos(\underset{\sim}{K}\cdot\underset{\sim}{v}_d\tau) \end{aligned} \tag{126}$$

The oscillatory behavior exhibited by Equation 126 indicates that a frequency shift is present. The exact form of the frequency spectrum is obtained by substituting Equation 126 into Equation 92,

$$\begin{aligned} I^{(2)}_{het}(\omega) &= I_L^2\,\delta(\omega) + \frac{2 I_L}{\pi} <I_S> \mathrm{Re}\int_0^\infty \cos(\underset{\sim}{K}\cdot\underset{\sim}{v}_d\tau)\, e^{-i\omega\tau} d\tau \\ &= I_L^2\,\delta(\omega) + I_L <I_S> [\delta(\omega - \underset{\sim}{K}\cdot\underset{\sim}{v}_d) + \delta(\omega + \underset{\sim}{K}\cdot\underset{\sim}{v}_d)] \end{aligned} \tag{127}$$

where, as in Equation 113, we have used the fact that the Fourier transform of unity is the delta function. Equation 127 shows that an infinitely narrow peak appears with a shift of $\Delta\omega = \underset{\sim}{K}\cdot\underset{\sim}{v}_d$ for either the velocity $\underset{\sim}{v}_d$ or $-\underset{\sim}{v}_d$. Perfectly uniform velocities are, of course, unattainable; and the distribution of velocities of the scattering particles will determine the width and shape of the shifted peak.

For application to real systems Equation 127 can be written as

$$I_{exp}(\omega) = \int P(v)\, I^{(2)}_{het}(\omega)\, dv \tag{128}$$

The experimental lineshape $I_{exp}(\omega)$ can then be used to determine the velocity distribution function P(v). Two important applications of this equation have been to blood flow in vessels and protoplasmic streaming in living cells. For example, Tanaka et al.[61] measured the flow rate in retinal vessels by focusing a low power laser into the eye and analyzing the back scattered light. Also, Tanaka and Benedek[62] measured blood flow in the femoral vein of a rabbit by inserting a fiber optic catheter with a beveled and polished end. Laser light passed through the fiber into the plasma, and the light scattered by moving particles (erythrocytes) was collected by the same fiber. In the first of a series of investigations Mustacich and Ware[63] scattered light from the protoplasm in a *Nitella* cell and used the frequency spectrum to map the flow pattern and to determine the velocity distribution function.

A final comment concerning the inability of the homodyne experiment to detect the uniform velocity $\underset{\sim}{v_d}$ is in order. The homodyne experiment does contain information about diffusion since the ensemble average of the phase factor $e^{i\underset{\sim}{K}\cdot\underset{\sim}{r}}$ produces the real function $F_s(\underset{\sim}{K},t)$ as shown in Equation 102. In a monodisperse solution each scattering particle exhibits the same average behavior, and each term on the rhs of Equation 82 makes the same contribution. A quite different result is expected for scattering from a gas at low pressure. The jth molecule has the velocity $\underset{\sim}{v_j}$ and contributes the term

$$< |A_j|^2 > e^{i\underset{\sim}{K}\cdot\underset{\sim}{v_j}\tau}\, E_o^2\, e^{-i\omega_o\tau}$$

to the sum in Equation 82. The contributions from the different molecules are not the same; however, a Maxwell-Boltzmann distribution of velocities is present and the summation in Equation 82 can be performed without difficulty (see Section III.E). The weighted sum of the phase factors $e^{i\underset{\sim}{K}\cdot\underset{\sim}{v_j}\tau}$ produces a real function which contains the mean square velocity $<\underset{\sim}{v^2}>$. The homodyne experiment, therefore, permits the measurement of $<\underset{\sim}{v^2}>$ even though no information would have been obtained if all of the molecules had the same velocity vector $\underset{\sim}{v}$. The conclusion is that the wave scattered from particles having the velocity v_i provides a reference for the wave scattered from particles having the velocity v_j. In fact, if only two velocities were present, the sum in Equation 82 would contain two terms and $|G^{(1)}(\tau)|^2$ would depend on the velocity difference $v_i - v_j$. This information, i.e., the quantity $<v^2>$, is obtained about the velocities only because more than one velocity is present. However, the average velocity $<v>$ is still inaccessible in the standard homodyne experiment.

2. Forced Diffusion

In a flowing solution of macromolecules or in electrophoresis where an electric field is applied, molecules undergo both directed flow and diffusion. The equation for forced diffusion can be used to describe this situation. The particle flux $\underset{\sim}{J}(\underset{\sim}{r},t)$ now contains the flow term $\underset{\sim}{v_d}C(\underset{\sim}{r},t)$ in addition to the diffusive term found in Equation 96. Thus

$$\underset{\sim}{J}(\underset{\sim}{r},t) = \underset{\sim}{v_d}\, C(\underset{\sim}{r},t) - D_T\, \underset{\sim}{\nabla}\, C(\underset{\sim}{r},t) \tag{129}$$

This expression for $\underset{\sim}{J}(\underset{\sim}{r},t)$ can be substituted into Equation 97 to obtain:

$$\frac{\partial C(\underset{\sim}{r},t)}{\partial t} + \underset{\sim}{\nabla}\cdot[\underset{\sim}{v_d}\, C(\underset{\sim}{r},t) - D_T\, \underset{\sim}{\nabla}\, C(\underset{\sim}{r},t)] = 0 \tag{130}$$

As in Section III.B, the conditional probability function $P(O|\underset{\sim}{r},t)$ is required so that the function $F_s(\underset{\sim}{K},t)$, which appears in $g^{(1)}(\tau)$, can be calculated. For dilute solutions $P(O|\underset{\sim}{r},t)$ obeys Equation 130, and we write

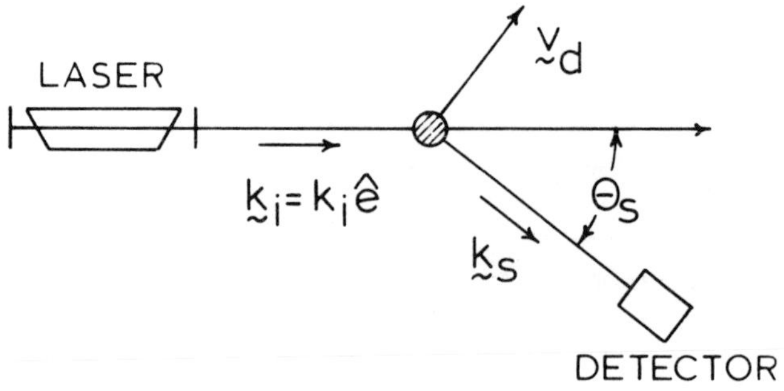

FIGURE 13. The Doppler scattering experiment for a particle having the constant velocity $\underset{\sim}{v}_d$.

$$\frac{\partial P(0|\underset{\sim}{r},t)}{\partial t} + \underset{\sim}{v}_d \cdot \underset{\sim}{\nabla} P(0|\underset{\sim}{r},t) - D_T \nabla^2 P(O|\underset{\sim}{r},t) = 0 \tag{131}$$

where the identity $\underset{\sim}{\nabla} \cdot \underset{\sim}{v}_d\, C = \underset{\sim}{v}_d \cdot \underset{\sim}{\nabla}\, C$ has been used. In solving for $F_s(\underset{\sim}{K},t)$ the same procedure is followed as in Equation 100. After multiplying Equation 131 by $e^{i\underset{\sim}{K}\cdot\underset{\sim}{r}}$ and integrating, we obtain

$$\frac{\partial F_S(\underset{\sim}{K},t)}{\partial t} + \underset{\sim}{v}_d \cdot \int_0^\infty e^{i\underset{\sim}{K}\cdot\underset{\sim}{r}}\, \underset{\sim}{\nabla} P(0|\underset{\sim}{r},t) d^3 r + D_T K^2 F_s(\underset{\sim}{K},t) = 0 \tag{132}$$

and using Equation 101 this becomes

$$\frac{\partial F_S(\underset{\sim}{K},t)}{\partial t} - i(\underset{\sim}{v}_d \cdot \underset{\sim}{K})\, F_s(\underset{\sim}{K},t) = -D_T K^2 F_S(\underset{\sim}{K},t) \tag{133}$$

The appropriate boundary condition is $F_s(\underset{\sim}{K},0) = 1$, and Equation 133 yields

$$F_S(\underset{\sim}{K},t) = e^{i\underset{\sim}{K}\cdot\underset{\sim}{v}_d t}\, e^{-D_T K^2 t} \tag{134}$$

This result is as simple as could be desired since the diffusion term from Equation 103 and the flow term from Equation 125 enter as independent factors.

In place of Equation 125 the first order correlation function for forced diffusion is

$$G^{(1)}(\tau) = \langle I_S \rangle\, e^{i\underset{\sim}{K}\cdot\underset{\sim}{v}_d \tau}\, e^{-D_T K^2 \tau}\, e^{-i\omega_0 \tau} \tag{135}$$

and for the heterodyne experiment Equations 111 and 135 give

$$G^{(2)}_{het}(\tau) = I_L^2 + 2\,I_L <I_S> \cos(\underset{\sim}{K}\cdot \underset{\sim}{v}_d\tau)\, e^{-D_T K^2 \tau} \tag{136}$$

The second term on the rhs is oscillatory, but decays to zero because of the damping factor exp $(-D_T K^2 \tau)$. This behavior is illustrated in Figure 14a where $G_{het}(\tau)$ is plotted vs. τ for $\underset{\sim}{K}\cdot\underset{\sim}{v}_d = 100\,\pi$ and $D_T K^2 = 5\,\pi$ which correspond to 50 Hz and 2.5 Hz, respectively. By comparison with the previous results for diffusion and flow in heterodyne experiments, we expect the power spectrum associated with Equation 136 to exhibit a line having the frequency shift $\underset{\sim}{K}\cdot\underset{\sim}{v}_d$ and the width (HWHH) $D_T K^2$. The calculation of $I_{het}(\omega)$, which bears this out, proceeds using Equation 92.

$$I^{(2)}_{het}(\omega) = I_L^2\,\delta(\omega) + \frac{2}{\pi} I_L <I_S> \mathrm{Re}\int_0^{\infty} \frac{(e^{i\underset{\sim}{K}\cdot\underset{\sim}{v}_d\tau} + e^{-i\underset{\sim}{K}\cdot\underset{\sim}{v}_d\tau})}{2}\, e^{-D_T K^2 \tau}\, e^{i\omega\tau}\, d\tau \tag{137}$$

The second term on the rhs of Equation 137 is easily evaluated to obtain

$$\frac{I_L<I_S>}{\pi}\left[\frac{DK^2}{(DK^2)^2 + (\omega + \underset{\sim}{K}\cdot\underset{\sim}{v}_d)^2} + \frac{DK^2}{(DK^2)^2 + (\omega - \underset{\sim}{K}\cdot\underset{\sim}{v}_d)^2}\right]$$

The function $I^{(2)}_{het}(\omega)$ is illustrated in Figure 14b for the same parameters which were used in part a. Equation 137 is very important since it shows that particles, e.g., molecules or cells, can be distinguished on the basis of their drift velocities, and that very small velocities can be measured.

3. Electrophoretic Light Scattering

The use of quasi-elastic light scattering to study molecules moving in an externally applied DC electric field is perhaps the most significant application of laser velocimetry in biology.[64] The basis of this experiment is that a charged molecule in an electrophoresis cell attains a characteristic drift velocity $\underset{\sim}{v}_d$ which is given by

$$\underset{\sim}{v}_d = u\,\underset{\sim}{E}_{dc} \tag{138}$$

where u is the electrophoretic mobility, characteristic of the molecule in a particular environment, and E_{dc} is the amplitude of the applied field. From simple arguments, similar to those used in Appendix H, the mobility is expected to be proportional to the net charge Ze of the particle and inversely proportional to the friction coefficient f_T. However, the exact dependence of u on these quantities is very complicated for charged macromolecules in electrolyte solutions. In general for spherical particles we can write[65]

$$u = \frac{Ze}{f_T} H(I, T, a) \tag{139}$$

where H is a screening function which depends at least on the ionic strength I, the temperature T, and the particle radius a. An approximate analytic equation is presented below. However, in discussing the experiment it suffices to say that the mobilities for particles of interest, i.e., macro-ions, cells, and viruses, usually lie in the range 1 to 5 μm $s^{-1}V^{-1}$cm. With electric fields in the neighborhood of 10^2 V cm^{-1}, which are easily obtained, the frequency shifts are in the range 10 to 100 Hz for the scattering angle $\theta_s \sim 5°$.

The experimental geometry is shown in Figure 15. The scattering plane is horizontal, i.e., the xy-plane, and the DC electric field is in the +x-direction. According to Equation 137, the frequency shift in radians/s is given by

$$\underset{\sim}{K} \cdot \underset{\sim}{v}_d = 2 k_i v_d \sin(\theta_s/2) \cos\phi = -k_i v_d \sin\theta_s \tag{140}$$

where ϕ is the angle between $\underset{\sim}{K}$ and $\underset{\sim}{v}_d$. Thus, the shift increases as sin θ_s increases, and it might appear that large scattering angles are desirable; however, this is not the case since the linewidth increases as $\sin^2(\theta_s/2)$. The important quantity in this experiment is the resolution which is determined by the ratio of the shift to the linewidth.

$$\left|\frac{\underset{\sim}{K} \cdot \underset{\sim}{v}_d}{D_T K^2}\right| = \frac{k_i u E_{dc} \sin\theta_s}{D_T (2 k_i)^2 [\sin(\theta_s/2)]^2} = \frac{\lambda_o u E_{dc}}{4\pi n D_T}\left[\frac{\cos(\theta_s/2)}{\sin(\theta_s/2)}\right] \tag{141}$$

For small scattering angles $\cos(\theta_s/2) \sim 1$ and $\sin(\theta_s/2) \sim \theta_s/2$. The resolution is then given by

$$\text{resolution} = \frac{\lambda_o u E_{dc}}{2\pi n D_T \theta_s} \tag{142}$$

It turns out that small scattering angles are desirable from the standpoint of resolution. Unfortunately, because of problems with stray light and the necessity of using finite collection apertures, about 1° is the practical lower limit on θ_s. An electrophoretic light scattering (ELS) spectrum taken from the work of Hass and Ware on carboxyhemoglobin is shown in Figure 16 to illustrate what can be done.[66] The scattering angle was

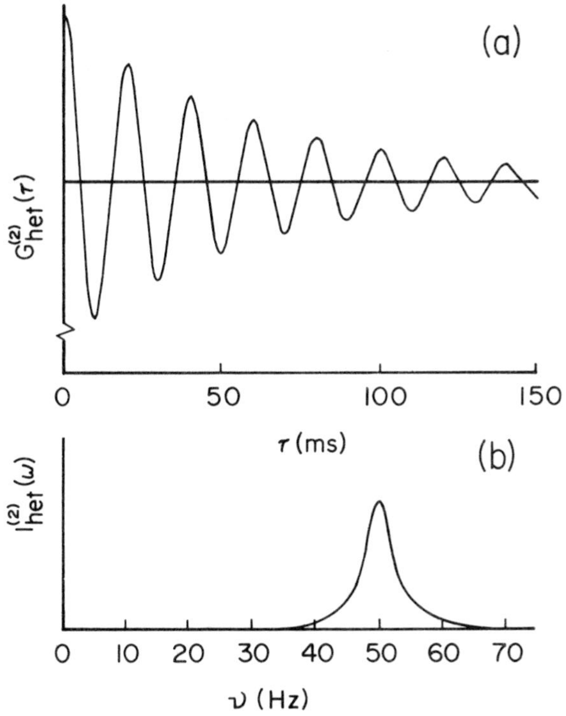

FIGURE 14. (a) Correlation function $G_{het}^{(2)}(\tau)$ calculated using Equation 136, (b) Power spectrum $I_{het}^{(2)}(\omega)$ calculated using Equation 137. The parameters used were $\underset{\sim}{K} \cdot \underset{\sim}{v_d} = 100\pi$ and $D_T K^2 = 5\pi$.

$$\underset{\sim}{k_i} = k_i \hat{j} \quad ; \quad \underset{\sim}{k_s} = k_i(\sin\Theta_s \hat{i} + \cos\Theta_s \hat{j})$$

$$\underset{\sim}{v_d} = v_d \hat{i} \quad ; \quad \underset{\sim}{K} = \underset{\sim}{k_i} - \underset{\sim}{k_s}$$

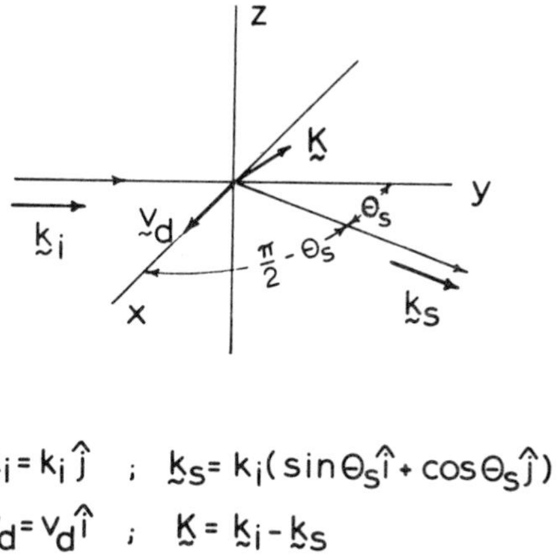

FIGURE 15. The geometry of an experiment in laser velocimetry with $\underset{\sim}{v_d}$ in the scattering plane.

$4.18°$, the electric field 88.8 V cm⁻¹, the ionic strength of 0.01 M, and the pH 9.5. From the peak position (shift) the electrophoretic mobility was calculated to be u_w^{20} = 2.74 μm s⁻¹ V⁻¹ cm. The experimental halfwidth (HWHH) was 11.5 ± 0.4 Hz, which is about 1 Hz larger than expected on the basis of measurements of D_T. The discrepancy was attributed to some experimental broadening effect.

ELS is an exciting development which is certain to be increasingly important in biological applications. The number of applications has thus far been surprisingly small. This probably results from technical problems which make the experiment far from routine with present techniques. For example, electrical currents passing through the solution lead to joule heating effects and to concentration polarization. Both of these effects can be reduced by turning the DC electric field on for only a fraction of the time and by alternating the polarity. With cell designs that place charged walls in contact with the scattering sample, electro-osmosis can also be a problem since it causes the solvent to flow. Special coatings on walls and windows can minimize this effect. These and other problems have motivated ingenious cell designs.[66,67] Even when the major problems have been solved there is still a limitation on the ionic strength which can be used, the typical value used being about 10^{-2} *M*.

The heterodyne experiment, which is normally used, also requires careful adjustment to get good mixing efficiency of the scattered and reference beams on the PMT. It should be mentioned that ELS can be done with a homodyne experiment if the laser cross-beam arrangement is used.[68] In this experiment the laser beam is split and then the two resulting beams are crossed in the sample with an intersection angle of about 5°. One interpretation of this experiment is that the crossed beams create a fringe intensity pattern and the scattering particles move through this pattern. Fluctuations in the scattered intensity are thus related to the separation of the fringes and the velocity of the particles. For an analysis of this experiment the reader is referred to the literature on laser anemometry.[69]

The ELS experiment is important because it permits several species in a sample to be resolved and studied simultaneously, and because it permits the electric mobilities to be determined. The mobilities are of interest since they are related to the net charges of the scattering particles. As previously mentioned, the relationship between the mobility and the charge is not simple. An approximate equation derived by Henry is often used for solutions having low ionic strengths. According to Henry's equation

$$u = \frac{Ze}{6\pi\eta a}\left[\frac{X_1(\kappa a)}{1 + \kappa a}\right] \tag{143}$$

where $X_1(\kappa a)$ is a complicated function which has the value 1.5 for $\kappa a >> 1$, monotonically decreases as κa decreases, and approaches 1.0 for $\kappa a << 1$. The parameter κ is the well-known Debye-Hückel constant

$$\kappa = \left[\frac{2N_A e^2 I}{10^3 \epsilon_0 k_e k_B T}\right]^{1/2} \tag{144}$$

The limitations of Henry's equation are discussed by Tanford and in specialized texts.[70]

Rapid *in situ* determination of electrophoretic mobility in a range of ionic strengths is of particular importance in the study of surface charges of cells. Table 4 gives a survey of the types of studies undertaken thus far. An example of the resolution of cell populations by means of ELS is shown in Figure 17 for lymphocytes and erythrocytes.[71] In this work an isoosmotic sucrose buffer solution of pH 7.3 and ionic strength

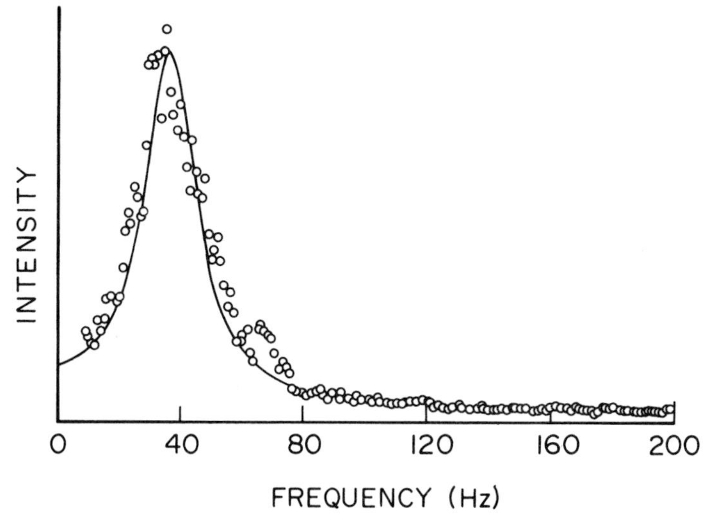

FIGURE 16. Electrophoretic light scattering spectrum of carboxyhemoglobin, 100 μM in heme. (From Haas, D. D. and Ware, B. R., *Anal. Biochem.*, 74, 175, 1976. With permission.)

I = 0.005 was used. The electrodes were platinum sheets, which had been platinized by electrodisposition and coated with bovine serium albumin to prevent reaction with the medium. The mobilities, calculated from the frequency shifts by means of Equation 140, had estimated accuracies of better than 5%, and were reproducible to better than 2% for different donors. The low ionic strength was selected in order to reduce heating and polarization effects, but it had the added advantage of giving larger cell mobilities and better resolution than that found at higher ionic strengths.

In another interesting study, mobilities for virus particles were used by Rimai et al. to calculate surface charge densities σ by means of the identity $Ze = \sigma(4\pi a^2)$.[76] For example with murine leukemia virus (MuLV), where the radius a = 72 nm was determined from diffusion measurements, Equation 143 was used to calculate a surface charge of 7.20×10^{-3} C.m^{-2} at pH 9 and 4.14×10^{-3} C.m^{-2} at pH 5. The value of $X_1(\kappa a)$ required in this calculation was obtained from Reference 1. The mobilities of all of the viruses studied were found to be quite similar in the high pH region.

Finally, we call attention to the study by Smith et al. of the electrophoretic mobilities of human peripheral blood lymphoblasts from normal donors and from patients with acute lymphocytic leukemia.[74] This study was motivated by the idea that surface charge density can be used to distinguish functionally different cells. An ELS cell similar to that described in Reference 66 was used in this work, and the individual averaged spectra required 1 to 5 min to collect. The major conclusion was that the mobilities corresponding to intensity maxima in ELS spectra were 7 to 28% lower for cells from nine leukemic patients than for normal cells. The standard deviation in the mobilities of the normal cells was only 2%. It was also confirmed that cryopreservation had no significant effect on the electrophoretic distribution. The potential clinical applications of this work are obvious.

D. Rotational Motion

1. Anisotropic Molecules

For molecules which are optically anisotropic it is possible to measure the rate of rotational motion by means of light scattering. The rotational diffusion coefficient D_R

Table 4
ELECTROPHORETIC MOBILITIES MEASURED BY MEANS OF ELECTROPHORETIC LIGHT SCATTERING

Macromolecule	Conditions	$\mu(\mu m\ s^{-1}\ V^{-1}\ cm)$	Ref.
Bovine serum albumin (BSA)	pH 9.55; I = 0.004	2.5	72
BSA dimers		1.7	72
Erythrocytes	0.29 *M* sucrose; 23.5°C	2.84	71
	0.145 *M* sucrose	3.40	71
	pH 7.2, 0.015 *M* NaCl; 4% sorbitol	1.10 ± 0.06	73
Lymphocytes	0.29 *M* sucrose, 23.5°C	2.4	71
	0.145 *M* sucrose	2.75	71
Normal lymphocytes	pH 7.1; 0.0145 *M* NaCl; 4% sorbitol	2.53 ± 0.05	74
Leukemic samples		2.13 ± 0.18	74
Rat thymus lymphocytes (RTL)		3.1	75
RTL stimulated with pokeweed mitogen		1.4	75
Calf thymus DNA	I = 0.004, 20°C	5.9	76
	0.01	5.0	
	0.02	4.4	
	0.05	3.3	
	0.1	2.8	
Tobacco mosaic virus (TMV)	I = 0.001, 20°C	5.2	76
	0.01	3.5	
	0.02	2.7	
	0.05	1.9	
	0.1	1.5	
Murine leukemia virus (MuLV)	pH 9, I = 0.005	2.8 ± 0.1	76
	7	2.7 ± 0.1	
	5	1.65 ± 0.1	
	3	1.02 ± 0.1	
Avian myeloblastosis virus (AMV)	pH 9, I = 0.005	2.64 ± 0.1	76
	7	2.78 ± 0.2	
	5	1.84 ± 0.05	
	3	1.63 ± 0.05	
Feline leukemia virus (FeLV)	pH 9, I = 0.005	3.19 ± 0.2	76
	7	2.71 ± 0.1	
	5	2.20 ± 0.1	
	3	2.60 ± 0.2	
Murine mammary tumor virus (MuMTV)	pH 9; I = 0.005	3.01 ± 0.2	76
	7	2.68 ± 0.2	
	5	2.36 ± 0.2	
Hamster fibroblasts	I = 0.15	1.0	76

often provides an adequate description of this motion for macromolecules. One reason for interest in D_R is that it is a sensitive function of the molecular radius. In 1928 Debye showed that[77]

$$D_R = \frac{k_B T}{f_R} \tag{145}$$

where f_R is the rotational friction coefficient. Stokes had previously shown that for a

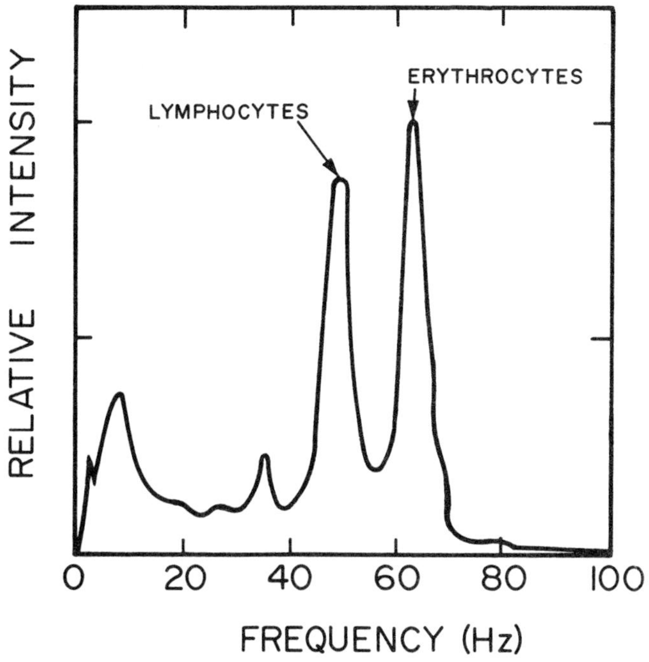

FIGURE 17. Electrophoretic light scattering spectrum of a mixture of erythrocytes and lymphocytes in 0.29 *M* sucrose buffer of ionic strength 0.005. The scattering angle was $\theta_s = 11.5°$. (From Uzgiris, E. E. and Kaplan, J. H., *Anal. Biochem.*, 60, 455, 1974. With permission.)

spherical particle of radius a in a solvent having the viscosity η, $f_R = 8\pi\eta a^3$. Thus D_R is seen to depend on the inverse cube of the radius.

A quantitative description of the scattering experiment requires that we specify the polarizability tensor for the scattering particle (see Appendix A). For simplicity we assume axial symmetry so that in the principal axes system the polarizability tensor has the form

$$\underset{\approx}{\alpha} = \begin{pmatrix} \alpha_\perp & 0 & 0 \\ 0 & \alpha_\perp & 0 \\ 0 & 0 & \alpha_{\parallel} \end{pmatrix} \qquad (146)$$

The molecular orientation is shown in Figure 18 where x, y, and z specify the laboratory fixed coordinate system, equivalent to that in Figure 1, and X, Y, and Z are the principal axes for the polarizability tensor. We take the unique axis to be Z, the orientation of which is described by the polar angles θ and ϕ with respect to the laboratory frame. The orientation of X is arbitrary, because of the axial symmetry, however, we choose to place X in the plane which contains both the z and Z axes.

We assume that the incident light propagates in the + y-direction and that it is polarized in the z-direction, i.e., the incident light is vertically polarized. The scattered light is collected at the scattering angle θ_s in the xy plane; and we are concerned with

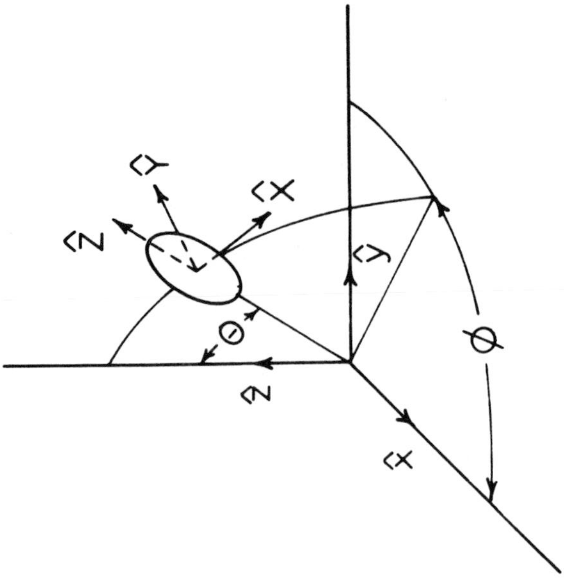

FIGURE 18. The laboratory coordinate system (xyz) and the principal axes system (XYZ) for a molecule having axial symmetry.

both the vertically and horizontally polarized components, the intensities of which are denoted by I_{VV} and I_{VH}, respectively. The induced dipole moments p_z and p_y are responsible for the scattered intensities, and according to Equation 210 these moments depend on the components α_{zz} and α_{yz}, respectively, of the polarizability tensor. For example $p_y = \alpha_{yz} E_z$. Our task then is to express α_{zz} and α_{yz} in terms of $\alpha_\perp$, $\alpha_\parallel$, and the molecular orientation. The transformation of $\underset{\sim}{\alpha}$ from the principal axes system to the laboratory frame is easily accomplished if the unit vectors $\hat{x}$, $\hat{y}$, and $\hat{z}$ are expressed in terms of the unit vectors, $\hat{X}$, $\hat{Y}$, and $\hat{Z}$.

By inspection of Figure 18 we see that

$$\hat{X} = \cos\theta\cos\phi\ \hat{x} + \cos\theta\sin\phi\ \hat{y} - \sin\theta\ \hat{z} \tag{147a}$$

$$\hat{Y} = -\sin\phi\ \hat{x} + \cos\phi\ \hat{y} \tag{147b}$$

$$\hat{Z} = \sin\theta\cos\phi\hat{x} + \sin\theta\sin\phi\hat{y} + \cos\theta\ \hat{z} \tag{147c}$$

The inverse of Equation 147, while not as easy to visualize, can be obtained by transposing the coefficients. Thus

$$\hat{x} = \cos\theta\cos\phi\ \hat{X} - \sin\phi\ \hat{Y} + \sin\theta\cos\phi\ \hat{Z} \tag{148a}$$

$$\hat{y} = \cos\theta\sin\phi\ \hat{X} + \cos\phi\ \hat{Y} + \sin\theta\sin\phi\ \hat{Z} \tag{148b}$$

$$\hat{z} = -\sin\theta\ \hat{X} + \cos\theta\ \hat{Z} \tag{148c}$$

Now armed with both $\underset{\approx}{\alpha}$ and the unit vectors, $\hat{x}$, $\hat{y}$, and $\hat{z}$ referred to the principal axes system, we can project out the desired components.

$$\alpha_{zz} = \hat{z}\cdot\underset{\approx}{\alpha}\cdot\hat{z}$$

$$\alpha_{zz} = (-\sin\theta, 0, \cos\theta)\begin{pmatrix} \alpha_\perp & 0 & 0 \\ 0 & \alpha_\perp & 0 \\ 0 & 0 & \alpha_{||} \end{pmatrix}\begin{pmatrix} -\sin\theta \\ 0 \\ \cos\theta \end{pmatrix} \tag{149}$$

$$\alpha_{zz} = \alpha_\perp \sin^2\theta + \alpha_{||}\cos^2\theta$$

Similarly

$$\alpha_{yz} = \hat{y}\cdot\underset{\approx}{\alpha}\cdot\hat{z} \tag{150}$$

$$= (\alpha_{||} - \alpha_\perp)\sin\theta\cos\theta\sin\phi$$

These equations are special cases of the application of Equation 212.

To see how α_{zz} and α_{yz} fit into the scattering theory, we return to Equation 83 where the scattering amplitude for the ℓth particle is specified by A_ℓ. In the VV experiment A_ℓ is proportional to α_{zz} and Equation 83 becomes

$$G^{(1)}_{VV}(\tau) = AN <\alpha^*_{zz}(0)\ \alpha_{zz}(\tau)> F_s(\underset{\sim}{K},\tau)\, e^{-i\omega_o\tau} \tag{151}$$

while in the VH experiment A_ℓ is proportional to α_{yz} and

$$G^{(1)}_{VH}(\tau) = AN<\alpha^*_{yz}(0)\ \alpha_{yz}(\tau)> F_s(\underset{\sim}{K},\tau)\, e^{-i\omega_o\tau} \tag{152}$$

Here A is a constant which includes E_o^2 and all of the other quantities from Equation 8 which determine the intensity. The evaluation of the correlation functions of elements of the polarizability tensor must be based on a particular model of the motion. In dealing with rotational diffusion it turns out to be much more convenient to express the angular dependences of α_{zz} and α_{yz} in terms of the spherical harmonics $Y_{\ell m}(\theta,\phi)$ rather than the trigonometric functions of Equations 149 and 150. The functions which appear in transformations of the polarizability tensor, i.e., spherical harmonics having $\ell = 2$, are listed in Table 5. By solving for $\cos^2\theta$ and $\sin^2\theta$ in terms of $Y_{2,0}(\theta,\phi)$ it is easy to show that

$$\alpha_{zz} = \alpha + \beta\sqrt{\frac{16\pi}{45}}\ Y_{2,0}(\theta,\phi) \tag{153}$$

Table 5

SPHERICAL HARMONICS FOR ℓ = 2

$$m = 2 \quad Y_{2,2}(\theta,\phi) = +\sqrt{\frac{15}{32\pi}} \sin^2\theta\, e^{2i\phi}$$

$$m = 1 \quad Y_{2,1}(\theta,\phi) = -\sqrt{\frac{15}{8\pi}} \sin\theta\cos\theta\, e^{i\theta}$$

$$m = 0 \quad Y_{2,0}(\theta,\phi) = +\sqrt{\frac{5}{16\pi}} (3\cos^2\theta - 1)$$

$$m = -1 \quad Y_{2,-1}(\theta,\phi) = +\sqrt{\frac{15}{8\pi}} \sin\theta\cos\theta\, e^{-i\phi}$$

$$m = -2 \quad Y_{2,-2}(\theta,\phi) = +\sqrt{\frac{15}{32\pi}} \sin^2\theta\, e^{-2i\phi}$$

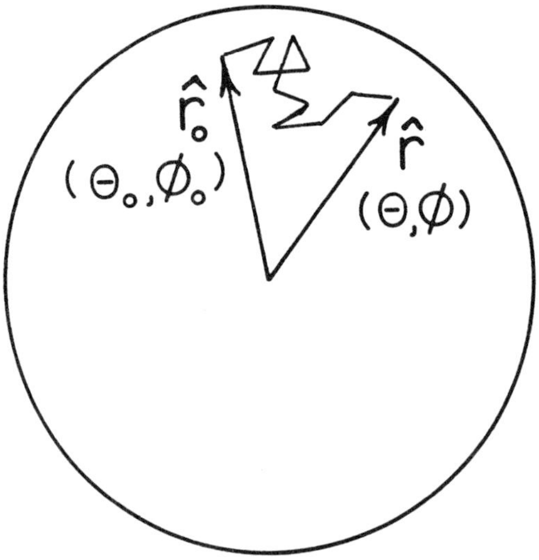

FIGURE 19. Rotational diffusion of the unit vector r̂. The t = O orientation is specified by r̂o.

Also, the combination $Y_{2,-1}(\theta,\phi) + Y_{2,1}(\theta,\phi)$ when used with Euler's theorem immediately gives

$$\alpha_{yz} = i\beta\sqrt{\frac{2\pi}{15}}\,[Y_{2,1}(\theta,\phi) + Y_{2,-1}(\theta,\phi)] \tag{154}$$

Here $\alpha = (\alpha_\parallel + 2\alpha_\perp)/3$ and $\beta = (\alpha_\parallel - \alpha_\perp)$, which are consistent with Equations 213 and 214. When Equations 153 and 154 are combined with Equations 151 and 152, it is clear that we must evaluate correlation functions of the form $<Y^*_{\ell m}(\theta_o,\phi_o)$

$Y_{\ell'm'}(\theta,\phi)>$ where (θ_o,ϕ_o) and (θ,ϕ) specify the molecular orientations at t = O and t = τ, respectively.

2. Rotational Diffusion

In rotational diffusion as described by Debye, a particle immersed in a viscous fluid is assumed to undergo numerous collisions with the solvent molecules. These collisions cause the particle to rotate through a sequence of small angular steps in random directions. If $\hat{r}$ is a unit vector fixed in the particle, we imagine that the tip of $\hat{r}$ undergoes a random walk on the surface of the unit sphere as shown in Figure 19.

The following probability functions are useful in describing the time evolution of $\hat{r}$ and functions of $\hat{r}$.

1. $P(\hat{r}_o) = P(\theta_o,\phi_o)$ = probability that the orientation is (θ_o,ϕ_o) at t = O.
2. $P(\hat{r}_o|\hat{r},\tau) = P(\theta_o,\phi_o|\theta,\phi,\tau)$ = conditional probability that the orientation is (θ,ϕ) at t = τ given that it was (θ_o,ϕ_o) at t = O.
3. $P(\hat{r}_o)\ P(\hat{r}_o|\hat{r},\tau)$ = probability that orientation is (θ_o,ϕ_o) at t = O and (θ,ϕ) at t = τ.

In calculating the correlation functions for the spherical harmonics $Y_{\ell m}(\theta,\phi)$ the function $P(\hat{r}_o)\ P(\hat{r}_o|\hat{r},\tau)$ is required. The first factor, $P(\hat{r}_o)$, is a constant since all initial orientations are equally likely. The probability that there is some initial orientation is, of course, unity. Thus

$$\int P(\hat{r}_0)\, d\Omega_0 = P(\theta_0,\phi_0) \int_{\phi_0 = O}^{2\pi} \int_{\theta_0 = O}^{\pi} \sin\theta_0\, d\theta_0\, d\phi_0 = 1 \tag{155}$$

and $P(\theta_o,\phi_o) = 1/4\pi$.
The second factor, $P(\hat{r}_o|\hat{r},\tau)$, obeys the rotational diffusion equation

$$\frac{\partial}{\partial\tau} P(\hat{r}_0|\hat{r},\tau) = D_R\, \nabla^2 P(\hat{r}_0|\hat{r},\tau) \tag{156}$$

This equation and its solutions are discussed in Appendix I where it is shown that

$$P(\hat{r}_0|\hat{r},\tau) = \sum_{\ell=O}^{\infty} \sum_{m=-\ell}^{+\ell} Y_{\ell m}(\theta_0,\phi_0)\, Y^*_{\ell m}(\theta,\phi)\, e^{-\ell(\ell+1)D_R\tau} \tag{157}$$

If $\hat{r}$ is identified with $\hat{Z}$ for an axially symmetric macromolecule, the correlation functions required in Equations 151 and 152 contain terms of the form

$$<Y^*_{\ell m}(\theta_o,\phi_o)\, Y_{\ell' m'}(\theta,\phi)> = \iint Y^*_{\ell m}(\hat{r}_0)\, Y_{\ell' m'}(\hat{r})\, P(\hat{r}_0)\, P(\hat{r}_0|\hat{r},\tau)\, d\Omega_0 d\Omega \tag{158}$$

Therefore, using Equations 151 and 157 the following integral must be evaluated

$$\iint Y^*_{\ell m}(\hat{r}_o)\, Y_{\ell' m'}(\hat{r})\, \frac{1}{4\pi} \sum_{\ell'',m''} Y^*_{\ell'' m''}(\hat{r}_o)\, Y_{\ell'' m''}(\hat{r})\, e^{-\ell''(\ell''+1)D_R\tau}\, d\Omega_o d\Omega$$

where for simplicity we have used $\hat{r}_o$ and $\hat{r}$ to specify the orientations (θ_o,ϕ_o) and (θ,ϕ), respectively. This integral is easily evaluated using the special property of spherical harmonics, namely

$$\int_{\phi=O}^{2\pi} \int_{\theta=0}^{\pi} Y^*_{\ell m}(\theta,\phi)\, Y_{\ell' m'}(\theta,\phi) \sin\theta\, d\theta\, d\phi = \delta_{\ell\ell'}\, \delta_{mm'} \tag{159}$$

The Kronecker delta δ_{ij} is defined so that $\delta_{ij} = 1$ for $i = j$ and $\delta_{ij} = 0$ for $i \neq j$. It is this property of spherical harmonics which justifies the introduction of the $Y\ell_m(\theta,\phi)$ in Equations 153 and 154. Equation 159 is used in both the integrations over (θ_o,ϕ_o) and (θ,ϕ) in 158 to obtain

$$< Y^*_{\ell m}(\theta_o,\phi_o)\, Y_{\ell' m'}(\theta,\phi)> = \frac{1}{4\pi}\, \delta_{\ell\ell'}\, \delta_{mm'}\, e^{-\ell(\ell+1)D_R\tau} \tag{160}$$

We are now in a position to express $G^{(1)}_{VV}(\tau)$ and $G^{(1)}_{VH}(\tau)$ in terms of the diffusion coefficients D_R and D_T. First Equations 153, 154, and 160 are combined to obtain

$$\begin{aligned} <\alpha^*_{zz}(0)\, \alpha_{zz}(\tau)> &= <\left[\alpha + \beta \sqrt{\frac{16\pi}{45}}\, Y^*_{2,0}(\theta_o,\phi_o)\right]\left[\alpha + \beta \sqrt{\frac{16\pi}{45}}\, Y_{2,0}(\theta,\phi)\right]> \\ &= \alpha^2 + \frac{4}{45}\beta^2 e^{-6D_R\tau} \end{aligned} \tag{161}$$

and

$$\begin{aligned} <\alpha^*_{yz}(0)\, \alpha_{yz}(\tau)> &= \frac{2\pi}{15}\beta^2 <\left[Y^*_{2,1}(\theta_o,\phi_o) + Y^*_{2,-1}(\theta_o,\phi_o)\right]\left[Y_{2,1}(\theta,\phi) + Y_{2,-1}(\theta,\phi)\right]> \\ &= \frac{1}{15}\beta^2 e^{-6D_R\tau} \end{aligned} \tag{162}$$

The first-order correlation functions can then be written as

$$G^{(1)}_{VV}(\tau) = AN\left[\alpha^2 + \frac{4}{45}\beta^2 e^{-6D_R\tau}\right] e^{-D_T K^2 \tau} e^{-i\omega_0 \tau} \tag{163}$$

$$G^{(1)}_{VH}(\tau) = AN \frac{\beta^2}{15} e^{-6D_R\tau} e^{-D_T K^2 \tau} e^{-i\omega_0 \tau} \tag{164}$$

In the VV experiment the correlation function contains an "isotropic" term identical to that found for spherical particles, but in addition contains a term, the amplitude of which depends on the anisotropy of the polarizability. The latter term decays with the time constant $\tau_c = (6D_R + D_T K^2)^{-1}$. In contrast to this, $G^{(1)}_{VH}(\tau)$ only contains an "anisotropic" term. It should be recalled that K is proportional to $\sin(\theta_s/2)$ and that the dependence on D_T can in principle be removed by extrapolating the decay rate to $\theta_s = O$. In a homodyne experiment of the VV type the relevant equation, obtained through use of Equation 163 and the Siegert relation, is

$$g^{(2)}_{VV}(\tau) = 1 + \frac{\left[B_0^2 e^{-2D_T K^2 \tau} + 2B_0 B_2 e^{-(2D_T K^2 + 6D_R)\tau} + B_2^2 e^{-2(D_T K^2 + 6D_R)\tau}\right]}{(B_0 + B_2)^2} \tag{165}$$

where $B_o = \alpha^2$ and $B_2 = 4\beta^2/45$.

The optical spectra associated with $G^{(1)}_{VV}(\tau)$ and $G^{(1)}_{VH}(\tau)$ are also of interest. The application of Equation 93 gives

$$I^{(1)}_{VV}(\omega) = \frac{\langle I^{ISO}\rangle}{\pi}\left[\frac{D_T K^2}{(D_T K^2)^2 + (\omega_0 - \omega)^2}\right] + \frac{4}{3}\frac{\langle I^{ANISO}\rangle}{\pi}\left[\frac{6D_R + D_T K^2}{(6D_R + D_T K^2)^2 + (\omega_0 - \omega)^2}\right] \tag{166}$$

$$I^{(1)}_{VH}(\omega) = \frac{\langle I^{ANISO}\rangle}{\pi}\left[\frac{6D_R + D_T K^2}{(6D_R + D_T K^2)^2 + (\omega_0 - \omega)^2}\right] \tag{167}$$

where $\langle I_s^{ANISO}\rangle = \langle I_{VH}\rangle$ and $\langle I_s^{ISO}\rangle$ is the scattering intensity resulting from α^2 alone. Typically $6D_R >> D_T K^2$, and for small globular proteins D_R is so large that the optical linewidths can be measured directly without the use of light-beating spectroscopy. Lysozyme provides a good example. For this molecule $6D_R(20°,w) \simeq 10^8$ s^{-1}, and for the scattering angle $\theta_s = 90°$ the quantity $D_T(20°,w)$ K^2 is only about 4×10^4 s^{-1}. The HWHH for the depolarized spectrum is between 10 and 20 MHz, which can be measured using a Fabry-Perot interferometer. Dubin et al. used this method to obtain $D_R(20°,w) = 16.7 \times 10^6$ s^{-1}.[51] By assuming that lysozyme can be approximated as a prolate ellipsoid of revolution, they were also able to use Perrin's expressions for f_T and f_R in terms of the ratio of the major to minor axes to show that the major and minor axes are (55 ± 1)Å and (33 ± 1)Å, respectively, for the hydrated molecule. For an ellipsoid having the semiaxes a,b, and b, the required expression for f_T is given in Appendix H. Perrin's equation for f_R for rotation of the symmetry axis is[78,79]

$$f_R = \frac{16\pi\eta a^3}{3}\left\{\frac{1-(b/a)^2}{[2-(b/a)^2]\,G(b/a)-1}\right\} \tag{168}$$

where G(b/a) was defined in connection with f_T in Appendix H.

Another simple case, which is important, arises for molecules which can be approximated as long thin rods of radius b and length L. If 20L ⩾ λ, the structure factor $S(\underset{\sim}{K})$ introduces an intensity fluctuation even if the molecule is composed of isotropic segments. The form of Equation 165 is approximately correct for KL<8, but the magnitudes of B_0 and B_2 must be looked up in tables.[80] For larger values of KL additional terms must be included. A well-studied example of a rod-shaped molecule is tobacco mosaic virus (TMV) for which $D_R \simeq 320\ s^{-1}$.[81] The linewidth corresponding to this value of the rotational diffusion coefficient is, of course, so small that light beating spectroscopy must be used.

E. Motility

Microorganisms often persist in their translational motion for distances which are much greater than $|\underset{\sim}{K}|^{-1}$. Except for the distribution of velocities, the situation is similar to that in a low pressure gas. The velocity distribution function $P(\underset{\sim}{v})$ is, in fact, the quantity of interest. The determination of $P(\underset{\sim}{v})$ by means of quasi-elastic light scattering was demonstrated by Nossal in 1971.[82] To show how this is done, we write Equation 82 as follows:

$$G^{(1)}(\tau) = \sum_{\ell=1}^{N} \langle A_\ell^*(O)\, A_\ell(\tau)\rangle \langle e^{i\underset{\sim}{K}\cdot[\underset{\sim}{r}_\ell(\tau)-\underset{\sim}{r}_\ell(O)]}\rangle E_o^2 e^{-i\omega_0\tau} \tag{169}$$

Thus far we have assumed that rotational and translational motions are independent and that the motions of different scattering particles are not correlated. If in addition we assume that either the particles are small and isotropic or that for large particles the scattering angle is small, then $\langle A^*_\ell(O)\, A_\ell(\tau)\rangle = \langle |A_\ell(O)|^2\rangle$ and Equation 169 can be written as

$$G^{(1)}(\tau) = \frac{\langle I_S\rangle}{N} \sum_{\ell=1}^{N} \langle e^{i\underset{\sim}{K}\cdot[\underset{\sim}{r}_\ell(\tau)-\underset{\sim}{r}_\ell(O)]}\rangle\, e^{-i\omega_o\tau} \tag{170}$$

The assumption of linear trajectories permits $r_\ell(\tau) - r_\ell(O)$ to be replaced with $v_\ell\tau$ where v_ℓ is the velocity for the ℓth particle or microorganism. Since the velocity distribution is isotropic, we can write

$$F_S(\underset{\sim}{K},\tau) = N^{-1}\sum_{\ell=1}^{N} \langle e^{i\underset{\sim}{K}\cdot\underset{\sim}{v}_\ell\tau}\rangle = \int P(v)\, e^{i\underset{\sim}{K}\cdot\underset{\sim}{v}\tau}\, d^3v \tag{171}$$

where P(v) depends only on the magnitude of $\underset{\sim}{v}$. For dilute gases $P(\underset{\sim}{v})$ is the Maxwell-Boltzmann distribution and Equation 171 is easily evaluated.[83]

In general the form of P(v) is unknown and must be obtained from the experimentally determined function $F_s(\underset{\sim}{K},t)$. For the purpose of evaluating the integral in Equation 171, we take the direction of $\underset{\sim}{K}$ to define the polar axis of a spherical polar coordinate system so that $\underset{\sim}{K}\cdot\underset{\sim}{v}\tau = \underset{\sim}{K}v(\cos\alpha)\tau$ where α is the angle between $\underset{\sim}{K}$ and $\underset{\sim}{v}$. Thus

$$
\begin{aligned}
F_S(K,\tau) &= \int_{\phi=O}^{2\pi}\int_{\alpha=0}^{\pi}\int_{v=O}^{\infty} P(v)\, e^{iKv(\cos\alpha)\tau}\, d\phi \sin\alpha d\alpha\, v^2 dv \\
&= 2\pi\int_{v=O}^{\infty} P(v)\left[\int_{\rho=-1}^{1} e^{iKv\rho\tau}\, d\rho\right] v^2 dv\ ;\ \rho = \cos\alpha \\
&= 4\pi\int_{v=O}^{\infty} P(v)\,\frac{\sin(Kv\tau)}{(kv\tau)}\, v^2 dv
\end{aligned}
\tag{172}
$$

If we define the swimming speed distribution $P_s(v) = 4\pi v^2 P(v)$, then Equation 172 shows that $(K\tau)\, F_s(\underset{\sim}{K},\tau)$ is the Fourier sine transform of $P_s(v)/v$, i.e.,

$$
(K\tau)\, F_S(\underset{\sim}{K},\tau) = \int_{0}^{\infty} \frac{P_S(v)}{v}\,\sin(Kv\tau)\, dv
$$

The inverse transform immediately gives

$$
P_S(v) = \frac{2v}{\pi}\int_{0}^{\infty} [(K\tau)\, F_S(\underset{\sim}{K},\tau)]\,\sin(Kv\tau)\, d(K\tau) \tag{173}
$$

which is the desired result. The result of applying this equation to an experimentally determined correlation function is shown in Figure 20. For details the reader is referred to the paper by Nossal et al:[84]

The simple treatment described here has been fairly successful for samples containing swimming bacteria. In certain cases corrections are required to take into account (1) distributions of particle sizes, (2) the presence of both motile and nonmotile bacteria, and (3) contributions from rotational motions.[85] Another quite interesting and useful effect is evident in some studies of bacteria. Namely, the number of bacteria (particles) in the scattering volume is so small that number fluctuations become important. The quantity $<\delta N(0)\delta N(\tau)>$, which is related to the mean square velocity and the mean free path, can be determined in these cases.[86] The general problem of number fluctuations is discussed in Sections III.F.

F. Number Fluctuations

In deriving an expression for $G^{(2)}(\tau)$, in Section III.A we assumed that the number

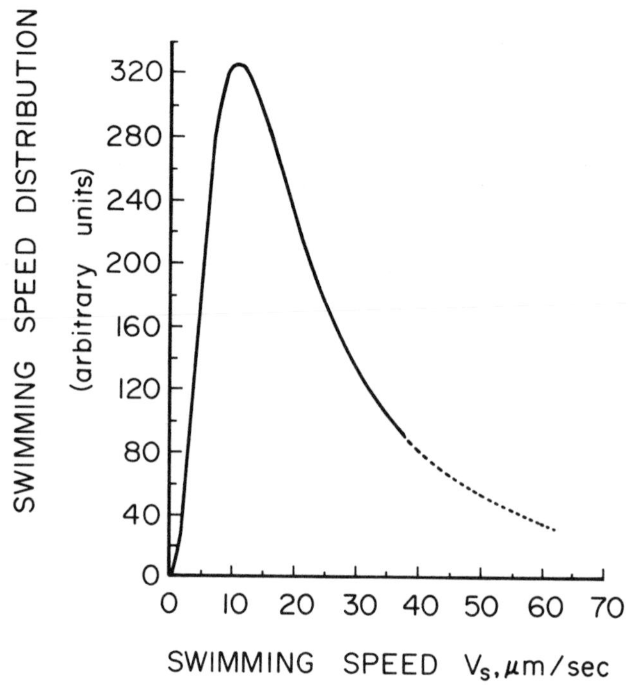

FIGURE 20. The swimming speed distribution $P_s(v)$ for motile bacteria (*Escherichia coli* K_{12}) at T = 25°C calculated from experimental data using Equation 173. (From Nossal, R., Chen, S.-H., and Lai, C.-C., *Opt. Commun.*, 4, 35, 1971. With permission.)

of particles in the scattering volume was $N = \langle N \rangle$. This is usually a very good approximation because N is normally so large that fractional fluctuations are negligible, and when N is small the intensity of the scattered light is very low. However, the situation changes when the scattering particles are very large. A small number of such particles may be in the scattering volume; and, since the scattered intensity is roughly proportional to the square of the mass, the scattered intensity may still be significant. The number of particles in the scattering volume at time t can be written as $N(t) = \langle N \rangle + \delta N(t)$ where $\langle N \rangle$ is the time average value of N(t). In this section we are concerned with the "number fluctuations" δN and their effects in quasi-elastic light scattering (QLS).[87]

Two features of number fluctuations should be noted at the beginning. First, the mean square fluctuations are usually very small. In fact, for an ideal solution $\langle(\delta N)^2\rangle / \langle N \rangle^2 = \langle(\delta C)^2\rangle / \langle C \rangle^2 = \langle N \rangle^{-1}$ and the fractional fluctuation in the scattered intensity is expected to roughly equal $\langle N \rangle^{-1/2}$. If $\langle N \rangle = 10^4$, the fluctuation will only be about 1%. The second feature is that the time scale of the fluctuations $\delta N(t)$ is usually very long compared to fluctuations in the phase factors which appear in $G^{(1)}(\tau)$. A fluctuation in N(t) occurs when a particle enters or leaves the scattering volume, and the characteristic time τ_N for the decay of the function $\langle \delta N(O) \delta N(\tau) \rangle$ depends on the dimensions of the scattering volume. For example, in diffusion the mean-square displacement which occurs in the time τ is given by $\langle x^2 \rangle = 2D_T\tau$. Therefore, if the diameter of the scattering volume is of order L, we estimate that $\tau_N \sim L^2/D_T$.

If number fluctuations are taken into account in the expression for E_S, new derivations must be given for $G^{(1)}(\tau)$ and $G^{(2)}(\tau)$. To anticipate the results, we will be able to show that $G^{(1)}(\tau)$ is unaffected by number fluctuations. In contrast to this, $G^{(2)}(\tau)$

is found to depend on $<\delta N(O)\delta N(\tau)>$. The previously stated relationship between $G^{(2)}(\tau)$ and $G^{(1)}(\tau)$ does not hold in this case since the number of particles is too small to permit the Gaussian assumption.

Following Berne and Pecora,[83] we rewrite Equation 77 as

$$E_S(t) = \sum_{j=1} A_j\, b_j(t)\, e^{i\underset{\sim}{K}\cdot\underset{\sim}{r}_j}\, E_o\, e^{-i\omega_o t} \tag{174}$$

where $b_j(t) = 1$ if the jth particle is in the scattering volume at time t and O otherwise. We assume that the scatterers are spherical and that their motions are statistically independently of $b_j(t)$. Equation 82 can now be rewritten as

$$G^{(1)}(\tau) = \sum_{j=1} |A_j|^2 <b_j(O)\, b_j(\tau)> <e^{i\underset{\sim}{K}\cdot[\underset{\sim}{r}_j(\tau) - \underset{\sim}{r}_j(O)]}> E_o^2\, e^{-i\omega_o \tau} \tag{175}$$

The crucial step is to realize that $\tau_N >> (D_T K^2)^{-1}$ so that $b_j(\tau)$ can be replaced by $b_j(O)$ without affecting the result, i.e., the factor containing $\underset{\sim}{K}$ decays to zero before $b_j(\tau)$ deviates much from $b_j(O)$ on average. Also, we noticethat $b_j^2(O) = b_j(O)$ and that $<\Sigma_{j=1} b_j(O)> = <N(O)>$. The conclusion is that

$$G^{(1)}(\tau) = <N> |A_j|^2\, F_S(\underset{\sim}{K},\tau)\, E_o^2\, e^{-i\omega_o \tau} \tag{176}$$

which is equivalent to Equation 94.

Since $G^{(2)}(\tau)$ cannot be derived from $G^{(1)}(\tau)$ by means of the Siegert relation in this case, we must return to the definition of $G^{(2)}(\tau)$. Thus from Equation 87 for the reduced second-order correlation function

$$\begin{aligned} g^{(2)}(\tau) &= <E_S^*(O)\, E_S(O)\, E_S^*(\tau)\, E_S(\tau)> / <I_S>^2 \\ &= \frac{1}{<N>^2} < \sum_{i,j,k,\ell} b_i(O)\, b_j(O)\, b_k(\tau)\, b_\ell(\tau)\, e^{i\underset{\sim}{K}\cdot[\underset{\sim}{r}_j(O) - \underset{\sim}{r}_i(O) + \underset{\sim}{r}_\ell(\tau) - \underset{\sim}{r}_k(\tau)]}> \end{aligned} \tag{177}$$

If any of the terms in this summation have one index which is unique, e.g., i is different from j,k, and ℓ, then the ensemble average of this term vanishes unless $\underset{\sim}{K} = O$. This occurs because the unique part can be factored out of the term and averaged separately. For example, if j is the unique index in the nth term, then one factor in this term is $<b_j e^{i\underset{\sim}{K}\cdot\underset{\sim}{r}_j}>$. The average here implies an integration over the scattering volume and since

$$V^{-1}\int e^{i\underset{\sim}{K}\cdot\underset{\sim}{r}_j}\, d^3 r_j \propto \delta(\underset{\sim}{K})$$

this term only produces forward scattering. Of the terms without unique indexes, there are only two types which can contribute to scattering. First we consider terms having $i=j$ and $k=\ell$ but no other restrictions. From Equation 177 we see that the exponent vanishes and we are left with

$$\frac{1}{<N>^2} \sum_{i,j} <b_i(O)\, b_j(\tau)> \tag{178}$$

where we have used the fact that $b_i^2 = b_i$. The other type of term which contributes has $i=\ell$ and $j=k$ but $i\neq j$. This gives

$$\frac{1}{<N>^2} \sum_{i\neq j} <b_i(O)\, b_i(\tau)\, b_j(O)\, b_j(\tau)> <e^{i\underset{\sim}{K}\cdot[\underset{\sim}{r}_i(\tau) - \underset{\sim}{r}_i(O)]}> <e^{-i\underset{\sim}{K}\cdot[\underset{\sim}{r}_j(\tau) - \underset{\sim}{r}_j(0)]}> \tag{179}$$

where we have used the fact that in a dilute solution the motion of particle j is uncorrelated with that of particle i. Since the factor containing $\underset{\sim}{r}_i(\tau) - \underset{\sim}{r}_i(O)$ decays rapidly to zero, we are justified in replacing $b_i(\tau)$ and $b_j(\tau)$ with $b_i(O)$ and $b_j(O)$, respectively. Finally, using $b_i^2 = b_i$ this type of term contributes

$$\frac{1}{<N>^2} \sum_{i\neq j} <b_i(O)\, b_j(O)> |\, F_S(\underset{\sim}{K},\tau)\, |^2 \tag{180}$$

Consideration of the terms having $i=k$ and $j=\ell$ but $i\neq j$ shows that they contribute only to forward scattering.

The summation which appears in Equation 178 can be written in terms of N as follows:

$$\sum_{i,j} <b_i(O)\, b_j(\tau)> = <\sum_{i=1} b_i(O) \sum_{j=1} b_j(\tau)> = <N(O)\, N(\tau)> \tag{181}$$

and using $N(t) = <N> + \delta N(t)$ this becomes

$$<N(O)N(\tau)> = <N>^2 + <\delta N(0)\, \delta N(\tau)> \tag{182}$$

since $<\delta N(0)> = O$. The summation in Equation 180 is given by

$$\sum_{i\neq j} <b_i(O)b_j(O)> = <N(N-1)> = <N^2> - <N> \tag{183}$$

The evaluation of the averages requires that the proper distribution function be selected. At low concentrations the probability of finding N particles in the scattering volume is given by the Poisson distribution

$$P(N) = \frac{<N>^N e^{-<N>}}{N!} \tag{184}$$

Using this function it can be shown that

$$<N^2> = \sum_N N^2 P(N) = <N>^2 + <N> \tag{185}$$

Then combining Equations 181, 183, and 185 we obtain

$$\begin{aligned} g^{(2)}(\tau) &= \frac{1}{<N>^2} [<N>^2 + <N>^2 |F_S(\underset{\sim}{K},\tau)|^2 + <\delta N(O)\, \delta N(\tau)>] \\ &= 1 + |F_S(\underset{\sim}{K},\tau)|^2 + \frac{<\delta N(O)\, \delta N(\tau)>}{<N>^2} \end{aligned} \tag{186}$$

The first two terms on the rhs of Equation 186 appear in the Siegert relation, which is based on the Gaussian approximation, while the last term is new. At $\tau = O$, $g^{(2)}(\tau)$ is equal to $2 + <N>^{-1}$, then for $\tau>(D_T K^2)^{-1}$ it drops to $1 + <N>^{-1}$, and finally for $\tau> \tau_N$ it approaches unity.

Number fluctuations in other types of experiments have been used in recent years to obtain kinetic information from fluctuations about equilibrium. For example fluctuations in the intensity of fluoresence have been analyzed to obtain the rate of binding of the dye ethydium bromide to DNA.[88] Also, fluctuations in conductance have been related to the rates of ionic association reactions.[89] In principal this kind of analysis can be carried out with any type of signal which is proportional to the number of molecules. However, success in a number fluctuation experiment demands a relatively small value of <N> and extreme sensitivity in detection. No one has reported measurements of fluctuations in absorption. Of course, these comments refer to fluctuations at thermal equilibrium. Nonequilibrium situations such as turbulance in gases produce much larger fractional fluctuations which can even be detected in Raman spectroscopy.[90]

G. Chemical Reactions

In chemical reactions the products differ from the reactants in their polarizabilities, diffusion coefficients, electrophoretic mobilities, and perhaps other properties. It is reasonable to expect that such changes will produce fluctuations in the intensity of scattered light. This idea has caused some excitement; however, experimental attempts to measure reaction rates by means of light scattering have not been successful. Apparently, in the systems studied, the changes in polarizabilities and diffusion coefficients have been quite small. The theoretical machinery for handling scattering from reacting mixtures is well developed, and the observation of reaction induced scattering may just depend on the judicious choice of systems.

As an example of polarizability fluctuation in a reaction, consider a conformational change in which form A converts into form B with the forward and reverse rates k_f and k_b, respectively, as shown below.[91]

$$A \underset{k_b}{\overset{k_f}{\rightleftharpoons}} B \tag{187}$$

The problem is similar to that encountered in Section III.D for rotational motion. Here the polarizability fluctuates between α_A and α_B because of the reaction rather than because of the rotation of an anisotropic molecule. For polarized scattering we can again use Equation 151 if the diffusion coefficients for forms A and B are similar and both conformations are optically isotropic. Suppose that P(A) is the probability that the polarizability is α_A, and $P(A|B,\tau)$ is the conditional probability that the polarizability is α_B at $t = \tau$ if it were α_A at $t = O$. The required correlation function can be written as

$$\langle\alpha(O)\alpha(\tau)\rangle = \sum_{i,j = A,B} P(i)\, P(i|j,\tau)\, \alpha_i \alpha_j \tag{188}$$

where $P(i|j,\tau)$ contains all of the information about the chemical reaction. In order to evaluate this function we assume that $\alpha(\tau)$ is a stationary Markov process. The conditional probabilities can be shown to obey the differential equation[92]

$$\frac{d\, P(i|j,\tau)}{d\tau} = \frac{-P(i|j,\tau)}{\tau_j} + \sum_k \frac{P(i|k,\tau)}{\tau_k} p_{kj} \tag{189}$$

The mean lifetime of the kth species is τ_k; and, when a jump occurs from the kth species, p_{kj} gives the probability that it will be the jth species. In the present case $p_{AB} = p_{BA} = 1$.

The equations, which must be solved are

$$\frac{d}{d\tau} P(A|A,\tau) = -k_f P(A|A,\tau) + k_b P(A|B,\tau) \tag{190a}$$

$$\frac{d}{d\tau} P(A|B,\tau) = -k_b P(A|B,\tau) + k_f P(A|A,\tau) \tag{190b}$$

and a similar set with A and B reversed. In Equation 190 we have introduced the definitions $k_f = \tau_A^{-1}$ and $k_b = \tau_B^{-1}$. The standard method for solving this pair of simultaneous differential equations is to let$P(A|A,\tau) = ae^{m\tau}$ and $P(A|B,\tau) = be^{m\tau}$. Substitution of these trial functions into Equation 190 yields the roots $m_1 = O$ and $m_2 = -(k_f + k_b)$. The solutions, taking into account the initial conditions $P(A|A,O) = 1$ and $P(A|B,O) = O$, are

$$P(A|A,\tau) = P(A) + P(B)\, e^{-(k_f + k_b)\tau} \tag{191a}$$

$$P(A|B,\tau) = P(B)\left[1 - e^{-(k_f + k_b)\tau}\right] \tag{191b}$$

$$P(B|A,\tau) = P(A)\left[1 - e^{-(k_f + k_b)\tau}\right] \tag{191c}$$

$$P(B|B,\tau) = P(B) + P(A)\, e^{-(k_f + k_b)\tau} \tag{191d}$$

Substituting these functions into Equation 188 and simplifying, we obtain

$$<\alpha(0)\alpha(\tau)> = (\bar{\alpha})^2 + P(A)\,P(B)\,(\alpha_A - \alpha_B)^2\; e^{-(k_f + k_b)\tau} \qquad (192)$$

where $\bar{\alpha} = P(A)\alpha_A + P(B)\alpha_B$. Now returning to Equation 151 we can obtain

$$G^{(1)}_{VV}(\tau) = AN\left[(\bar{\alpha})^2 + X_A X_B(\alpha_A - \alpha_B)^2\; e^{-(k_f + k_b)\tau}\right] e^{-D_T K^2 \tau} e^{-i\omega_0 \tau} \qquad (193)$$

The mole fractions X_A and X_B are equal to the probabilities P(A) and P(B), respectively, and the equilibrium constant is

$$K_{eq} = X_B/X_A = k_f/k_b$$

The first term on the rhs of Equation 193 gives the "normal" scattering while the second term results from the chemical reaction. If $(\alpha_A - \alpha_B)^2/(\bar{\alpha})^2 << 1$, the reaction term will, of course, be difficult to detect. Notice that an extrapolation to $\theta_s = O$ should permit the rate constant to be extracted.

In transparent, i.e., nonabsorbing samples, polarizability differences such as $\alpha_A - \alpha_B$ are usually quite small. If, on the other hand, the products have significantly different absorption frequencies than the reactants, it is possible to increase the polarizability difference by tuning the exciting light into an absorption band.[6,93] This depends on the well known dispersion of the index of refraction. Consider for example the reaction of an indicator dye In^- with an H^+ ion to give the product HIn.

$$\begin{array}{ccccc} In^- & + & H^+ & \underset{k_b}{\overset{k_f}{\rightleftharpoons}} & HIn \\ \bar{C}_1 & & \bar{C}_2 & & \bar{C}_3 \end{array} \qquad (194)$$

where $\bar{C}_i$ is the mean concentration of the ith species. The H^+ ion will not be detected and we assume that $D_{H^+} >> D_1 = D_3 = D_T$. The equation for $G^{(1)}_{VV}(\tau)$ in this case can be obtained either by treating the A + B ⇌ C reaction by the methods of Reference 12 or by analogy with the treatment given above. Either method gives an equation of the same form as Equation 192, but with A and B replaced by 1 and 3, respectively, and $k_f + k_b$ replaced by $\bar{C}_2 k_f + k_b$. Also, for reaction Equation 194 the mole fractions are given by

$$X_1 = \frac{k_b}{\bar{C}_2 k_f + k_b}\;;\quad X_3 = \frac{\bar{C}_2 k_f}{\bar{C}_2 k_f + k_b} \qquad (195)$$

Because of the color change in this type of reaction it may be possible to choose a wavelength that will permit the reaction term to be measured.

It has been suggested that changes in diffusion coefficients of more than 30% may be found in macromolecular dimerization and isomerization reactions, and that this could lead to detectable changes in quasi-elastic light scattering.[94] An even more likely possibility is that the reacting molecules will have different electrophoretic mobilities and the ELS spectrum will depend on the reaction rate.[11] Calculated curves for different reaction rates have been presented, but as yet no experimental examples have been reported. A formulation of the problem which includes all of these effects is discussed in detail in Chapter 6 of *Dynamic Light Scattering.*[12]

H. Experimental Capabilities and Limitations

1. Light Sources and Detectors

Collimated beams of continuous wave (CW), monochromatic radiation are required in QLS, and lasers are the only practical sources of such radiation. Contrary to common belief, conventional sources could be used in light-beating spectroscopy (LBS) if sufficient intensity were available.[95-97] Restrictions on the bandwidths of the sources are, in fact, not very severe. For example, dye lasers with tuning elements but without high-resolution etalons are satisfactory sources even though the bandwidths are of the order of 10^{10} Hz.[98] The bandwidth of the source would clearly dominate in optical spectroscopy, but in intensity fluctuation spectroscopy (IFS) extremely high resolution can, in fact, be obtained with fairly broadband sources.

A simple analysis of light scattering using a quasi-monochromatic source reveals the requirements and limitations. The electric field of the laser light at the sample can be written as

$$E_L(t) = E_L^0 \, e^{-i[\omega_0 t + \phi_L(t)]} \tag{196}$$

where $\theta_L(t)$ is a time-dependent phase fluctuation, which is characteristic of instabilities in the source. It is this phase fluctuation which is responsible for most of the optical line width. In IFS $\theta_L \cdot (t)$ is of no consequence since the measured quantity is the intensity $I = E^*E$. The important factor turns out to be the amplitude fluctuation in the source, i.e., fluctuations in $|E_L{}^0|$. The electric field of the scattered light can be expressed as

$$E_S(t) = E_L^0(t) \, e^{-i[\omega_0 t + \phi_L(t)]} \, f_M(t) \tag{197}$$

where $f_M(t)$ describes the modulation resulting from the scattering medium. If fluctuations in the source and the scattering medium are independent and Gaussian distributed, the reduced second-order correlation function can be factored as follows:

$$\begin{aligned} g_S^{(2)}(\tau) &= g_L^{(2)}(\tau) \, g_M{}^{(2)}(\tau) \\ &= \left[1 + |g_L^{(1)}(\tau)|^2\right] \left[1 + |g_M^{(1)}(\tau)|^2\right] \end{aligned} \tag{198}$$

Now suppose that $|g_L{}^{(1)}(\tau)| \sim e^{-\tau/\tau_L}$ and $|g_M{}^{(1)}(\tau)| \sim e^{-\tau/\tau_M}$. If only low frequency amplitude fluctuations are present in the laser, then $\tau_L >> \tau_M$, and we find that $g_S^{(2)}(\tau) \simeq 2$

$g_M^{(2)}(\tau)$. At the opposite limit high frequency amplitude fluctuations give $\tau_L << \tau_M$ and $g_S^{(2)}(\tau) \simeq g_M^{(2)}(\tau)$. Therefore, in principle $g_M^{(2)}(\tau)$ can be obtained using either narrow or broadband sources. In general when laser intensity fluctuations are present the scattered light is not Gaussian.

In practice two difficulties arise. First, the intensity fluctuations in the laser may have frequency components in the same range as the fluctuations of interest in the sample, i.e., τ_L may be roughly equal to τ_M. Usually source fluctuations of up to 2% root-mean-square (rms) can be tolerated. Some lasers are inherently noisy while others have noise which results from malfunctions in their power supplies. These fluctuations are unimportant in many applications but are catastrophic in IFS. This kind of problem must be solved at the source or at least before the beam reaches the sample. It is an important consideration in selecting a laser system, but one should be aware that the noise level for a given laser may change with time. An effective but expensive remedy is to insert an electro-optic noise reduction system, i.e., a "noise eater", in the optical path. For example, the Coherent Associates Model 307 provides a 30 to 40 dB reduction in noise below 100 kHz.

The second problem limits the laser bandwidths which can be used. Recall that in nonideal situations the Siegert relation is given by

$$g^{(2)}(\tau) = 1 + \beta\,|\,g^{(1)}(\tau)\,|^2 \qquad (105')$$

where β can become very small for samples which are not coherently illuminated. The effective scattering volume depends on the coherence length $\ell_c = c/(n\ \Delta\nu)$ of the incident light where n is the index of refraction of the medium and $\Delta\nu$ is the laser band width. In order to have a coherence length of 1 mm the laser bandwidth must be less than 8 cm^{-1}. Of course, a narrow band filter can be inserted if necessary, but this may reduce the useable intensity to an unsatisfactory level.

In Table 6 we list some commercially available CW lasers which may be useful in IFS. Most experimental studies to date have made use of He-Ne lasers, which are quiet and reliable though limited in power. Interest in the other lasers arises because other wavelengths are sometimes required in order to avoid absorption or to search for resonance enhancement. Dye lasers have the advantage of tunability, but with a given dye and the corresponding set of mirrors the range is somewhat limited. Also, unacceptable intensity fluctuations often result from bubbles in the dye jet and other mechanical instabilities, especially when operating near threshold.

The incident beam in a light scattering experiment is expected to be quasi-monochromatic, well-defined in space, and polarized in a known direction. Lasers are less than perfect in these characteristics. For example, gas lasers emit light at frequencies corresponding to plasma oscillations as well as at the primary beam frequency ω_o, and they often have secondary output beams in various directions. Dye lasers always produce some background fluorescence. Therefore, it is advisable to prepare the beam by using narrow band filters, apertures, polarization rotators, and polarizing prisms. A vibration isolation table also may be required if there is evidence of effects of mechanical oscillations in the power spectra or correlation functions.

The light detectors are essentially always photomultiplier tubes (PMT). These are photo emissive devices coupled with electron multipliers. When a photon strikes the photocathode, an electron is emitted with a quantum efficiency of as high as 0.3. The multiplier sections typically have gains of about 10^6 so that a measurable current is produced at the anode. In a photon counting system the pulses resulting from single photons are selected, amplified, and counted. In analog systems the average anode

Table 6
CONTINUOUS LASERS THAT MAY BE SUITABLE FOR INTENSITY FLUCTUATION SPECTROSCOPY

Laser	Wavelength range (nm)	Comments
Argon ion (gas)	454.5-528.7	Discrete lines with maximum power at 488 and 514.5;
	351.1-363.8, 1090	requires special optics.
Krypton ion[a] (gas)	476.2-799.3	Discrete lines with maximum power at 647.1;
	350.7, 364.4	requires special optics.
Helium-Cadmium[a]	441.6	
(gas)	325	Requires special optics.
Helium-Neon (gas)	632.8	Power <75 mW but stable and reliable;
	611.8; 640.1, 1084, 1152	requires special optics.
Dye laser (liquid)	430-900	Requires about ten different dyes and several sets of mirrors to cover this range; also, intense pump lasers in UV and visible (both green and red) are required.
Semiconductor diode laser[a]	820-870	Single line somewhere in this region; a linewidth of ∼0.1 nm at 15 mW can be obtained.
Neodymium Yag[a]	1060	

[a] Use in intensity fluctuation experiments has not been reported.

current i(t) is measured. Spectrum analyzers usually operate on i(t) to determine the power spectrum $I^{(2)}(\omega)$. In this case the requirements on the PMT are not very severe, and inexpensive tubes such as the RCA 7265® can be used. With autocorrelators however, there are advantages to using photon counting electronics, and the response time of the PMT becomes important. Favorite tubes for photon counting applications have been the ITT FW130® and the Channeltron BX-7500®. Both of these tubes have S20 response and, therefore, have adequate quantum efficiency at 632.8 nm; but their sensitivities decrease rapidly with increasing wavelength. The newer RCA C31034(A)® has a GaAs photocathode that permits a spectral response range of 200 to 950 nm. In addition, it is superior in quantum counting efficiency and dead time.[99] This appears to be the tube of choice in spite of its relatively high price and the requirement that it be cooled to −20°C to reduce the dark count. At wavelengths longer than 950 nm one of the Varian InGaAsP® tubes such as the VPM-159A2 would probably be the best detector for photon counting applications.

2. Spectrum Analyzers and Correlators

Spectrum analyzers are commonly used in laser velocimetry where frequency shifts appear and several frequency components may be present, e.g., in ESL. The simplest spectrum analyzer consists of a narrow band amplifier, the center frequency of which can be scanned across the frequency range of interest to obtain $I^{(2)}(\omega)$. This was the basis of the old HP-302A wave analyzers which used an electric motor to scan the amplifier at rates up to 1 kHz/min. The same idea was implemented in the fast scanning cathode ray tube (CRT) type spectrum analyzers such as the Singer Panaramic MF-5®. Much greater efficiency and resolution is now available in "time compression" type spectrum analyzers which record the incoming signal and then play it back at different rates to pick out the various frequency components. An example of this type of analyzer is the 400 channel Honeywell SAICOR® which provides 11 frequency ranges from 0 to 20 Hz to 0 to 1 MHz and has built-in signal averaging capability. For very high resolution in the frequency range below 50 kHz the realtime FFT spectrum analyzers offer the best performance. These microprocessor based units perform a fast Fourier transformation on an input record containing n points to produce a power spectrum having n/2 points. Examples of FFT analyzers are the PAR 4512 and 4513 and the recently announced HP 3582 A.

In experiments where the correlation function $G^{(2)}(\tau)$ is a monotonically decreasing function of time it is most efficient to determine this function directly from the PMT output. Digital correlators calculate approximations to $G^{(2)}(\tau)$ as discussed in Section III.A.3. In photocurrent autocorrelation an analog to digital (A/D) converter is required in the correlator while with photon counting systems the intensity $I_s(t)$ of the incident light is proportional to the number of counts $n(t,\Delta T)$ in the time interval ΔT at t, and the autocorrelation function is given by $G^{(2)}(\tau) = \langle n(O,\Delta T)\, n(\tau,\Delta T)\rangle$ except for a proportionality constant. A correlator calculates and stores in the kth channel the function

$$C(k\Delta T) = \sum_{i=1}^{N_s} n(t_i,\Delta T)\, n(t_i + \tau,\Delta T) \tag{199}$$

where $\tau = k\Delta T$ and $n(t_i,\Delta T)$ is the number of photocounts measured in the time interval ΔT starting at time t_i. The Siegert relation is usually assumed and the equation

$$C(\tau) = A[1 + \beta \mid g^{(1)}(\tau) \mid^2] \tag{200}$$

must be used to obtain $|g^{(1)}(\tau)|$. In photocurrent autocorrelation both A and β must be determined from a least squares fit unless special filtering has been introduced to remove the DC component. One of the advantages of photocount autocorrelation is that A can be calculated directly. This is possible since the correlator keeps up with the total number of counts.

$$N_T = \sum_{i=1}^{N_s} n(t_i,\Delta T) \tag{201}$$

and the number of terms N_s in the summation. Therefore, $\langle n\rangle = N_T/N_S$, $C(\tau) = N_sG^{(2)}(\tau)$, and Equation 200 can be written as

$$N_S G^{(2)}(\tau) = C(k\Delta T) = \frac{N_T^2}{N_S} [1 + \beta \mid g^{(1)}(\tau) \mid^2] \tag{202}$$

High speed digital correlators are now available which permit sample times ΔT as short as 100 nsec. The units manufactured by Langley-Ford and by Malvern are real time in the sense that processing occurs simultaneously for all of the channels, and the processing efficiency does not change significantly when the shorter time intervals are selected. A typical number of channels is 64 and expansion is offered as an option. A very important feature is that the last few channels can be shifted out to greater time delays. This permits A in Equation 200 to be determined independently of the calculation described above. If the two values differ, contamination of the sample or intensity fluctuations in the laser should be considered.

Thus far we have assumed that the full correlation function $C(k\Delta T)$ is calculated by the correlator. In practice this means that 4-bit shift registers are used and $n(t,\Delta T)$ is stored as an integer from 0 to 15. If more than 15 counts occur in the interval ΔT, a correction is required. This scheme is called 4 × 4 or full processing. Furthermore, some correlators gain speed by operating in the single or double clipped mode. This means that $n(t,\Delta T)$ is stored as 1 or 0 depending on whether its magnitude exceeds some predetermined level, i.e., the "clipped" count rate is defined as

$$n_k(t,\Delta T) = 1 \text{ for } n(t,\Delta T) > k$$

$$= 0 \text{ for } n(t,\Delta T) \leq k$$

The single and double clipped correlation functions are defined by Equations 203 and 204, respectively, where ΔT has been suppressed.

$$g_k^{(2)}(\tau) = \langle n_k(t)\, n(t+\tau)\rangle / \langle n_k\rangle \langle n\rangle \qquad (203)$$

$$g_{kk'}^{(2)}(\tau) = \langle n_k(t)\; n_{k'}(t+\tau)\rangle / \langle n_k\rangle \langle n_{k'}\rangle \qquad (204)$$

If $\langle n\rangle \ll 1$, both $g_k^{(2)}(\tau)$ and $g_{kk}^{\prime(2)}(\tau)$ are equivalent to $g^{(2)}(\tau)$, but as $\langle n\rangle$ increases distortions can occur. However, in the special case of Gaussian light it can be shown that

$$g_k^{(2)}(\tau) = 1 + \frac{(1+k)}{(1+\langle n\rangle)} \,|\, g^{(1)}(\tau)\,|^2 \qquad (205)$$

The time dependence is not distorted and for $k = \langle n\rangle$, $g_k^{(2)}(\tau)$ is identical to $g^{(2)}(\tau)$. At higher count rates the double clipped function will always introduce some distortion.

3. Special Requirements

a. Coherence Areas

A major difference between the experimental requirements for classical light scattering and intensity (interference) fluctuation spectroscopy (IFS) is the importance of <u>coherence areas</u> in the latter. The following facts characterize the IFS experiment:

1. Light scattered from a coherently illuminated volume gives a characteristic speckle pattern, i.e., an array of bright nonoverlapping spots.
2. The temporal fluctuations of interest appear in one speckle, i.e., in one coherence area, and the signal-to-noise ratio (S/N) does not increase when more than one speckle is detected.

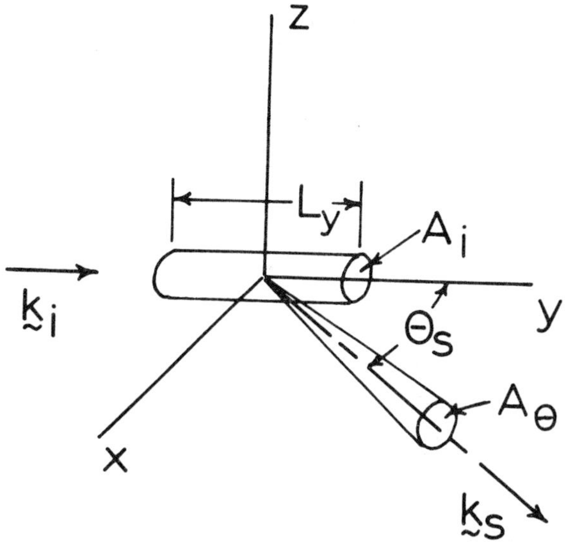

FIGURE 21. The approximate shape of an illuminated volume having the cross-sectional area A_i and the length L_y.

If a detector collects light in the solid angle $\Delta\Omega$ at the distance R, the area of the detector is given by $A_\theta = R^2\,\Delta\Omega$. The coherence area and the coherence angle are denoted by A_θ and $\Delta\Omega_{coh}$, respectively, and the number of coherence areas collected is $N_{coh} = A_\theta/A_{coh}$. Since S/N does not increase as N_{coh} increases for $N_{coh} > 1$, and may in fact decrease, the important quantity is the amount of power scattered into one coherence area. From Equation 16 we see that the power P_θ in the solid angle $\Delta\Omega$ is given by

$$P_\theta = \mathcal{R}_\theta\, L_y\, P_i\, \Delta\Omega \tag{206}$$

where L_y is the length of the scattering volume. The Rayleigh ratio R_θ is characteristic of the sample, but the other quantities are, to a certain extent, under our control.

Consider the geometry shown in Figure 21. The illuminated volume has the dimensions L_x, L_y, and L_z and the cross-sectional area $A_i \sim L_x\, L_z$ so that the magnitude of the volume is given by $V = A_i\, L_y$. The calculation of the coherence angle is not trivial, but a detailed treatment has been given by Lastovka.[100] A summary can also be found in Chu's book.[13] The basic idea is that the speckle pattern is an instantaneous Fourier expansion of the distribution of scatterers in the illuminated volume. Only discrete values of $\underset{\sim}{K}$ are permitted in this expansion, and these correspond to the speckles. The speckles turn out to have finite size since the Fourier integrals span a volume which is finite in extent. The complete analysis gives

$$\Delta\Omega_{coh} = \frac{\lambda^2}{L_y\, L_z[\sin\theta_s + (L_x/L_y)\cos\theta_s]} \tag{207}$$

It is interesting and significant that the denominator in Equation 207 is equal to R^2

times the solid angle Ω subtended by the illuminated volume at the detector, i.e., $A_{coh} = \lambda^2/\Omega$. When Equations 206 and 207 are combined, we obtain an expression for the power scattered into one coherence area.

$$P_\theta(\Delta\Omega_{coh}) = \frac{\mathcal{R}_\theta P_i \lambda^2}{L_z[\sin\theta_s + (L_x/L_y)\cos\theta_s]} \tag{208}$$

Equation 208 suggests that L_x and L_z should be minimized, but that L_y can be increased without penalty. Also, the power in $\Delta\Omega_{coh}$ is seen to increase as θ_s decreases. It turns out that the wavelength dependence of $P_\theta(\Delta\Omega_{coh})$ is much less than expected since the λ^2 in $\Delta\Omega_{coh}$ partially cancels the λ^{-4} which appears in R_θ. In fact, a complete analysis, taking into account the wavelength dependence of the characteristic decay time of $G^{(2)}(\tau)$ and the energy per photon hc/λ, shows that the statistical accuracy in a photon correlation spectroscopy (PCS) experiment of fixed duration is independent of the wavelength.[37] This, of course, assumes that the quantum efficiency of the PMT is constant.

The precise dependence of S/N on N_{coh} in a real experiment is not known. Experience indicates that it is best to collect a few coherence areas, e.g., somewhere in the range 2 to 5. It is clear that large values of N_{coh} are not beneficial.

b. Stray Light

Uncontrolled light can mix with the light from the scattering volume to give a heterodyne component. This may, of course, have a large effect on the measured line width. Also, back scattered light from the exit wall of the sample cell can have sufficient intensity to give secondary scattering at the angle $\theta'_s = \pi - \theta_s$. These effects are sufficient to indicate that a well-characterized scattering experiment requires that stray light be eliminated as much as possible. Since each air-glass interface can scatter about 4% of the incident light, it is important to keep these interfaces far from the scattering volume. A cell holder designed for this purpose has been described by Jolly and Eisenberg.[101] The important feature is that their cylindrical sample cell (10 mm or 29 mm diameter) is suspended in an index of refraction matching fluid which is contained in a cylindrical tank having a diameter of 125 mm. The incident beam is defined by an aperture in the bath, and after the sample cell there is a light trap, also in the bath.

Another solution to the problem of stray light, especially in connection with low angle scattering, has been discussed by Kaye et al., and incorporated into the Chromatix KMX-6® photometer.[17,102] Their idea is to contain the sample inside a spacer between two thick fused-silica windows. The exiting light on axis ($\theta_s = O$) is trapped while the light scattered into the cone between θ_s and $\theta_s + \Delta\theta_s$ is collected through a defining annulus. It is reported that scattering angles from 2° to 7° can be covered and that the sample volume is only 150 $\mu\ell$. With this type of optical arrangement the angular factor $(1 + \cos^2\theta_s)/2$ must be included in the expression for the scattered intensity regardless of the polarization state of the incident light (see Figure 17). Also, the effect of the number of coherence areas collected must be considered as previously discussed.

c. Particulate Contamination

Dust and other large particles can be a severe problem in QLS since the scattering intensity depends on the square of the mass of the scattering particle. Closed loop recirculation systems containing millipore or other types of filters are fairly effective in removing dust, but these systems must be rigorously sealed. The performance of such a system can be monitored by observing the illuminated volume with a small microscope. Another convenient test is to compare the scattered intensities at $\theta_s = 45°$ and 135° using polarized incident light since for small particles the intensity should be independent of the scattering angle. If the sample of interest is known to be monodisperse, the variance $\mu_2/\langle\Gamma\rangle^2$, obtained from an analysis of the measured correlation function, can also be used as a criterion of sample quality. Still another effective test is to compare the normalization constant A in Equation 200 obtained from the baseline at large τ with the calculated value N_T^2/N_s. If large, slowly moving particles are present, the correlation function will contain slowly decaying components which may contribute to $G^{(2)}(\tau)$ at large but not infinite τ; and effectively establish a baseline at a higher level than that calculated by Equation 202. In such cases the measured baseline should be used; however, the only good solution to the dust problem is to make sure that the dust is removed.

d. Absorbing Samples

When a sample has significant absorption at the wavelength of the incident light, the efficiency of the PCS experiment is greatly reduced and effects appear which may complicate the interpretation of the results. The most obvious problem is the loss of signal, which results from the attenuation of both the incident and the scattered light. If the scattering volume is at the center of a cell having a diameter of 1.0 cm, an absorption coefficient (absorbance/cm) of 1.0 reduces the signal by a factor of 10. It is not permissible to compensate for this effect by increasing the incident power since this causes heating of the sample and leads to additional problems. The best approach is to minimize the path lengths for the incident and scattered light by judicious cell design. A fairly satisfactory arrangement is to use a cylindrical jacketed cell, e.g., the Hellma cell no. 165, and to pass the focused laser beam parallel to the axis of the cell but only ~ 1 mm from the wall. The scattered light can then be collected through the side of the water jacketed at fairly large scattering angles or at smaller angles through the flat front face of the water jacket. The total path length in the sample can be held to about 2.5 mm in this way.

Heating effects can be serious at even moderate power levels. The diffusion coefficient for an aqueous solution increases roughly 3% per degree centrigrade, so the temperature rise alone is significant. It is not easy to calculate the temperature rise in the scattering volume for a given power level, but an estimate can be based in the equations of Gordon et al.[103-105] For a typical experiment involving an aqueous solution with an absorption coefficient of unity and a laser beam having a diameter of 0.1 mm, the temperature rise turns out to be 0.8°C at 2 mm from the entrance face for an incident power of 10 mW. This crude estimate only serves to emphasize that power levels should be kept low, and that measured diffusion coefficients should be extrapolated to zero power.[117]

For aqueous samples the temperature rise is usually the major effect; however, convection and thermal lensing must also be considered. The increased temperature in the laser beam results in reduced density which in turn gives an upward buoyant force. As discussed in Section III.C.1, uniform velocity cannot be detected in homodyne experiments; however, convection may produce a distribution of velocities. Again, low incident power is recommended. Also, because of viscous forces the use of a tightly focused laser beam will minimize velocity gradients in the scattering volume. In any event it is important to keep the scattering plane horizontal so that the dot product $\underset{\sim}{K} \cdot \underset{\sim}{v}_d$ will be zero. This removes the effect of the vertical component of the convection velocity in both homodyne and heterodyne experiments, even when a distribution of velocities is present.

Thermal lensing results from gradients in the index of refraction which exist because of the steady-state temperature distribution. Since the index is usually lowest along the axis of the laser beam, the sample behaves as a diverging lens. The result is that the beam may increase in diameter as it passes through the sample and a distribution of scattering angles may contribute to the light scattered in a given direction. At higher power levels severe divergence and even aberrations may occur.

APPENDIXES

A. The Polarizability Tensor

The need for tensors arises when the induced dipole is not in the same direction as the electric field. The tensor $\underset{\approx}{\alpha}$ permits both the magnitude and the direction of the induced dipole to be calculated when the direction and magnitude of the electric field are specified. The equation $\underset{\sim}{p} = \underset{\approx}{\alpha} \cdot \underset{\sim}{E}$ is shorthand notation for the set of three equations:

$$
\begin{aligned}
P_x &= \alpha_{xx} E_x + \alpha_{xy} E_y + \alpha_{xz} E_z \\
P_y &= \alpha_{yx} E_x + \alpha_{yy} E_y + \alpha_{yz} E_z \\
P_z &= \alpha_{zx} E_x + \alpha_{zy} E_y + \alpha_{zz} E_z
\end{aligned} \tag{209}
$$

where the nine quantities α_{ij} (i,j = x,y,z) are elements of the polarizability tensor. It is common to express Equation 209 in matrix form as:

$$
\begin{pmatrix} P_x \\ P_y \\ P_z \end{pmatrix} = \begin{pmatrix} \alpha_{xx} & \alpha_{xy} & \alpha_{xz} \\ \alpha_{yx} & \alpha_{yy} & \alpha_{yz} \\ \alpha_{zx} & \alpha_{zy} & \alpha_{zz} \end{pmatrix} \cdot \begin{pmatrix} E_x \\ E_y \\ E_z \end{pmatrix} \tag{210}
$$

In transparent media these tensors are usually real and symmetric, i.e., $\alpha_{ij} = \alpha_{ji}$ and $\alpha_{ij} = \alpha_{ij}^*$.

The form of a tensor depends on the coordinate system chosen for its representation. Since both E and p are specified in the laboratory fixed coordinate system, we chose xyz in Equations 209 and 210 as axes in this frame. Suppose, for example, that the scatterer has $\alpha_{xy} = \alpha_{xz} = O$. If the incident light is polarized in the z-direction, as shown in Figure 1, the induced dipole will be given by $p = (\alpha_{yz}\,\hat{j} + \alpha_{zz}\,\hat{k})\,E_z$. It is always possible to choose a coordinate system so that the matrix $[\alpha]$ has the diagonal form, i.e.,

$$
[\alpha] = \begin{pmatrix} \alpha_{XX} & 0 & 0 \\ 0 & \alpha_{YY} & 0 \\ 0 & 0 & \alpha_{ZZ} \end{pmatrix} \tag{211}
$$

Here XYZ, which are called the principal axes for the polarizability tensor, have the special property that a field directed along one of the axes will induce a dipole along the same axis (see Figure 18).

The principal axes for a molecule can usually be determined without difficulty by considering molecular symmetry, e.g., the six-fold rotational axis of benzene is one of the principal axes of the polarizability tensor. The elements of the polarizability tensor in the laboratory frame can then be related to the elements in the molecule fixed frame by means of the equation:[106]

$$\alpha_{ij} = \sum_{i',j'} \alpha_{i'j'} \ell_{i'i} \ell_{j'j} \tag{212}$$

where $\ell_{i',i}$ is the cosine of the angle between i′ and i axes. However, regardless of the coordinate system chosen, certain properties of the matrix $[\alpha]$ are unchanged. The invariants of interest here are the mean polarizability α defined by[107]

$$\alpha = \frac{1}{3}(\alpha_{XX} + \alpha_{YY} + \alpha_{ZZ}) \tag{213}$$

and the anisotropy β defined by

$$\beta^2 = \frac{1}{2}[(\alpha_{XX} - \alpha_{YY})^2 + (\alpha_{YY} - \alpha_{ZZ})^2 + (\alpha_{ZZ} - \alpha_{XX})^2 + 6(\alpha^2_{XY} + \alpha^2_{YZ} + \alpha^2_{ZX})] \tag{214}$$

In liquids and gases the situation is somewhat more complicated since the molecules are randomly oriented with respect to the laboratory coordinate system. Any measurement on a bulk system which depends on combinations of components of the polarizability tensor must in fact determine only averages of the combination over all orientations. The intensity of the scattered radiation is an important example. According to Equation 9, the intensity of light scattered by an isotropic scatterer depends on α^2. When anisotropy is present, the scattered light contains both polarized and depolarized components. Consider the geometry shown in Figure 1. Light polarized in the z-direction is said to be vertically polarized and is denoted by the index V. Similarly, light polarized in the x-direction is horizontally polarized and is denoted by H. For simplicity let $\chi = \theta_s = \pi/2$. If the incident light is vertically polarized, the intensity of the scattered light which is polarized vertically is proportional to $\overline{\alpha^2}_{zz}$ where the subscripts refer to the laboratory coordinate system and the bar indicates an average over molecular orientations. Similarly, the intensity of the scattered light which is polarized horizontally is proportional to $\overline{\alpha^2}_{zy}$. When vertically polarized incident light is used, the depolarization ratio is defined as

$$\rho_V = \frac{I_{VH}}{I_{VV}} = \frac{\overline{\alpha^2_{zy}}}{\overline{\alpha^2_{zz}}} \tag{215}$$

When naturally polarized incident light is used, the depolarization ratio is given by

$$\rho_n = \frac{I_{VH} + I_{HH}}{I_{VV} + I_{HV}} = \frac{\overline{\alpha_{zy}^2} + \overline{\alpha_{xy}^2}}{\overline{\alpha_{zz}^2} + \overline{\alpha_{xz}^2}} \tag{216}$$

When two indexes appear on the scattered intensity, the first indicates the direction of polarization of the incident light and the second refers to the scattered light.

The problem of calculating averages for products of tensor components for rotating molecules arises in many branches of physical chemistry, and the solution has been given by many authors. A complete derivation can be found in Reference 106. The averages required in light scattering are the following:[107]

$$\overline{\alpha_{xx}^2} = \overline{\alpha_{yy}^2} = \overline{\alpha_{zz}^2} = \alpha^2 + \frac{4}{45}\beta^2 \tag{217}$$

$$\overline{\alpha_{xy}^2} = \overline{\alpha_{yz}^2} = \overline{\alpha_{zx}^2} = \frac{\beta^2}{15} \tag{218}$$

$$\overline{\alpha_{xx}\alpha_{yy}} = \overline{\alpha_{yy}\alpha_{zz}} = \overline{\alpha_{zz}\alpha_{xx}} = \alpha^2 - \frac{2}{45}\beta^2 \tag{219}$$

Therefore, depolarization ratios can be expressed in terms of the invariants α and β, and the intensity of the depolarized component of the scattered light can be used to determine β. Using Equations 217 and 218, depolarization ratios can be written as:

$$\rho_V = \frac{3\beta^2}{45\alpha^2 + 4\beta^2} \tag{220}$$

$$\rho_n = \frac{6\beta^2}{45\alpha^2 + 7\beta^2} \tag{221}$$

In many scattering experiments the incident light is vertically polarized, but the scattered light at $\theta_s = \pi/2$ is collected without the use of a polarization analyzer. If the anisotropy $\beta \neq O$, the total scattered intensity is greater than that resulting from the mean polarizability. It is sometimes necessary to derive the Rayleigh ratio expected for isotropic scatterers having the polarizability α from the measured total intensity. This can be done as follows. For the total intensity we have

$$I_{TOTAL} = I_{VV} + I_{VH} \propto \overline{\alpha_{zz}^2} + \overline{\alpha_{zy}^2} = \frac{45\alpha^2 + 7\beta^2}{45}$$

Since the Rayleigh ratio is defined by $R_\theta = \overline{I}_s R^2/\overline{I}_i$, we can write

$$\frac{\mathcal{R}_\theta\ (\beta \neq 0)}{\mathcal{R}_\theta\ (\beta = 0)} = \frac{45\,\alpha^2 + 7\,\beta^2}{45\,\alpha^2} = \frac{1 + \rho_v}{1 - \frac{4}{3}\rho_v} \qquad (222)$$

This ratio, which is well known in classical light scattering, is known as the Cabannes factor. When naturally polarized incident light is used, the Cabannes factor has the form[108]

$$\left[\frac{6 + 6\,\rho_n}{6 - 7\,\rho_n}\right]$$

Thus, the Rayleigh ratio $\mathcal{R}_\theta(\beta = O)$ for scattering by concentration fluctuations alone can easily be derived from the experimental Rayleigh ratio by using Equation 222 if the depolarization ratio is known.

B. Electromagnetic Waves

The purpose of this appendix is to show how the susceptibility χ, the dielectric constant k_e, the polarizability α, and the index of refraction n are related.[109] For simplicity we consider a plane wave propagating in the +y-direction with the velocity v. The evolution of the amplitude U(y,t) for this wave obeys the equation:

$$\frac{\partial^2 U(y,t)}{\partial y^2} = \frac{1}{v^2}\frac{\partial^2 U(y,t)}{\partial t^2} \qquad (223)$$

For light propagating in the +y-direction in a vacuum Maxwell's equations show that the amplitude of the electric field is described by

$$\frac{\partial^2 E(y,t)}{\partial y^2} = \mu_o \epsilon_o \frac{\partial^2 E(y,t)}{\partial t^2} \qquad (224)$$

where $\mu_o = 4\pi \times 10^{-7}$ H/m is the permeability of the vacuum and $\varepsilon_o = 8.854 \times 10^{-12}$ F/m is the permittivity of the vacuum. A comparison of Equations 223 and 224 indicates that the speed of light in vacuum is given by

$$c = \frac{1}{\sqrt{\epsilon_o \mu_o}} = 2.997 \times 10^8 \text{ m/s} \qquad (225)$$

The solution of Equation 224 has the form

$$E(y,t) = E_o \cos(k_o y - \omega_o t) = E_o \text{ Re}\left[e^{i(k_o y - \omega_o t)}\right] \qquad (226)$$

where $k_o = 2\pi/\lambda_o$, λ_o is the wavelength in vacuum, and Re means "the real part of". Substitution of Equation 226 into 224 immediately gives $k_o = \omega_o/c$ as expected.

In nonconducting media, Equation 224 takes the form:

$$\frac{\partial^2 E(y,t)}{\partial y^2} = \mu_0 \frac{\partial^2}{\partial t^2}(\epsilon_0 E + P) \tag{227}$$

The quantity $\varepsilon_0 E + P$ is known as the electric displacement and is assigned the symbol D. The displacement is also written as

$$D = \epsilon E = \epsilon_0 E + P \tag{228}$$

where ε is the permittivity of the medium. Assuming that E is sufficiently small that the response of the medium is linear, the polarization P is proportional to E and can be written as:

$$P = \epsilon_0 \chi E = N \alpha E \tag{229}$$

Equation 229 serves to define the susceptibility χ per unit volume and the polarizability α per scatterer. N is the number of scatterers per unit volume. At this point we combine Equations 228 and 229 to define the dielectric constant k_e, which is also called the relative permittivity.

$$k_e = \frac{\epsilon}{\epsilon_0} = (1 + \chi) = \left(1 + \frac{N\alpha}{\epsilon_0}\right) \tag{230}$$

With these definitions Equation 227 can be written as

$$\frac{\partial^2 E(y,t)}{\partial y^2} = \mu_0 \epsilon_0 (1 + \chi) \frac{\partial^2 E(y,t)}{\partial t^2} = \frac{k_e}{c^2} \frac{\partial^2 E(y,t)}{\partial t^2} \tag{231}$$

By considering only the electric polarization of the medium we have already made the assumption that the permeability μ does not differ from its value μ_o in vacuum. This is a reasonable assumption for nonmagnetic materials.

The solution of Equation 231 has the same form as that for Equation 224 except that we now must allow for both retardation and absorption of the incident light. These features can be incorporated by permitting k_e and hence χ to have both real and imaginary parts. With this change the solution still has the form shown in Equation 226, and the substitution of 226 into 231 gives $k^2c^2/\omega^2_o = k_e$. It is consistent with common usage to define the index of refraction n and the extinction coefficient $\varkappa$ through the equation

$$kc/\omega_0 = n + i\kappa \tag{232}$$

Then the wave amplitude in Equation 226 becomes

$$\begin{aligned} E(y,t) &= E_0 \,\mathrm{Re}\left\{ e^{i\left[\frac{\omega_0}{c}(n + i\kappa)y - \omega_0 t\right]} \right\} \\ &= E_0\, e^{-\omega_0 \kappa y/c}\, \mathrm{Re}\left[e^{-(nk_0 y - \omega_0 t)} \right] \end{aligned} \tag{233}$$

The intensity I, which is proportional to E^2, is seen to decay as $e^{-2\omega_0 \kappa y/c}$; and the wavelength λ is given by $\lambda = \lambda_0/n$ or $k = nk_0$. Notice also that Equation 231 indicates that the velocity v for light in the medium is related to the speed of light c in vacuum by $k_e/c^2 = 1/v^2$ or $c/v = \sqrt{k_e}$. By definition c/v is equal to n, and therefore $n^2 = k_e$ in the absence of absorption.

For macromolecules in solution it is convenient to write

$$\begin{aligned} k_e &= (\chi_{solvent} + \chi_{solute}) + 1 \\ &= (\chi_{solvent} + 1) + \chi_{solute} \end{aligned} \tag{234}$$

Then

$$k_e - (k_e)_{solvent} = \chi_{solute}$$

$$n^2 - n_0^2 = N\alpha/\epsilon_0$$

where N is again the number of scatterers per unit volume.

C. Thermodynamic Relations

1. The Relationship of $(\partial^2 A/\partial C^2)_{T,V}$ to $(\partial\mu_1/\partial C)_{T,V}$

Since V is constant, the number of moles of components 1 and 2 are related by

$$\delta V = n_1\overline{V}_1 + n_2\overline{V}_2 \tag{235}$$

where $\overline{V}_1$ and $\overline{V}_2$ are the partial molar volumes of components 1 and 2, respectively. By convention the solvent is denoted by 1 and the solute by 2. Therefore, a change in the concentration of one species will be reflected in a change in the other according to

$$dn_1 = -\frac{\overline{V}_2}{\overline{V}_1}\, dn_2 \tag{236}$$

The change in Helmholtz free energy associated with the composition change at constant volume and temperature is given by

$$dA = \mu_1 dn_1 + \mu_2 dn_2 \qquad (237)$$

where μ_1 and μ_2 are the chemical potentials of the solvent and solute, respectively. By combining Equations 236 and 237 we obtain

$$dA = \left[\mu_2 - \frac{\bar{V}_2}{\bar{V}_1}\mu_1\right] dn_2 \qquad (238)$$

It is also apparent that dn_2 can be expressed in terms of the concentration C since $(n_2/\delta V) = C/M$ and $dn_2 = (\delta V/M)dC$. Substituting for dn_2 in Equation 238 gives

$$\left(\frac{\partial A}{\partial C}\right)_{T,V} = \left[\mu_2 - \frac{\bar{V}_2}{\bar{V}_1}\mu_1\right]\frac{\delta V}{M} \qquad (239)$$

Differentiation of Equation 239 with respect to C gives

$$\left(\frac{\partial^2 A}{\partial C^2}\right)_{T,V} = \left[\left(\frac{\partial \mu_2}{\partial C}\right)_{T,V} - \frac{\bar{V}_2}{\bar{V}_1}\left(\frac{\partial \mu_1}{\partial C}\right)_{T,V}\right]\frac{\delta V}{M} \qquad (240)$$

It should also be noted that the partial molar volumes depend on the concentration, but the fluctuations are too small for this to be important. We also note that the differentials of chemical potentials are related by the Gibbs-Duhem equation

$$n_1 d\mu_1 + n_2 d\mu_2 = 0 \qquad (241)$$

so $d\mu_2 = -n_1/n_2\, d\mu_1$. Substitution for $d\mu_2$ in Equation 240 gives

$$\left(\frac{\partial^2 A}{\partial C^2}\right)_{T,V} = -\frac{\delta V}{M}\left[\frac{n_1\bar{V}_1 + n_2\bar{V}_2}{n_2\bar{V}_1}\right]\left(\frac{\partial \mu_1}{\partial C}\right)_{T,V} \qquad (242)$$

and recalling that $C = Mn_2/(n_1\bar{V}_1 + n_2\bar{V}_2)$ allows Equation 242 to be simplified to the desired relation,

$$\left(\frac{\partial^2 A}{\partial C^2}\right)_{T,V} = -\frac{\delta V}{C\bar{V}_1}\left(\frac{\partial \mu_1}{\partial C}\right)_{T,V} \qquad (243)$$

2. Virial Expansion for the Chemical Potential

In the limit of infinite dilution, solutions tend to become ideal and the following expression for the concentration dependence of the chemical potential can be used.

$$\mu_1 - \mu_1^0 = RT \ln X_1 \tag{244}$$

Here μ_1 is the chemical potential of species 1, μ_1° is the standard chemical potential of species 1, X_1 is the mole fraction of species 1, and R and T have their usual meaning. For a binary system it is clear that $X_1 = 1 - X_2$ so that

$$\mu_1 - \mu_1^0 = RT \ln (1 - X_2) \tag{245}$$

where X_2 is the mole fraction of the solute and can be approximated by $C\,\overline{V}_1/M$. Since the expansion of $\ln(1 - X_2)$ is given by

$$\ln(1 - X_2) = -X_2 - \frac{1}{2} X_2^2 - \ldots \tag{246}$$

Equation 245 can be rewritten as

$$\mu_1 - \mu_1^0 = -RTV_1^0\, C \left[\frac{1}{M} + \frac{V_1^0 C}{2M^2} \right] \tag{247}$$

It should be noted that at the limit of $C = 0$, the partial molar volume $\overline{V}_1$ becomes the molar volume V°_1. Equation 247 is of the form of the well-known virial expansion which is usually written as

$$\mu_1 - \mu_1^0 = -RT\, V_1^0\, C \left[\frac{1}{M} + B_2 C + B_3 C^2 + \ldots \right] \tag{248}$$

where B_n is the nth virial coefficient. Equation 248 is differentiated with respect to solute concentration to obtain

$$\left(\frac{\partial \mu_1}{\partial C} \right)_{T,V} = -RT\, \overline{V}_1 \left[\frac{1}{M} + 2B_2 C + 3B_3 C^2 + \ldots \right] \tag{249}$$

Using the fact that $R = N_A k_B$ and rearranging, the standard form of the concentration dependence of the chemical potential is obtained.

$$-\frac{1}{k_BT\bar{V}_1}\left(\frac{\partial\mu_1}{\partial C}\right)_{T,V} = N_A\left[\frac{1}{M} + 2B_2C + 3B_3C^2 + \ldots\right] \tag{250}$$

D. Number, Weight, and z-Averages

In the study of macromolecules polydisperse samples are often encountered which present distribution of values of properties such as molecular weight, radius of gyration, and the degree of polymerization.[110] Different analytical methods report different types of averages for these properties. For example, in osmotic pressure measurements, the equilibrium across the membrane is influenced by the concentration of particles. The molecular weight determined by this method is the number average molecular weight, which is defined by

$$M_n = \frac{\sum_i N_iM_i}{\sum_i N_i} \tag{251}$$

where M_i is the molecular weight of the ith species and N_i is the number of particles of the ith type per unit volume.

On the other hand, analytic methods such as light scattering and ultracentrifugation give averages that depend on the weight of the macromolecule. The weight average molecular weight is defined by

$$M_w = \frac{\sum_i (N_iM_i)M_i}{\sum N_iM_i} \tag{252}$$

That the weight average molecular weight is obtained from light scattering data when the scattering intensities are plotted as $(KC/\mathcal{R}_\theta)_{c=0}$ vs. $\sin^2(\theta_s/2)$ can be seen from the following. First, assume that each species present has the same optical constant K and that the total concentration is given by

$$C = \sum_i C_i$$

At very low concentrations we can rewrite Equation 32 as

$$\mathcal{R}_\theta = K\sum_i C_iM_i \tag{253}$$

Then using the identity $C_i = M_iN_i/N_A$, we find that

$$\left(\frac{KC}{\mathcal{R}_\theta}\right)_{C=0} = \frac{\sum_i C_i}{\sum_j C_i M_i} = M_w^{-1} \tag{254}$$

The weight average molecular weight tends to emphasize the higher molecular weight species.

The molecular weights obtained from different averages may be quite different, depending on the distribution of the solute mass. Of course for a monodisperse solution the averages will be identical. This fact leads to an important property of these averages in that the degree of polydispersity of the system can be accessed. When a distribution of molecular weights is present, the various averages will assume a definite numerical order, i.e.,

$$M_n < M_w < M_z < M_{z+1} \cdots$$

where M_z, the z-average molecular weight, is defined as

$$M_z = \frac{\sum_i (N_i M_i^2) M_i}{\sum_i N_i M_i^2} \tag{255}$$

For a polydisperse sample the initial slope of a plot of $(K\,C/\mathcal{R}_\theta)$ vs. $\sin^2(\theta_s/2)$ turns out to give the z-average radius of gyration.

$$S(\underset{\sim}{K})_{exp} = \frac{\mathcal{R}_\theta}{\mathcal{R}_\theta\,(\text{at } \theta_s = O)} = \frac{\sum_i S(\underset{\sim}{K})_i\, C_i M_i}{\sum_i C_i M_i} \tag{256}$$

where $S(\underset{\sim}{K})_i$ is the structure factor for the ith species. Using Equation 57 the inverse of the experimental structure factor can be expressed as

$$S(\underset{\sim}{K})^{-1} = \sum_i C_i M_i \left[1 - \frac{16\pi^2 n^2 (R_G^2)_i \sin^2(\theta_s/2)}{3\lambda_o^2}\right] \Bigg/ \sum_i C_i M_i \tag{257}$$

Then, since

$$<R_G^2>_z = \frac{\sum_i C_i M_i (R_G^2)_i}{\sum_i C_i M_i} = \frac{\sum_i N_i M_i^2 (R_G^2)_i}{\sum_i N_i M_i^2} \tag{258}$$

and the second term in the brackets is small

$$S(\underset{\sim}{K})^{-1} = 1 + \frac{16\pi^2 n^2}{3\lambda_o^2} <R_G^2>_z \sin^2 (\theta_s/2) \tag{259}$$

Equation 259 can be combined with Equation 58 to give the desired result

$$\left(\frac{Kc}{\mathcal{R}_\theta}\right)_{C=0} = \frac{1}{M_w}\left[1 + \frac{16\pi^2 n^2 <R_G^2>_z}{3\lambda_o^2} \sin^2\left(\frac{\theta_s}{2}\right)\right] \tag{260}$$

E. Correlation Functions and Spectra of Scattered Light

Suppose the x(t) is a random variable, i.e., a function of t with an allowed set of values, each of which occurs with a definite probability. The power associated with x(t) is $|x(t)|^2$, and the power spectrum $J(\omega)$ is defined so that $J(\omega)d\omega$ gives the fraction of the average power which lies in the frequency range from ω to $\omega + d\omega$. The correlation function $C(\tau)$ for x(t) is written as[111,112]

$$C(\tau) = <x^*(t)x(t+\tau)> = \lim_{T\to\infty} \frac{1}{T} \int_{-T/2}^{+T/2} x^*(t)x(t+\tau)dt \tag{261}$$

For example x(t) might be the noise on an electrical signal, e.g., current fluctuations in a resistor. In this case $J(\omega)$ would be proportional to the signal from a narrow band amplifier tuned to detect the component of x(t) at ω. The purpose of this appendix is to derive the Wiener-Khintchine theorem which relates $J(\omega)$ to $C(\tau)$.

The record of x(t) in the time interval $-T/2 \leqslant t \leqslant T/2$ can be Fourier analyzed to obtain the frequency components of x(t). For this purpose we define $x_T(t)$ so that

$$x_T(t) = x(t) \quad \text{for} \quad |t| \leqslant T/2$$

$$x_T(t) = 0 \quad \text{otherwise}$$

The variable $x_T(t)$ and its Fourier transform $\hat{x}_T(\omega)$ are related by the transform pair

$$x_T(t) = \int_{-\infty}^{+\infty} \hat{x}_T(\omega)e^{-i\omega t}\,d\omega \tag{262}$$

$$\hat{x}_T(\omega) = \frac{1}{2\pi}\int_{-T/2}^{+T/2} x_T(t)e^{+i\omega t}dt \tag{263}$$

First, we obtain an expression for $J(\omega)$ in terms of $\hat{x}_T(\omega)$ which is consistent with the identity

$$<|x_T(t)|^2> = \int_{-\infty}^{+\infty} J(\omega)d\omega \tag{264}$$

The average power can be written as

$$<|x_T(t)|^2> = \lim_{T\to\infty} \frac{1}{T} \int_{-T/2}^{+T/2} x_T^*(t)x_T(t)dt \tag{265}$$

and this provides a starting point. By inserting the rhs of Equation 262 for $x_T(t)$ into the integral of Equation 265 we obtain

$$\int_{-T/2}^{+T/2} x_T^*(t)x_T(t)dt = \int_{-T/2}^{+T/2} x_T^*(t) \left[\int_{-\infty}^{+\infty} \hat{x}_T(\omega)e^{-i\omega t}\, d\omega \right] dt \tag{266}$$

Then by exchanging the order of integration and using the complex conjugate of Equation 263, we find that

$$\int_{-\infty}^{+\infty} \hat{x}_T(\omega) \left[\int_{-T/2}^{+T/2} x_T^*(t)e^{-i\omega t}\, dt \right] d\omega = \int_{-\infty}^{+\infty} \hat{x}_T(\omega)[2\pi x_T^*(\omega)]\, d\omega \tag{267}$$

A comparison of Equations 267 and 265 then shows that

$$<|x_T(t)|^2> = \int_{-\infty}^{+\infty} \lim_{T\to\infty} \left(\frac{2\pi|\hat{x}_T(\omega)|^2}{T} \right) d\omega \tag{268}$$

and Equation 264 requires that

$$J(\omega) = \lim_{T\to\infty} \frac{(2\pi)\,|\hat{x}_T(\omega)|^2}{T} \tag{269}$$

Using the form of $J(\omega)$ given in Equation 269, it is easy to show that the power spectrum is equal to the Fourier transform of the correlation function. With $\hat{x}_T(\omega)$ from Equation 263 we obtain

$$J(\omega) = \lim_{T\to\infty} \frac{2\pi}{T}\left[\frac{1}{2\pi}\int_{-T/2}^{+T/2} x_T^*(t)e^{-i\omega t}dt\right]\left[\frac{1}{2\pi}\int_{-T/2}^{+T/2} x_T(t')e^{+i\omega t'}dt'\right]$$

$$= \lim_{T\to\infty} \frac{1}{2\pi T}\int_{-T/2}^{+T/2}\int_{-T/2}^{+T/2} x_T^*(t)x_T(t')e^{-i\omega(t-t')}dtdt' \qquad (270)$$

Changing the variables of integration from t, t′ to t, τ where τ = t′ − t gives

$$J(\omega) = \frac{1}{2\pi}\int_{-T/2}^{+T/2}\left[\lim_{T\to\infty}\frac{1}{T}\int_{-T/2}^{+T/2} x_T^*(t)x_T(t+\tau)dt \quad e^{i\omega\tau}\,d\tau\right] \qquad (271)$$

Since in cases of interest C(τ) vanishes for τ > T/2, this is equivalent to

$$J(\omega) = \frac{1}{2\pi}\int_{-\infty}^{+\infty} C(\tau)e^{i\omega\tau}d\tau \qquad (272)$$

and the reverse transform is

$$C(\tau) = \int_{-\infty}^{+\infty} J(\omega)e^{-i\omega\tau}d\omega \qquad (273)$$

Equation 272 can be written in a form involving only integration over positive values of τ by considering the properties of the correlation function C(τ). Since x(t) represents a stationary random variable, any time origin can be used in the calculation of C(τ). Therefore

$$C(\tau) = \langle x^*(\tau)\;X(t+\tau)\rangle = \langle x^*(t-\tau)\;X(\tau)\rangle = C^*(-\tau) \qquad (274)$$

The identity C(τ) = C*(−τ) could also have been derived from the requirement that J(ω) be a real function, i.e., that J(ω) = J*(ω). In order to use this condition, we write Equation 272 as

$$J(\omega) = \frac{1}{2\pi}\left[\int_{-\infty}^{0} C(\tau)\,e^{i\omega\tau}d\tau + \int_{0}^{\infty} C(\tau)\,e^{i\omega\tau}d\tau\right] \qquad (275)$$

The first integral on the rhs of Equation 275 can be combined with 274 to obtain:

$$\int_{-\infty}^{0} C^*(-\tau)e^{i\omega\tau}d\tau = \int_{0}^{\infty} C^*(\tau)\,e^{-i\omega\tau}d\tau \tag{276}$$

Therefore Equation 275 can be written as

$$J(\omega) = \frac{1}{2\pi}\int_{0}^{\infty} [C^*(\tau)e^{-i\omega\tau} + C(\tau)e^{+i\omega\tau}]\,d\tau \tag{277}$$

and

$$J(\omega) = \frac{1}{\pi}\,\mathrm{Re}\int_{0}^{\infty} C(\tau)\,e^{i\omega\tau}d\tau \tag{278}$$

The last step follows from the identity Re z = (z + z*)/2 where z is an arbitrary random variable.

F. The Heterodyne Correlation Function

If I_{det} is the intensity of light at the surface of a PMT, the photocurrent or the photon count rate can be used to calculate an approximation to $G^{(2)}(\tau) = \langle I_{det}(t)\, I_{det}(t + \tau)\rangle$. In "homodyne" or "self-beat" experiments $I_{det}(t) = I_s$ and $G^{(2)}(\tau) = \langle I_s(t)\, I_s(t + \tau)\rangle$. As pointed out in Section III.A.3, this function does not in general permit the electric field correlation function $G^{(1)}(\tau)$ to be determined. Even with Gaussian light, only $|G^{(1)}(\tau)|$ can be derived from $G^{(2)}(\tau)$, as discussed in Section III.A.3, and important oscillatory factors may be lost. It is possible, however, to arrange the experiment so that Re $G^{(1)}(\tau)$ is obtained directly. A factor such as $e^{i\omega_1 t}$, which would be lost if the absolute magnitude were taken, will then appear as $\cos(\omega_1 t)$. For this to be possible, a coherent reference beam of light must be mixed with the scattered light on the surface of the PMT. An analysis of this experiment follows.[37]

Suppose that a reference beam is obtained either by picking off a fraction of the incident beam before it reaches the scattering volume or by inserting an object into the scattering volume which will scatter incident light without a frequency shift. The latter method might be realized by simply collecting some light scattered from the wall of the sample cell in addition to light scattered from the sample. By analogy to radio frequency techniques, this is called a heterodyne experiment, and the reference beam is referred to as the local oscillator. The electric field of the light at the PMT is then given by

$$E_{det}(t) = E_S(t) + E_L(t) \tag{279}$$

where

$$E_S(t) = f(t)\, E_o e^{-i\omega_o t}$$

$$E_L(t) = E_{LO}\, e^{-i[\omega_o t - \phi_L]}$$

Here f(t) specifies the modulation resulting from the scattering process, and ϕ_L takes into account a phase shift for E_L if the reference beam has a different path length to the detector than does the scattered beam. The autocorrelation function for I_{det} in this case is

$$G^{(2)}(\tau) = <[E_L^*(t) + E_S^*(t)]\; [E_L(t) + E_S(t)]\; [E_L^*(t+\tau) + E_S^*(t+\tau)]\, [E_L(t+\tau) + E_S(t+\tau)]>$$

This function can be multiplied out and simplified to obtain

$$G^{(2)}(\tau) = I_L^2 + 2\, I_L <I_S> + <I_S>^2 + I_L <E_S^*(t)\, E_S(t+\tau)>\, e^{+i\omega_o \tau} + I_L <E_S(t)\, E_S^*(t+\tau)>\, e^{-i\omega_o \tau} \qquad (280)$$

where we have used the following results

$$<E_L(t)> = <E_S(t)> = <E_S(t)\, E_S(t+\tau)> = <E_L(t)\, F_L(t+\tau)> = 0$$

and similar equations for the complex conjugates. All of these averages vanish because

$$<e^{i\omega_o t}> = <\cos \omega_o t> + i <\sin \omega_o t> = 0$$

As in Section III, we suppress the factor of ($\varepsilon_o c/2$) so that $I_L = |E_L|^2$ and $<I_s> = <|E_s(t)|^2>$. In the experiment, as usually performed, the intensity of the reference beam is adjusted so that $I_{LO} >> <I_s>$. Therefore, Equation 280 can be written as

$$G^{(2)}(\tau) = I_L^2 + 2\, I_L\, \mathrm{Re}[G_s^{(1)}(\tau) e^{+i\omega_o \tau}] \qquad (281)$$

According to Equation 273, the power spectrum associated with I_{det} in the heterodyne experiment is

$$I^{(2)}(\omega) = I_L^2\, \delta(\omega) + 2\, I_L \frac{\mathrm{Re}}{\pi} \int_0^{+\infty} e^{i\omega\tau}\, \mathrm{Re}[G_s^{(1)}(\tau) e^{i\omega_o \tau}]\, d\tau \qquad (282)$$

It is also possible to use a local oscillator having a frequency ω_L different from ω_o. In this case ω_o appearing in Equations 280, 281, and 282 would be replaced with ω_L. A discussion of this experiment and other details can be found in References 37 and 113.

G. Cumulant Analysis

The homodyne light scattering experiment yields $\beta|g^{(1)}(\tau)|^2$, and for monodisperse scatterers $|g^{(1)}(\tau)| = \text{xp}(-\Gamma\tau)$ where $\Gamma = D_T K^2$. However, samples of macromolecules are often polydisperse because of aggregation or contamination, and a distribution of Γ values must be considered. If $G(\Gamma)$ is the normalized distribution function for Γ, then

$$|g^{(1)}(\tau)| = \int_0^\infty G(\Gamma)\, e^{-\Gamma\tau}\, d\Gamma \tag{283}$$

and

$$\int_0^\infty G(\Gamma)\, d\Gamma = 1$$

Discrete particle sizes can be treated by writing $G(\Gamma)$ in the form

$$G(\Gamma) = \frac{\sum_i N_i m_i^2\, \delta\,(\Gamma - \Gamma_i)}{\sum_i N_i m_i^2} = \frac{\sum_i C_i M_i\, \delta\,(\Gamma - \Gamma_i)}{\sum_i C_i M_i} \tag{284}$$

The notation is consistent with that used in Section II.1. For the ith species N_i is the number of scatterers per unit volume, m_i is the mass of a particle, C_i is the weight concentration (weight/unit volume), and M_i is the molecular weight. The form of $G(\Gamma)$ in Equation 284 shows that the average specified in Equation 283 is a z-average.

One method for obtaining the cumulant expansion of $\ell n|g^{(1)}(\tau)|$ is to expand $\exp(-\Gamma\tau)$ about $\exp(-<\Gamma>\tau)$. Thus[40]

$$|g^{(1)}(\tau)| = e^{-<\Gamma>\tau} \int_0^\infty G(\Gamma)\, e^{-(\Gamma - <\Gamma>)\tau}\, d\Gamma$$

$$= e^{-<\Gamma>\tau} \int_0^\infty G(\Gamma)\, [1 - (\Gamma - <\Gamma>)\tau + (\Gamma - <\Gamma>)^2 \frac{\tau^2}{2!} + \ldots]\, d\Gamma \tag{285}$$

where

$$<\Gamma> = \int_0^\infty \Gamma G(\Gamma) d\Gamma$$

The moments of the distribution are defined by

$$\mu_n = \int_0^\infty (\Gamma - <\Gamma>)^n G(\Gamma) d\Gamma \qquad (286)$$

and Equation 285 can be written as

$$|g^{(1)}(\tau)| = e^{-<\Gamma>\tau}\left[1 + \frac{\mu_2}{2!}\tau^2 - \frac{\mu_3}{3!}\tau^3 + \frac{\mu_4}{4!}\tau^4 + \ldots\right] \qquad (287)$$

The desired result is obtained using the Taylor's series

$$\ell n(1 + x) = x - \frac{x^2}{2} + \frac{x^3}{3} - \frac{x^4}{4} + \ldots , \quad (-1 < x < +1) \qquad (288)$$

and collecting terms so that

$$\ell n|g^{(1)}(\tau)| = -<\Gamma>\tau + \frac{\mu_2}{2!}\tau^2 - \frac{\mu_3}{3!}\tau^3 + \frac{(\mu_4 - 3\mu_2^2)}{4!}\tau^4 + \ldots \qquad (289)$$

The coefficients in this series, which are known as cumulants, describe some of the properties of $G(\Gamma)$. The standard deviation is $\sqrt{\mu_2}$, the skewness is $\mu_3/\mu_2^{3/2}$, and the kurtosis is $[(\mu_4/\mu_2^2) - 3]/2$. Another quantity, which is often used to specify the degree of polydispersity, is the normalized variance defined as $\mu_2/<\Gamma>^2$. Since $\Gamma = D_T K^2$, the variance can also be expressed as

$$\frac{<(\Gamma - <\Gamma>)^2>}{<\Gamma>^2} = \frac{<D_T^2> - <D_T>^2}{<D_T>^2} \qquad (290)$$

H. The Diffusion Coefficient

An expression for D_T can easily be obtained by considering the particle flux $\underset{\sim}{J}(\underset{\sim}{r},t)$.[42] From the definition of the flux as mass through unit area in unit time we can write

$$\underset{\sim}{J}(\underset{\sim}{r},t) = \underset{\sim}{v}(\underset{\sim}{r},t)\, C(\underset{\sim}{r},t) \qquad (291)$$

where $\underset{\sim}{v}$ is the average velocity of the solute molecules at the position $\underset{\sim}{r}$ at time t, and

C is again the mass concentration of the solute. We assume that $\underset{\sim}{v}$ is a steady-state velocity which is attained by a particle under the influence of an applied force $\underset{\sim}{F}$ and a frictional force $-f_T\underset{\sim}{v}$ where f_T is the translational friction coefficient. For a sphere in a viscous medium Stokes' equation gives

$$f_T = 6\pi\eta a_h \tag{292}$$

where η is the coefficient of viscosity and a_h is the hydrodynamic radius of the sphere. Perrin and others have extended Stokes' treatment to cover ellipsoids of revolution. If the semi-axes are denoted by a, b, and b, Equation 292 can be rewritten as[114]

$$f_T = \frac{6\pi\eta a}{G(b/a)} \tag{293}$$

where

$$G(b/a) = \frac{1}{\sqrt{1-(b/a)^2}}\,\ell n\left[\frac{1+\sqrt{1-(b/a)^2}}{(b/a)}\right] \quad ; \ a > b \text{ (prolate)}$$

and

$$G(b/a) = \frac{\tan^{-1}\sqrt{(b/a)^2-1}}{\sqrt{(b/a)^2-1}} \quad ; \ a < b \text{ (oblate)}$$

According to Newton's second law for a particle of mass m

$$m\frac{d^2\underset{\sim}{r}}{dt^2} = \text{net force} = \underset{\sim}{F} - f_T\underset{\sim}{v} \tag{294}$$

and under steady-state conditions, where the acceleration vanishes, we find that

$$\underset{\sim}{v} = \underset{\sim}{F}/f_T \tag{295}$$

In translational diffusion the effective force per particle results from the gradient in the chemical potential. Therefore, the driving force for diffusion can be written as

$$\underset{\sim}{F} = -\frac{1}{N_A}\underset{\sim}{\nabla}\mu_2 \tag{296}$$

where μ_2 is the chemical potential of the solute. Equations 291, 295, and 296 can be combined to obtain

$$\underset{\sim}{J}(\underset{\sim}{r},t) = -\frac{C(\underset{\sim}{r},t)}{f_T N_A}\underset{\sim}{\nabla}\mu_2 \qquad (297)$$

In order to obtain an expression for D_T by comparing Equations 297 and 96 we must express $\underset{\sim}{\nabla}\mu_2$ in terms of the concentration. The chemical potential μ_2 is related to the mass concentration by the equation

$$\mu_2 = \mu_2^{\circ} + RT\, \ell n\, [\gamma C(\underset{\sim}{r}, t)/M] \qquad (298)$$

where μ°_2 is the standard chemical potential, which is a constant, and γ is the activity coefficient. Thus

$$\underset{\sim}{\nabla}\mu_2 = \frac{RT}{C(r,t)}\underset{\sim}{\nabla}C(\underset{\sim}{r},t) \qquad (299)$$

and

$$\underset{\sim}{J}(\underset{\sim}{r},t) = -\frac{RT}{N_A f_T}\underset{\sim}{\nabla}C(\underset{\sim}{r},t) \qquad (300)$$

Equations 300 and 96 can now be combined to obtain the Stokes-Einstein equation

$$D_T = \frac{k_B T}{f_T} = \frac{k_B T}{6\pi\eta a_h} \qquad (301)$$

We choose to retain the form of the diffusion equation as the concentration increases and simply to let D_T become a function of C. The concentration dependence of D_T enters the theory both through the chemical potential μ_2 and the friction coefficient f_T. From Appendix C we have the virial expansion for the chemical potential μ_1 of the solvent

$$\mu_1 = \mu_1^{\circ} - RT\, V_1^{\circ}\, C\left[\frac{1}{M} + B_2\, C + B_3\, C^2 + \ldots\right]$$

where V°_1 is the molar volume of the solvent and the B_i are virial coefficients. Considering only the x-direction, we have

$$\frac{\partial \mu_1}{\partial x} = -RT\, V_1^{\circ}\left(\frac{1}{M} + 2B_2\, C + 3B_3\, C^2 + \ldots\right)\frac{\partial C}{\partial x} \qquad (302)$$

The desired expression for the gradient of μ_2 is obtained by using the Gibbs-Duhem relation $n_1 d\mu_1 = -n_2 d\mu_2$ where n_1 and n_2 are the number of moles of solvent and solute, respectively. Equation 302 then becomes

$$\frac{\partial \mu_2}{\partial x} = \frac{RT}{C}\,(1 + 2B_2\,MC + 3B_3\,MC^2 + \ldots)\,\frac{\partial C}{\partial x} \qquad (303)$$

where we have used the fact that $(n_1\,V°_1/n_2\,M) = C^{-1}$. The friction coefficient can also be expanded in terms of C as follows

$$f_T = f_0\,(1 + B'C + \ldots) \qquad (304)$$

When Equations 303 and 304 are combined with Equation 297, the resulting expression for D_T is

$$\begin{aligned} D_T(C) &= \frac{k_BT}{f_0}\,\frac{(1 + 2\,B_2\,MC + 3\,B_3\,MC^2 + \ldots)}{(1 + B'C + \ldots)} \\ &= \frac{k_BT}{f_0}\,[1 + (2\,M\,B_2 - B')\,C + \ldots] \end{aligned} \qquad (305)$$

I. The Rotational Diffusion Equation

The rotational random walk of a unit vector is depicted in Figure 19. By considering the probability of rotations into and out of an element of solid angle $d\Omega$, Debye was able to derive an equation for rotational diffusion similar to the translational diffusion equation. The conditional probability that the orientation is along r at time τ given that it was along $\hat{r}_o$ at t = O is denoted by $P(\hat{r}_o|\hat{r},\tau)$, which is also known as the transition probability from $\hat{r}_o$ to $\hat{r}$. This function obeys the rotational diffusion equation[77]

$$\frac{\partial P(\hat{r}_o|\hat{r},\tau)}{\partial \tau} = D_R\,\nabla^2 P(\hat{r}_o|\hat{r},\tau) \qquad (306)$$

where in spherical polar coordinates

$$\nabla^2 = \frac{1}{\sin\theta}\,\frac{\partial}{\partial\theta}\left(\sin\theta\,\frac{\partial}{\partial\theta}\right) + \frac{1}{\sin^2\theta}\,\frac{\partial^2}{\partial\phi^2} \qquad (307)$$

In the following we take the polar coordinates of $\hat{r}_o$ and $\hat{r}$ to be (θ_o,ϕ_o) and (θ,ϕ), respectively.

Students of atomic physics will recognize that the angular momentum operator $\hat{L}^2$ is equal to the negative of ∇^2. In addition it is well known that the spherical harmonics $Y_{\ell m}(\theta,\phi)$ are eigenfunctions of $\hat{L}^2$ and satisfy the equation[115]

$$\hat{L}^2 Y_{\ell m}(\theta,\phi) = \ell(\ell + 1) Y_{\ell m}(\theta,\phi) \quad (308)$$

where $\ell = 0,1,2,3 \ldots, m = \ell, \ell-1, \ell-2, \ldots, -\ell + 1, -\ell$. Therefore,

$$\nabla^2 Y_{\ell m}(\theta,\phi) = -\ell(\ell + 1) Y_{\ell m}(\theta,\phi) \quad (309)$$

Using this result, it is easy to verify that the following is a solution of Equation 306.

$$P(\hat{r}_o|\hat{r},\tau) = Y^*_{\ell m}(\theta,\phi)\, e^{-\ell(\ell + 1) D_R \tau} \quad (310)$$

A linear combination of solutions is also a solution, and any solution can be written as

$$P(r_o|r,\tau) = \sum_{\ell=O}^{\infty} \sum_{m=-\ell}^{\ell} c_{\ell m} Y^*_{\ell m}(\theta,\phi)\, e^{-\ell(\ell + 1) D_R \tau} \quad (311)$$

where the $c_{\ell m}$ are coefficients which must be determined. The functions $Y_{\ell m}(\theta,\phi)$ are a complete set of normalized and orthogonal functions. Thus

$$\int_{\phi=O}^{2\pi} \int_{\theta=0}^{\pi} Y^*_{\ell m}(\theta,\phi) Y_{\ell' m'}(\theta,\phi) \sin\theta \, d\theta \, d\phi = \delta_{\ell\ell'} \delta_{mm'} \quad (312)$$

where the δ_{ij} are Kronecker delta functions which are unity for $i = j$ and are zero otherwise. Therefore, $c_{\ell m}$ can be determined by setting $\tau = 0$ and using the usual trick of multiplying through Equation 311 by $Y_{\ell m}(\theta,\phi)$ and integrating over the coordinates. We recognize that

$$P(\hat{r}_o|\hat{r},O) = \delta(\hat{r}_o - \hat{r}) = \delta(\theta - \theta_o)\,\delta(\phi - \phi_o) \quad (313)$$

where $\delta(x - x_o)$ is the Dirac delta function which is normalized but equals zero if $x \neq x_o$. The integration then gives

$$\int Y_{\ell m}(\theta,\phi) P(\hat{r}_o|\hat{r},0)\, d\Omega = c_{\ell m} = Y_{\ell m}(\theta_o,\phi_o) \quad (314)$$

Now returning to Equation 311 we can write

$$P(r_0|r,\tau) = \sum_{\ell=O}^{\infty} \sum_{m=-\ell}^{\ell} Y_{\ell m}(\theta_0,\phi_0)\, Y_{\ell m}^*(\theta,\phi)\, e^{-\ell(\ell+1) D_R \tau} \tag{315}$$

It is this probability function which is required in the calculation of correlation functions of the components of the polarizability tensor.

REFERENCES

1. **Tanford, C.,** *Physical Chemistry of Macromolecules,* John Wiley & Sons, New York, 1961.
2. **Timasheff, S. N. and Townsend, R.,** Light scattering, *Physical Principles and Techniques of Protein Chemistry Part B,* Leach, S. J., Ed., Academic Press, New York, 1970, 147.
3. **Huglin, M. B.,** *Light Scattering from Polymer Solutions,* Academic Press, New York, 1972.
4. **Fabelinskii, I. L.,** *Molecular Scattering of Light,* Plenum Press, New York, 1968.
5. **Kerker, M.,** *The Scattering of Light,* Academic Press, New York, 1969.
6. **Long, D. A.,** *Raman Spectroscopy,* McGraw-Hill, New York, 1977.
7. **Schurr, J. M.,** Dynamic light scattering of biopolymers and biocolloids, *CRC Crit. Rev. Biochem.,* 4, 371, 1977.
8. **Pusey, P. N. and Vaughan, J. M.,** Light scattering and intensity fluctuation spectroscopy, *Dielectric and Related Molecular Processes,* Vol. 2, Davies, M., Ed., The Chemical Society, London, 1975, 48.
9. **Carlson, F. D.,** The application of intensity fluctuation spectroscopy to molecular biology, *Ann. Rev. Biophys. and Bioeng.,* 4, 243, 1975.
10. **Ware, B. R.,** Applications of laser velocimetry in biology and medicine, *Chemical and Biochemical Applications of Lasers,* Vol. 2, Moore, C. B., Ed., Academic Press, New York, 1977, chap. 5.
11. **Ware, B. R.,** Electrophoretic light scattering, *Adv. Colloid Interface Sci.,* 4, 1, 1974.
12. **Berne, B. J. and Pecora, R.,** *Dynamic Light Scattering,* John Wiley & Sons, New York, 1976.
13. **Chu, B.,** *Laser Light Scattering,* Academic Press, New York, 1974.
14. **Cummins, H. Z. and Pike, E. R.,** *Photon Correlation and Light Beating Spectroscopy,* Plenum Press, New York, 1974.
15. **Loudon, R.,** *The Quantum Theory of Light,* Clarendon Press, Oxford, 1973.
16. **Rayleigh, Lord,** On the light from the sky, its polarization and color, *Phil. Mag.,* XLI, 4th series, 107, 1871.
17. **Kaye, W. and Havlik, A. J.,** Low angle laser light scattering — absolute calibration, *Appl. Opt.,* 12, 541, 1973.
18. **Smoluchowski, M.,** Molekular-kinetische Theorie der Opaleszenz von Gasen im kritischen Zustande sowie einiger verwandter Erscheinungen, *Ann. Phys.,* 25, 205, 1908.
19. **Einstein, A.,** Theory of the opalescence of homogeneous liquids and mixtures of liquids in the vicinity of the critical state, English translation, *Colloid Chemistry,* Vol. I, Alexander, J., Ed., Reinhold, New York, 1926, 323.
20. **Neugebauer, T.,** Berechnung der Lichtzerstreuung von Fadenkettenlosungen, *Ann. Phys.,* 42, 509, 1943.
21. **Kratochvil, P.,** Particle scattering functions, *Light Scattering from Polymer Solutions,* Huglin, M. B., Ed., Academic Press, New York, 1972, chap. 7.
22. **Gunier, A.,** La diffraction des rayons X aux tres petits angles: application a l'etude de phenomenes ultramicroscopique, *Ann. Phys.,* 12, 161, 1939.
23. **Tanford, C.,** *Physical Chemistry of Macromolecules,* John Wiley & Sons, New York, 1961, chap. 3.
24. **Zimm, B. H.,** Apparatus and methods for measurement and interpretation of angular variation of light scattering; preliminary results on polystyrene solutions, *J. Chem. Phys.,* 16, 1099, 1948.
25. **Morris, V. J., Coles, H. J., and Jennings, B. R.,** Infrared plots for macromolecular characterization, *Nature,* 249, 240, 1974.
26. **Benoit, H.,** On the effect of branching and polydispersity on the angular distribution of light scattering by Gaussian coils, *J. Polym. Sci.,* 11, 507, 1953.
27. **Benoit, H., Holtzer, A. M., and Doty, P.,** An experimental study of polydispersity by light scattering, *J. Phys. Chem.,* 58, 635, 1954.

28. **Holtzer, A. M., Benoit, H., and Doty, P.,** The molecular configuration and hydrodynamic behavior of cellulose trinitrate, *J. Phys. Chem.*, 58, 624, 1954.
29. **Holtzer, A.,** Interpretation of the angular distribution of the light scattered by a polydisperse system of rods, *J. Polym. Sci.*, 17, 432, 1955.
30. **Goldstein, M.,** Scattering factors for certain polydisperse systems, *J. Chem. Phys.*, 21, 1255, 1953.
31. **Rice, S. A.,** Particle scattering factors in polydisperse systems, *J. Polym. Sci.*, 16, 94, 1955.
32. **Kratochvil, P.,** Some remarks on light scattering by solutions of highly polydisperse polymers, *J. Polym. Sci.*, Part C, 23, 143, 1968.
33. **Carpenter, D. K.,** Light scattering study of the molecular weight distribution of polypropylene, *J. Polym. Sci.*, Part A-2, 4, 923, 1966.
34. **Cummins, H. Z., Knable, N., and Yeh, Y.,** Observation of diffusion broadening of Rayleigh scattered light, *Phys. Rev. Lett.*, 12, 150, 1964.
35. **Mandel, L.,** Fluctuations in light beams, *Progress in Optics*, Vol. 2, Wolf, E., Ed., Wiley-Interscience, New York, 1963, 181.
36. **Pecora, R.,** Doppler shifts in light scattering from pure liquids and polymer solutions, *J. Chem. Phys.*, 40, 1604, 1964.
37. **Jakeman, E.,** Photon correlation, *Photon Correlation and Light Beating Spectroscopy*, Cummins, H. Z. and Pike, E. R., Eds., Plenum Press, New York, 1974, 75.
38. **Jones, C. R.,** Photon Correlation Spectroscopy of Hemoglobin: Diffusion of Oxy-HbA and Oxy-HbS, Ph.D. thesis, University of North Carolina, Chapel Hill, 1977.
39. **Wilson, W. W.,** Light Beating Spectroscopy Applied to Biochemical Systems, Ph.D. thesis, University of North Carolina, Chapel Hill, 1973.
40. **Koppel, D. E.,** Analysis of macromolecular dispersity in intensity correlation spectroscopy: the method of cumulants, *J. Chem. Phys.*, 57, 4814, 1972.
41. **Bezot, P., Ostrowsky, N., and Hesse-Bezot, C.,** Light scattering data analysis for samples with large polydispersities, *Optics Commun.*, 25, 14, 1978.
42. **Einstein, A.,** *Investigations on the Theory of the Brownian Movement*, Dover Publications, New York, 1926.
43. **Cummins, H. Z.,** Applications of light beating spectroscopy in biology, *Photon Correlation and Light Beating Spectroscopy*, Cummins, H. Z. and Pike, E. R., Eds., Plenum Press, New York, 1974, 285.
44. **Kitchen, R. G., Preston, B. N., and Wells, J. D.,** Diffusion and sedimentation of serum albumin in concentrated solutions, *J. Polymer Sci. Symp.*, 55, 39, 1976.
45. **Keller, K. H., Canales, E. R., and Yum, S. I.,** Tracer and mutual diffusion coefficients of proteins, *J. Phys. Chem.*, 75, 379, 1971.
46. **Herbert, T. J. and Carlson, F. D.,** Spectroscopic study of the self-association of myosin, *Biopolymers*, 10, 2231, 1971.
47. **Palmer, G., Fritz, O. G., and Hallett, F. R.,** Quasi-elastic light scattering on human fibrinogen. I. Fibrinogen, *Biopolymers*, 18, 1647, 1979.
48. **Cohen, R. J., Jedziniak, J. A., and Benedek, G. B.,** Study of the aggregation and allosteric control of bovine glutamate dehydrogenase by means of quasi-elastic light scattering spectroscopy, *Proc. R. Soc. London, Ser. A*, 345, 73, 1975.
49. **Jones, C. R., Johnson, C. S., Jr., and Penniston, J. T.,** Photon correlation spectroscopy of hemoglobin: diffusion of oxy-HbA and oxy-HbS, *Biopolymers*, 17, 1581, 1978.
50. **Dubin, S. B., Lunacek, J. H., and Benedek, G. B.,** Observation of the spectrum of light scattered by solutions of biological macromolecules, *Proc. Natl. Acad. Sci. U.S.A.*, 57, 1164, 1967.
51. **Dubin, S. B., Clark, N. A., and Benedek, G. B.,** Measurement of the rotational diffusion coefficient of lysozyme by depolarized light scattering: configuration of lysozyme in solution, *J. Chem. Phys.*, 54, 5158, 1971.
52. **Rimai, L., Hockmott, J. T., Jr., Cole, T., and Carew, E. B.,** Quasi-elastic light scattering by diffusional fluctuations in RNase solutions, *Biophys. J.*, 10, 20, 1970.
53. **Bellamy, A. R., Gillies, S. C., and Harvey, J. D.,** Molecular weight of two oncornavirus geomes: derivation from particle molecular weights and RNA content, *J. Virol.*, 14, 1388, 1974.
54. **Ware, B. R., Raj, T., Flygare, W. H., Lesnaw, J. A., and Reichman, M. E.,** Molecular weights of vesicular stomatitis virus and its defective particles by laser light-beating spectroscopy, *J. Virol.*, 11, 141, 1973.
55. **Schaefer, D. W., Benedek, G. B., Schofield, P., and Bradford, E.,** Spectrum of light quasielastically scattered from tobacco mosaic virus, *J. Chem. Phys.*, 55, 3884, 1971.
56. **Dubos, P., Hallett, R., Kells, D. T. C., Sorensen, O., and Rowe, D.,** Biophysical studies of infectious pancreatic necrosis virus, *J. Virol.*, 22, 150, 1977.

57. **Camerini-Otero, R. D., Pusey, P. N., Koppel, D. E., Schaefer, D. W., and Franklin, R. M.,** Intensity fluctuation spectroscopy of laser light scattered by solutions of spherical viruses: R17, Qβ, BSV, PM2, and T7. II. Diffusion coefficients, molecular weights, solvation, and particle dimensions, *Biochemistry,* 13, 960, 1974.
58. **Koppel, D. E.,** Study of *Escherichia coli* ribosomes by intensity fluctuation spectroscopy of scattered laser light, *Biochemistry,* 13, 2712, 1974.
59. **Tanford, C.,** *Physical Chemistry of Macromolecules,* John Wiley & Sons, New York, 1961, 356.
60. **Tanford, C.,** *Physical Chemistry of Macromolecules,* John Wiley & Sons, New York, 1961, 379.
61. **Tanaka, T., Riva, C., and Ben-Sira, I.,** Blood velocity measurements in human retinal vessels, *Science,* 186, 830, 1974.
62. **Tanaka, T. and Benedek, G. B.,** Measurement of the velocity of blood flow (in vivo) using a fiber optic catheter and optical mixing spectroscopy, *Appl. Opt.,* 14, 189, 1975.
63. **Mustacich, R. V. and Ware, B. R.,** Observation of protoplasmic streaming by laser-light scattering, *Phys. Rev. Lett.,* 33, 617, 1974.
64. **Ware, B. R. and Flygare, W. H.,** The simultaneous measurement of the electrophoretic mobility and diffusion coefficient in Bovine Serum albumin solutions by light scattering, *Chem. Phys. Lett.,* 12, 81, 1971.
65. **Tanford, C.,** *Physical Chemistry of Macromolecules,* John Wiley & Sons, New York, 1961, 414.
66. **Haas, D. D. and Ware, B. R.,** Design and construction of a new electrophoretic light scattering chamber and applications to solutions of hemoglobin, *Anal. Biochem.,* 74, 175, 1976.
67. **Moran, R., Steiner, R., and Kaufmann, R.,** Laser Doppler spectroscopy as applied to electrophoresis of protein solutions, *Anal. Biochem.,* 70, 506, 1976.
68. **Josefowicz, J. and Hallett, F. R.,** Homodyne electrophoretic light scattering of polystyrene spheres by laser cross-beam intensity correlation, *Appl. Optics,* 14, 740, 1975.
69. **Chu, B.,** *Laser Light Scattering,* Academic Press, New York, 1974, 283.
70. **Tanford, C.,** *Physical Chemistry of Macromolecules,* John Wiley & Sons, New York, 1961, 414.
71. **Uzgiris, E. E. and Kaplan, J. H.,** Study of lymphocyte and erythrocyte electrophoretic mobility by laser Doppler spectroscopy, *Anal. Biochem.,* 60, 455, 1974.
72. **Ware, B. R. and Flygare, W. H.,** Light scattering in mixtures of BSA, BSA dimers, and fibrinogen under the influence of electric fields, *J. Colloid Interface Sci.,* 39, 670, 1972.
73. **Luner, S. J., Szklarek, D., Knox, R. J., Seaman, G. V. F., Josefowicz, J. Y., and Ware, B. R.,** Red cell charge is not a function of cell age, *Nature (London),* 269, 719, 1977.
74. **Smith, B. A., Ware, B. R., and Weiner, R. S.,** Electrophoretic distributions of human peripheral blood mononuclear white cells from normal subjects and from patients with acute lymphocyte leukmia, *Proc. Natl. Acad. Sci. U.S.A.,* 73, 2388, 1976.
75. **Josefowicz, J. and Hallett, F. R.,** Cell surface effects of pokeweed observed by electrophoretic light scattering, *FEBS Lett.,* 60, 62, 1975.
76. **Rimai, L., Salmeen, I., Hart, D., Liebes, L., Rich, M. A., and McCormick, J. J.,** Electrophoretic mobilities of RNA tumor viruses. Studies by Doppler-shifted light scattering spectroscopy, *Biochemistry,* 14, 4621, 1975.
77. **Debye, P.,** *Polar Molecules,* Dover Publications, New York, 1929, chap. 5.
78. **Berne, B. J. and Pecora, R.,** *Dynamic Light Scattering,* John Wiley & Sons, New York, 1976, 143.
79. **Perrin, F.,** Movement Brownien d'un ellipsoide (II). Rotation libre et depolarization des fluorescences translation et diffusion de molecules ellipsoidales, *J. Phys. Radium,* 7, 1, 1936.
80. **Pecora, R.,** Spectral distribution of light scattered by monodisperse rigid rods, *J. Chem. Phys.,* 48, 4126, 1968.
81. **Cummins, H. Z., Carlson, F. D., Herbert, T. J., and Woods, G.,** Translational and rotational diffusion constants of tobacco mosaic virus from Rayleigh linewidths, *Biophys. J.,* 9, 518, 1969.
82. **Nossal, R.,** Spectral analysis of laser light scattered from motile microorganisms, *Biophys. J.,* 11, 341, 1971.
83. **Berne, B. J. and Pecora, R.,** *Dynamic Light Scattering,* John Wiley & Sons, New York, 1976, chap. 5.
84. **Nossal, R., Chen, S.-H., and Lai, C.-C.,** Use of laser scattering for quantitative determinations of bacterial motility, *Opt. Commun.,* 4, 35, 1971.
85. **Schaefer, D. W., Banks, G., and Alpert, S. S.,** Intensity fluctuation spectroscopy of motile microorganisms, *Nature (London),* 248, 162, 1974.
86. **Schaefer, D. W.,** Dynamics of number fluctuations: motile microorganisms, *Science,* 180, 1293, 1973.
87. **Schaefer, D. W. and Berne, B. J.,** Light scattered from non-Gaussian concentration fluctuations, *Phys. Rev. Lett.,* 28, 475, 1972.
88. **Magde, D., Elson, E., and Webb, W. W.,** Thermodynamic fluctuations in a reacting system-measurement by fluorescence correlation spectroscopy, *Phys. Rev. Lett.,* 29, 705, 1972.

89. **Feher, G. and Weissman, M.,** Fluctuation spectroscopy: determination of chemical reaction kinetics from the frequency spectrum of fluctuations, *Proc. Natl. Acad. Sci. U.S.A.*, 70, 870, 1973.
90. **Birch, A. D., Brown, D. R., Dodson, M. G., and Thomas, J. R.,** The determination of gaseous turbulent concentration fluctuations using Raman photon correlation spectroscopy, *J. Phys. D.*, 8, L167, 1975.
91. **Berne, B. J., Deutch, J. M., Hynes, J. T., and Frisch, H. L.,** Light scattering from chemically reactive mixtures, *J. Chem. Phys.*, 49, 2864, 1968.
92. **Feller, W.,** *An Introduction by Probability Theory and Its Applications,* Vol. 1, 3rd ed., John Wiley & Sons, New York, 1968, chap. 17.
93. **Bauer, D. R., Hudson, B., and Pecora, R.,** Resonance enhanced depolarized Rayleigh scattering from diphenylpolyenes, *J. Chem. Phys.*, 63, 588, 1975.
94. **Bloomfield, V. A. and Benbasat, J. A.,** Inelastic light-scattering study of macromolecular reaction kinetics. I. The reactions A ⇌ B and 2A ⇌ A_2, *Macromolecules*, 4, 609, 1971.
95. **Jakeman, E., Pusey, P. N., and Vaughan, J. M.,** Intensity fluctuation light-scattering spectroscopy using a conventional light source, *Optics Commun.*, 17, 305, 1976.
96. **Cummins, H. Z. and Swinney, H. L.,** Light beating spectroscopy, *Progress in Optics,* Vol. 8, Wolf, E., Ed., North-Holland, Amsterdam, 1970, 133.
97. **Mandel, L.,** Correlation properties of light scattered from fluids, *Phys. Rev.*, 181, 75, 1969.
98. **Jones, C. R. and Johnson, C. S., Jr.,** Photon correlation spectroscopy using a jet stream dye laser, *J. Chem. Phys.*, 65, 2020, 1976.
99. **Gulari, E. and Chu, B.,** Photon correlation in the nanosecond range and its application to the evaluation of RCA C31034 photomultiplier tubes, *Rev. Sci. Instrum.*, 48, 1560, 1977.
100. **Lastovka, J. B.,** Light Mixing Spectroscopy and the Spectrum of Light Scattered by Thermal Fluctuations in Liquids, Ph.D. thesis, Massachusetts Institute of Technology, Cambridge, 1967.
101. **Jolly, D. and Eisenberg, H.,** Photon correlation spectroscopy, total intensity light scattering with laser radiation, and hydrodynamic studies of a well-fractionated DNA sample, *Biopolymers,* 15, 61, 1976.
102. **Kaye, W. and McDaniel, J. B.,** Low-angle laser light scattering-Rayleigh factors and depolarization ratios, *Appl. Optics,* 13, 1934, 1974.
103. **Gordon, J. P., Leite, R. C. C., Moore, R. S., Porto, S. P. S., and Whinnery, J. R.,** Long-transient effects in lasers with inserted liquid samples, *J. Appl. Phys.*, 36, 3, 1965.
104. **Whinnery, J. R.,** Laser measurement of optical absorption in liquids, *Acc. Chem. Res.*, 7, 225, 1974.
105. **Whinnery, J. R., Miller, D. T., and Dabby, F.,** Thermal convection and spherical aberration distortion of laser beams in low loss liquids, *IEEE J. Quantum Electron,* QE-3, 382, 1967.
106. **Carrington, A. and McLachlan, A. D.,** *Introduction to Magnetic Resonance,* Harper & Row, New York, 1967, 260.
107. **Long, D. A.,** *Raman Spectroscopy,* McGraw-Hill, New York, 1977, 46.
108. **Stacey, K. A.,** *Light Scattering in Physical Chemistry,* Academic Press, New York, 1956, 21.
109. **Loudon, R.,** *The Quantum Theory of Light,* Clarendon Press, Oxford, 1973, chap. 2.
110. **Tanford, C.,** *Physical Chemistry of Macromolecules,* John Wiley & Sons, New York, 1961, 145.
111. **Flygare, W. H.,** *Molecular Structure and Dynamics,* Prentice-Hall, Englewood Cliffs, N.J., 1978, chap. 1.
112. **McQuarrie, D. A.,** *Statistical Mechanics,* Harper & Row, New York, 1976, chap. 22.
113. **Chu, B.,** *Laser Light Scattering,* Academic Press, New York, 1974, 101.
114. **Perrin, F.,** Mouvement Brownien d'un ellipsoide (I.) Dispersion dielectrique pour des molecules ellipsoidales, *J. Phys. Radium,* 5, 497, 1934.
115. **Johnson, C. S., Jr. and Pedersen, L. G.,** *Problems and Solutions in Quantum Chemistry and Physics,* Addison-Wesley, Reading, Mass., 1974, chap. 5.
116. **Hall, R. S. and Johnson, C. S., Jr.,** Experimental evidence that mutual and tracer diffusion coefficients for hemoglobin are not equal, *J. Chem. Phys.*, 72, 4251, 1980.
117. **Hall, R. S., Oh, Y. S., and Johnson, C. S., Jr.,** Photon correlation spectroscopy in strongly absorbing and concentrated samples and applications to unliganded hemoglobin, *J. Phys. Chem.*, 84, 756, 1980.

INDEX

A

β-(2-furyl) acryloyl phosphate, I: 23, 27
glucose oxidase, I: 36

B

C

D

E

F

G

H

I

J

K

L

M

N

O

P

Q

R

S

T

U

V

W

X

Y

Z